NOT FOR LOAN

WITHDRAWN

ST. MARY'S HOSPITAL
LIBRARY

Since 1973 the Royal College of Obstetricians and Gynaecologists has regularly convened Study Groups to address important growth areas within obstetrics and gynaecology. An international group of eminent scientists and clinicians from various disciplines is invited to present the results of recent research and take part in in-depth discussion. The resulting volume containing the papers presented and also edited transcripts of the discussions is published within a few months of the meeting and provides a summary of the subject that is both authoritative and up to date.

Previous Study Group publications available from Springer-Verlag:

Early Pregnancy Loss
Edited by R. W. Beard and F. Sharp

AIDS in Obstetrics and Gynaecology
Edited by C. N. Hudson and F. Sharp

Fetal Growth
Edited by F. Sharp, R. B. Fraser and R. D. G. Milner

Micturition
Edited by J. O. Drife, P. Hilton and S. L. Stanton

HRT and Osteoporosis
Edited by J. O. Drife and J. W. W. Studd

The Royal College of Obstetricians and Gynaecologists gratefully acknowledges the sponsorship of this Study Group by the Department of Health.

Antenatal Diagnosis of Fetal Abnormalities

Edited by
J. O. Drife and D. Donnai

With 44 Figures

Springer-Verlag
London   Berlin   Heidelberg   New York
Paris   Tokyo   Hong Kong
Barcelona   Budapest

James O. Drife, MD, FRCSEd, FRCOG
Professor of Obstetrics and Gynaecology, Clarendon Wing, Leeds General Infirmary, Leeds LS2 9NS, UK

Dian Donnai, MBBS, FRCP, DCH, DObst RCOG
Consultant Clinical Geneticist, Department of Medical Genetics, St Mary's Hospital, Whitworth Park, Manchester M13 0JH, UK

ISBN 3-540-19673-0 Springer-Verlag Berlin Heidelberg New York
ISBN 0-387-19673-0 Springer-Verlag New York Berlin Heidelberg

British Library Cataloguing in Publication Data
Antenatal diagnosis of fetal abnormalities.
1. Humans. Foetuses. Diagnosis
I. Drife, J. O. (James Owen) 1947– II. Donnai, D. (Dian) 1945–
618.32075
ISBN 3-540-19673-0

Library of Congress Cataloguing-in-Publication Data
Antenatal diagnosis of fetal abnormalities/edited by J. O. Drife and D. Donnai.
p. cm.   Includes index. Report of a study group called by the Royal College of Obstetricians and Gynaecologists.
ISBN 3-540-19673-0.—ISBN 0-387-19673-0
1. Prenatal diagnosis. 2. Fetus—Abnormalities—Diagnosis.
3. Genetic counseling. I. Drife, James O., 1947– . II. Donnai, D. (Dian), 1945– .
III. Royal College of Obstetricians and Gynaecologists (Great Britain)
[DNLM: 1. Abnormalities—diagnosis—congresses. 2. Fetal Diseases—diagnosis—congresses. 3. Genetic Counseling—congresses.
4. Prenatal Diagnosis—congresses. WQ 209 A6258 1991]
RG628.A555 1991 618.3'2042—dc20
DNLM/DLC
for Library of Congress   91-4698
                                                                        CIP

Apart from any fair dealing for the purpose of research or private study, or criticism or review, as permitted under the Copyright, Designs and Patents Act 1988, this publication may only be reproduced, stored or transmitted, in any form or by any means, with the prior permission in writing of the publishers, or in the case of reprographic reproduction in accordance with the terms of licences issued by the Copyright Licensing Agency. Enquiries concerning reproduction outside those terms should be sent to the publishers.

© Springer-Verlag London Limited 1991
Printed in Great Britain

The use of registered names, trademarks etc. in this publication does not imply, even in the absence of a specific statement, that such names are exempt from the relevant laws and regulations and therefore free for general use.

Product liability: The publisher can give no guarantee for information about drug dosage and application thereof contained in this book. In every individual case the respective user must check its accuracy by consulting other pharmaceutical literature.

Typeset by Wilmaset, Birkenhead, Wirral
Printed by The Alden Press, Osney Mead, Oxford
2128/3830-543210 Printed on acid-free paper

Preface

In few areas of medicine is progress more spectacular than in the field of prenatal diagnosis. New clinical techniques such as chorion villus sampling, detailed ultrasound scanning and cordocentesis are being evaluated by obstetricians, and refinement of biochemical testing is widening the scope of maternal serum screening. In the laboratory, dramatic advances in molecular biology are occurring: families at risk of genetic disease can be investigated with gene probes, and preimplantation diagnosis of the embryo is now becoming a reality.

These technical advances have important ethical and practical implications, among which will be a further increase in public expectations of the standards required of antenatal services. Clinicians will need a high degree of skill to inform healthy women about the options for screening normal pregnancies, and to counsel high-risk women about the benefits and limitations of prenatal diagnosis. Obstetricians, scientists and health service managers will face the difficult task of deciding how prenatal diagnosis can be made available to women in a caring and cost-effective way.

Recognising the rapid progress in this field, the Royal College of Obstetricians and Gynaecologists made prenatal diagnosis the subject of its 23rd Study Group. An international panel of leading researchers, whose expertise ranged from molecular biology to philosophy, was invited to participate in a three day workshop, with time for in-depth discussion as well as the presentation of papers. Sessions covered the epidemiology of fetal abnormality, routine screening, special diagnostic techniques, DNA and biochemical analysis, psychological and ethical considerations, and the economic and practical aspects of service provision. The result is an authoritative, comprehensive and up-to-date review of this important subject.

The rapid publication of the proceedings of the Study Group would not have been possible without the efficiency and charm of Miss Sally Barber, Postgraduate Secretary of the RCOG. The Editors wish to thank her for her skilful co-ordination of this project. We also wish to express our warm thanks to the participants who willingly gave up their time for this meeting and who communicated their considerable expertise so effectively. The group put much effort into the formulation of constructive recommendations, which conclude this volume. We hope that these will be of benefit to obstetricians and others working in this field, and to our patients.

December 1990
J. O. Drife
D. Donnai

Contents

Preface v

Participants xi

SECTION I – EPIDEMIOLOGY AND ROUTINE SCREENING

1   **Trends in Prevalence of Congenital Abnormalities**
N. C. Nevin 3

2   **The Northern Regional Fetal Abnormality Survey**
A. F. J. Atkins and E. N. Hey 13
Discussion 30

3   **Routine Fetal Anomaly Screening**
M. J. Whittle 35

4   **Some Practical Issues in the Antenatal Detection of Neural Tube Defects and Down's Syndrome**
N. Wald and H. Cuckle 45
Discussion 54

5   **Heterozygote Screening for Cystic Fibrosis**
D. J. H. Brock, M. E. Mennie, I. McIntosh, C. Jones and
A. E. Shrimpton 59
Discussion 67

SECTION II – SPECIAL TECHNIQUES: 1

6   **Chorion Villus Sampling: The MRC European Trial**
T. W. Meade 73

7 Invasive Diagnostic Procedures in the First Trimester
R. J. Lilford 79
Discussion 91

8 Cardiac Ultrasound Scanning
L. D. Allan 97

9 Doppler Ultrasound Studies and Fetal Abnormality
B. J. Trudinger 113
Discussion 122

SECTION III – DNA ANALYSIS

10 Overview of Linkage and Probes
M. E. Pembrey 129
Discussion 135

11 Diagnosis of Genetic Defects in Eggs and Embryos
M. Monk 137
Discussion 148

SECTION IV – CYTOGENETIC AND BIOCHEMICAL DISORDERS

12 Fetal Karyotyping Using Chorionic Villus Samples
C. M. Gosden 153

13 Prenatal Diagnosis of the Fragile-X Syndrome
T. Webb 169
Discussion 179

14 Advances in Diagnosis of Biochemical Disorders
H. Galjaard 183
Discussion 197

SECTION V – SPECIAL TECHNIQUES: 2

15 Cordocentesis
K. H. Nicolaides 201

16 Intrauterine Therapy
C. H. Rodeck and N. M. Fisk 217

17 Magnetic Resonance Imaging (MRI) Scanning
I. R. Johnson 229
Discussion 238

SECTION VI – COUNSELLING, ECONOMICS AND ETHICAL ISSUES

18 Psychological Implications of Prenatal Diagnosis
T. M. Marteau 243

19 Counselling after Prenatal Diagnosis
D. Donnai and L. Kerzin-Storrar 255
Discussion 265

20 Economic Aspects of Prenatal Diagnosis
J. B. Henderson 269

21 Ethical Aspects of Prenatal Diagnosis
J. Harris 279
Discussion 289

SECTION VII – SERVICE PROVISION

22 Organisation of Genetic Services in the Netherlands
H. Galjaard 297

23 Genetic Services
R. Harris 311

24 National Coordination of Molecular Genetic Services
A. P. Read 321
Discussion 329

25 Provision of Service: The Obstetrician's View
R. H. T. Ward 335
Discussion 348

Conclusions and Recommendations 353

Subject Index 357

Participants

Dr L. D. Allan
Senior Lecturer, Department of Perinatal Cardiology, Guy's Hospital, St Thomas Street, London SE1 9RT, UK

Mr A. F. J. Atkins
Consultant Obstetrician, Maternity Unit, South Cleveland Hospital, Marton Road, Middlesborough, Cleveland TS4 3BW, UK

Professor D. J. H. Brock
Human Genetics Unit, University of Edinburgh, Western General Hospital, Crewe Road, Edinburgh EH4 2XU, UK

Professor S. Campbell
Department of Obstetrics & Gynaecology, King's College School of Medicine & Dentistry, Denmark Hill, London SE5 9RS, UK

Dr D. Donnai
Consultant Clinical Geneticist, Department of Medical Genetics, St Mary's Hospital, Whitworth Park, Manchester M13 0JH, UK

Professor J. O. Drife
Department of Obstetrics & Gynaecology, The University of Leeds, Leeds General Infirmary, D Floor, Clarendon Wing, Belmont Grove, Leeds LS2 9NS, UK

Professor Dr H. Galjaard
Professor of Cell Biology, Chairman, Department of Genetics, University Hospital, Westreedijk 112, Rotterdam 3016 AH, The Netherlands

Professor C. M. Gosden
MRC Human Genetics Unit, Western General Hospital, Crewe Road, Edinburgh EH4 2XU, UK

Professor J. Harris
Professor of Applied Philosophy, The Centre for Social Ethics & Policy, University of Manchester, Oxford Road, Manchester M13 9PL, UK

Professor R. Harris
Department of Medical Genetics, St Mary's Hospital, Hathersage Road, Manchester M13 0JH, UK

Mr J. B. Henderson
Economic Adviser, Economic Adviser's Office, Department of Health, 719 Friars House, 157–168 Blackfriars Road, London SE1 8EU, UK

Professor I. R. Johnson
Department of Obstetrics & Gynaecology, City Hospital, Hucknall Road, Nottingham NG5 1PB, UK

Professor R. J. Lilford
Department of Obstetrics & Gynaecology, St James's University Hospital, Beckett Street, Leeds LS9 7TF, UK

Dr T. M. Marteau
Lecturer in Health Psychology & Honorary Principal Clinical Psychologist, Royal Free Hospital School of Medicine, Pond Street, London NW3 2QG, UK

Dr T. W. Meade
Director, MRC Epidemiology & Medical Care Unit, Northwick Park Hospital, Harrow HA1 3UJ, UK

Dr W. J. Modle
Senior Medical Officer, Department of Health, Room B1103, Elephant and Castle, London SE1 6BY, UK

Dr M. Monk
Senior Scientist, MRC Mammalian Development Unit, Wolfson House, (University of London), 4 Stephenson Way, London NW1 2HE, UK

Professor N. C. Nevin
Department of Medical Genetics, The Queen's University of Belfast, Floor A, Tower Block, Belfast City Hospital, Lisburn Road, Belfast BT9 7AB, UK

Mr K. H. Nicolaides
Director, Harris Birthright Research Centre for Fetal Medicine, Department of Obstetrics & Gynaecology, King's College School of Medicine & Dentistry, Denmark Hill, London SE5 8RX, UK

Professor M. E. Pembrey
Mothercare Professor of Paediatric Genetics, Institute of Child Health, 30 Guilford Street, London WC1N 1EH, UK

Dr A. P. Read
Senior Lecturer in Medical Genetics, St Mary's Hospital, Hathersage Road, Manchester M13 0JH, UK

Professor C. H. Rodeck
Department of Obstetrics & Gynaecology, University College & Middlesex School of Medicine, Chenies Mews, London WC1E 6HX, UK

Professor B. J. Trudinger
Department of Obstetrics & Gynaecology, The University of Sydney at Westmead Hospital, Westmead, New South Wales 2145, Australia

Professor N. Wald
Department of Environmental & Preventive Medicine, The Medical College of St Bartholomew's Hospital, Charterhouse Square, London EC1M 6BQ, UK

Mr R. H. T. Ward
Consultant Obstetrician, University College Hospital, London, 86–96 Chenies Mews, London WC1E 6HX, UK

Dr T. Webb
Lecturer in Clinical Genetics, Birmingham Maternity Hospital, Queen Elizabeth Medical Centre, Birmingham B15 2TG, UK

Dr M. J. Whittle
Consultant Obstetrician/Perinatologist, The Queen Mother's Hospital, Yorkhill, Glasgow G3 8SH, UK

Professor R. Williamson
Surgical Unit, St Mary's Hospital Medical School, Praed Street, London W2 1NY, UK

Additional Contributors

Dr H. Cuckle
Reader in Preventive Medicine, CRC Fellow, Department of Environmental & Preventive Medicine, The Medical College of St Bartholomew's Hospital, London, EC1M 6BQ, UK

Dr N. M. Fisk
Department of Fetal Medicine, King George V Hospital, Missenden Road, Camperdown, New South Wales 2050, Australia

Dr E. N. Hey
Consultant Paediatrician, Princess Mary Maternity Hospital, Great North Road, Newcastle-upon-Tyne, NE2 3BD, UK

Dr C. Jones
Human Genetics Unit, University of Edinburgh, Western General Hospital, Crewe Road, Edinburgh EH4 2XU, UK

Mrs L. Kerzin-Storrar
Genetic Associate, Department of Medical Genetics, St Mary's Hospital, Whitworth Park, Manchester, M13 0JH, UK

Dr I. McIntosh
Human Genetics Unit, University of Edinburgh, Western General Hospital, Crewe Road, Edinburgh EH4 2XU, UK

Mrs M. E. Mennie
Human Genetics Unit, University of Edinburgh, Western General Hospital, Crewe Road, Edinburgh EH4 2XU, UK

Dr A. E. Shrimpton
Human Genetics Unit, University of Edinburgh, Western General Hospital, Crewe Road, Edinburgh EH4 2XU, UK

Back row: Professor C. M. Gosden, Professor R. Harris, Professor M. E. Pembrey, Dr T. Webb, Mr J. B. Henderson, Mr R. H. T. Ward, Mr K. H. Nicolaides, Dr M. Monk, Professor S. Campbell, Professor R. J. Lilford.
Middle row: Dr W. J. Modle, Dr L. D. Allan, Professor J. Harris, Dr T. M. Marteau, Dr A. P. Read, Professor C. H. Rodeck, Dr M. J. Whittle, Professor D. J. H. Brock, Professor I. R. Johnson.
Front row: Professor B. J. Trudinger, Professor Dr H. Galjaard, Dr D. Donnai, Professor J. O. Drife, Professor N. C. Nevin, Mr A. F. J. Atkins.

Section I
Epidemiology and Routine Screening

Chapter 1

Trends in Prevalence of Congenital Abnormalities

N. C. Nevin

In Western societies, during the present century, infant and neonatal mortality has declined steadily due to the effective control of infectious diseases, better obstetric care and the improvement in social and environmental conditions. Congenital abnormalities, with a few exceptions, have not shared this amelioration. In 1978, the Medical Research Council, London, in a review of clinical genetics, emphasised that "handicaps due to a genetic disorder or congenital malformation are the major child health problem today" [1]. As a community health problem, genetic disease and congenital abnormalities have become increasingly important [2].

For the family, congenital abnormalities have important implications. Some couples refrain from having another pregnancy because they have had an infant with a congenital abnormality or because some relative suffers from a genetic disorder. Other couples, in a similar situation, embark on another pregnancy unaware of the risk involved. Developments in clinical genetics and in obstetrics provide new approaches in the diagnosis, management and prevention of congenital abnormalities. Today, no couple planning to have a child can afford to ignore these new advances.

The recognition of the risk of having a child with a congenital abnormality or genetic disorder begins with the establishment of an accurate diagnosis in an affected child, fetus or relative and involves not only detailed clinical examination but also extensive laboratory investigation including cytogenetic, biochemical and molecular tests. It will also require post-mortem examination of a deceased infant or fetus. It may involve also extensive family study and perhaps tests for the detection of abnormal genes or balanced structural chromosomal abnormalities in relatives. Once a disorder has been diagnosed and the risk to future children assessed, the couple must be informed of the risk and of the reproductive choices available for avoiding the birth of an affected child. These choices may include accepting an increased risk of an abnormal child or refraining from having children (or further children) or considering antenatal screening and diagnosis with selective abortion.

In order to assess the impact of parental decisions, it is essential to have accurate baseline data on the prevalence rate of congenital abnormalities and genetic disease in the community. This paper discusses the prevalence rates of congenital abnormalities, the trends in prevalence rates and the impact of antenatal diagnosis with selective abortion on these rates.

Prevalence of Congenital Abnormalities and Genetic Disease

Congenital abnormalities can be classified into four main groups: those associated with abnormalities of the chromosomes (chromosomal abnormalities); those due to a single abnormal gene or abnormal gene pair (genic disorders); those caused by the interaction of abnormal genes, each with small detrimental effects and environmental influences (multifactorial disorders) and those due solely to environmental factors (environmental abnormalities). The total contribution which genetic factors make to human disease and abnormality is of the order of 1.6% [3]. However, this is a conservative estimate for chromosomal abnormalities and genic disorders. If congenital abnormalities and disease with a genetic component (multifactorial) are included, the genetic contribution to disease can be estimated at greater than 5% [4]. Table 1.1 shows the estimated prevalence per 1000 for human disease with a genetic factor(s) [5].

In Northern Ireland, approximately 1 in 50 infants born has a major congenital abnormality. The prevalence of some of the common lethal or handicapping congenital abnormalities are shown in Table 1.2. The major congenital abnormalities are Down's syndrome with a prevalence rate per 1000 livebirths of 1.41, neural tube defects (spina bifida, anencephaly and encephalocele) with a

Table 1.1. Estimation of genetic factors on human disease and congenital abnormalities

Disorder	Frequency per 1000
Chromosomal abnormalities	6.9
Disorders due to single genes:	
Autosomal dominant	2.1–2.8
Autosomal recessive	2.3–2.6
X-linked	3.0–4.8
Congenital abnormalities	32.3–37.0
Complex disorders	20.0–29.0
Total genetic contribution	66.6–76.2

Modified from Tables in reference [5].

Table 1.2. Prevalence rate (per 1000) of congenital anomalies in Northern Ireland, 1980–1986[a]

Anomaly	Number	Prevalence rate
Congenital heart defect	612	3.14
Spina bifida	315	1.62
Anencephaly	189	0.97
Encephalocele	45	0.23
Hydrocephaly	87	0.45
Microcephaly	90	0.46
Eye anomalies	149	0.76
Ear anomalies	204	1.05
Cleft lip	159	0.82
Cleft palate	149	0.76
Oesophageal atresia	55	0.28
Atresia intestinal tract	110	0.56
Hypospadias	147	0.75
Renal anomalies	112	0.58
Limb reduction anomalies	155	0.79
Omphalocele/gastroschisis	78	0.40
Down's syndrome	274	1.41
Other chromosome anomalies	44	0.23
	2944	15.10

[a]Total births (live and stillbirths) 195 012.

prevalence rate of 2.82 per 1000 total births and congenital heart abnormalities of 3.14 per 1000 total births. However, the prevalence rate of congenital heart abnormalities is at least double that in Table 1.2, as many congenital heart defects are not recognised within the first 7 days of life, the period of notification of congenital abnormalities in the Northern Ireland register.

Neural Tube Defects

Neural tube defects (NTDs), anencephaly and/or spina bifida, are among the most common and the most serious of congenital anomalies in man. The precise cause(s) still remains unknown. A high prevalence rate of NTDs has been reported for Northern Ireland [6–8]. In Belfast, for the period 1964–1968, the prevalence rate of anencephalus, which also included infants with both anencephalus and spina bifida, was 4.2 and of spina bifida alone was 4.5 per 1000 total births, giving an overall prevalence rate of NTDs of 8.7 per 1000 total births. In Northern Ireland, for the period 1974–1976, the prevalence rate of anencephalus was 3.1 and of spina bifida, 4.0 per 1000 total births, giving an overall prevalence rate of 7.1 per 1000 total births. In the period 1980–1986, among 195 012 total births, there were 315 with spina bifida and 189 anencephalus infants. The prevalence rate of anencephalus was 0.97 and of spina bifida 1.62 per 1000 total births, giving an overall prevalence rate of NTDs of 2.59 per 1000 total births (Table 1.3). The table summarises the changing prevalence rates for neural tube defects from 1964–1986.

Table 1.3. Prevalence of neural tube defects: Northern Ireland

	1964–1968[a]	1974–1976[b]	1980–1986[c]
Births (live and still)	41 351	79 783	195 012
Anencephalus	175 (4.2)	245 (3.1)	189 (1.0)
Spina bifida	185 (4.5)	319 (4.0)	315 (1.6)
Anencephalus and spina bifida	360 (8.7)	564 (7.1)	504 (2.6)

[a]Elwood and Nevin [6].
[b]Nevin et al. [8].
[c]Nevin (unpublished data).

Neural tube defects show marked geographical variations in prevalence rates which steadily decrease from north and west to the south and east of the British Isles and Europe. Table 1.4 shows this geographical gradient.

Table 1.4. Neural tube defects: study population, total births and prevalence rate (per 1000 births)

Population	Total births	Number of NTDs	Rate
Northern Ireland	195 012	671	3.44
Ireland	161 861	559	3.45
Mainland UK	221 116	634	1.20

Trend in Prevalence Rates of NTDs

The prevalence of NTDs also varies with time. In the Boston area, a peak prevalence at birth in 1930 ended in about 1950 [9]. In New York, there was a continued decreasing rate from 1945–1971 [10]. In other parts of the world, a decreasing rate of NTDs has been described during the past few decades. A declining rate has been seen in England and Wales [11], Northern Ireland [12], Ireland [13], Hungary [14], Liverpool [15], Netherlands [16], New South Wales and Australia [17]. In Sweden, 1947–1981, the prevalence rate approximately halved but the decline was not smooth and occurred in three separate waves [18]. In Northern Ireland, for the period 1964–1986, the overall prevalence rate of anencephalus declined from 4.2 to 1.0 per 1000 total births. The corresponding rates for spina bifida were 4.5 and 1.6 per 1000 total births respectively.

At present, prevalence rates of NTDs in many communities may reflect a worldwide decline. It is important that trends in the prevalence rates of neural tube defects are accurately monitored in order to assess the effectiveness of intervention programmes.

Impact of Antenatal Screening and Diagnosis on the Prevalence of Neural Tube Defects

With the establishment of maternal screening with maternal serum alphafetoprotein (MSAFP) for open NTDs, it is important to assess the impact of such a programme on the prevalence rates. Using information derived from the voluntary system of notification of congenital abnormalities in England and Wales, Cuckle and Wald [19] found that the birth prevalence of NTDs had declined by 80% from 3.15 to 0.62 per 1000 between 1964–1972 and 1985. Over the same period, notifications of terminations of pregnancies with a suspected NTD increased from less than 1% to 56% of neural tube defect births. Overall, it was estimated that 31% of the prevalence rate reduction could be attributed to antenatal detection and selective abortion.

A similar finding was observed in Scotland where, between 1974 and 1982, there was a 75% reduction in birth prevalence of which 49% was considered to be attributed to antenatal detection and selective abortion [20]. However, in both studies [19,20], it was clear that the reporting of terminations of abnormal pregnancies was incomplete.

The European Registration of Congenital Abnormalities and Twins, (EUROCAT) is a system for the registration of congenital abnormalities in a number of regions of the member states of the European Economic Community [21]. Similar reductions in prevalence rates of NTDs have been noted in several United Kingdom and European regions. Table 1.5 shows the impact of antenatal

Table 1.5. Impact of antenatal diagnosis and induced abortion on birth prevalence of anencephaly, 1980–1986

Centre	Induced abortion		Birth prevalence rate (B)	Proportion of birth prevalence prevented (A/B)
	Number	Rate (A)		
Belfast	89	0.46	1.48	32%
Liverpool	66	0.54	0.87	62%
Glasgow	119	1.31	1.46	90%
Paris	65	0.30	0.41	73%
Marseilles	11	0.46	0.59	78%
Strasbourg	18	0.27	0.32	84%
Firenze	18	0.28	0.44	64%
Groningen	9	0.18	0.66	27%

Table 1.6. Impact of antenatal diagnosis and induced abortion on birth prevalence of spina bifida, 1980–1986

Centre	Induced abortion		Birth prevalence rate (B)	Proportion of birth prevalence prevented (A/B)
	Number	Rate (A)		
Belfast	8	0.04	1.66	2%
Liverpool	43	0.35	1.32	27%
Glasgow	73	0.79	1.78	44%
Paris	33	0.15	0.49	31%
Marseilles	3	0.13	0.42	31%
Strasbourg	16	0.25	0.69	36%
Firenze	6	0.09	0.55	16%
Groningen	2	0.04	0.64	6%

...is on the prevalence rate for anencephaly. The proportion of the ...ience rate attributable to antenatal diagnosis and selective abortion ranged ... 32% (Belfast) to 90% (Glasgow). Table 1.6 shows the proportion of the ...ina bifida prevalence rate prevented by antenatal diagnosis and selective abortion. This ranged from 2% (Belfast) to 44% (Glasgow).

Prevalence of Down's Syndrome

Down's syndrome is the most important chromosomal abnormality, with a birth prevalence rate of one to two per 1000 livebirths. It is the most common single cause of mental retardation and, as such, is a major community health problem. Table 1.7 shows the birth prevalence rate of Down's syndrome in Western European areas for the years 1980–1986 [21]. A variety of chromosome aberrations result in Down's syndrome. About 93%–96% of patients have a regular trisomy 21, about 2%–5% have unbalanced translocations and a further 2%–4% have mosaicism. Double trisomies and rarer chromosomal abnormalities occur in less than 1% [22]. Data from amniocentesis indicates that maternal age specific rates of Down's syndrome are greater than those derived from livebirth rates [23]. Of the cases diagnosed at 16 weeks' gestation, 20%–30% end as spontaneous abortion or fetal death [24]. Results of chorion villus sampling at 8–10 weeks' gestation show a prevalence rate nearly double that at 16 weeks' gestation [25]. Spontaneous abortion in the first trimester has a ten times higher prevalence rate than that observed at birth [26]. At conception, the prevalence rate of trisomy 21 is extremely high with a continued loss throughout pregnancy.

Table 1.7. Down's syndrome: study population, total births and prevalence rate (per 1000 livebirths)

Population	Total livebirths	Number with Down's syndrome	Rate
Northern Ireland	195 012	269	1.38
Ireland	183 278	321	1.75
Mainland UK	213 710	210	0.98
Europe	823 258	904	1.10
Total	1 415 254	1704	1.20

Trends in Prevalence Rate of Down's Syndrome

The prevalence of Down's syndrome may be altered by such factors as changes in the age structure of women of reproductive age, restriction in family size and antenatal screening and diagnosis with selective abortion. The most significant factor in the aetiology of Down's syndrome is advanced maternal age. Changes in the age structure of the female population can produce a profound effect on the birth prevalence rate. In Denmark, where almost complete ascertainment of individuals with Down's syndrome is available, comparison of the periods 1960–1971 and of 1980–1985, showed that the proportion of mothers aged 25–29 years rose from 29.2% to 38.9% and the proportion of patients with Down's syndrome rose from 18.7% to 27.4%. Similar findings were observed in mothers aged 30–34 years [27,28]. However, in several other countries, no changes in the birth prevalence rates have been observed [29]. In a survey of publications on prevalence rate of Down's syndrome, none supported either an increase or a decrease with time [30]. Also in the 20 registries for chromosomal abnormalities in EUROCAT for 1980–1986, none reported an alteration in the trend of prevalence rate. Interestingly, in Sweden an increase was observed for women aged 25–34 years [31].

Impact of Antenatal Diagnosis on Prevalence rates of Down's Syndrome

Antenatal screening and diagnosis has had a significant impact on the prevalence rates of Down's syndrome. Currently, the indications for antenatal diagnosis of Down's syndrome are advanced maternal age, usually mothers over 35 years, mothers who have had a child with a chromosomal abnormality, a maternal or paternal balanced chromosomal translocation, and low serum alphafetoprotein. The main factor in the inefficiency of antenatal diagnosis is the poor utilisation of the antenatal diagnostic facilities by older mothers. The uptake of antenatal facilities varies from one country to another, and indeed within countries. In Denmark, the overall utilisation of antenatal diagnosis by mothers over 36 years was 70%. In the period 1980–1985, 15 333 pregnancies in women over 35 years were screened and 123 fetuses with Down's syndrome were aborted [27]. Antenatal diagnosis with selective abortion resulted in marked changes in the birth prevalence rate. For the period 1960–1971, 6.5% of mothers were over 40 years, whereas the corresponding figure for the period 1980–1985 was 6.8%. However, the percentage of Down's syndrome infants born to mothers over 40 years for these periods fell from 28.1% to 8.6%. Indeed, the prevalence rate per 1000 births fell from 5.13 to 1.05 [27]. Data of the effectiveness of antenatal diagnosis in reducing the prevalence rate of Down's syndrome in other countries is less impressive. Table 1.8 shows the effect of antenatal diagnosis from several EUROCAT centres. The proportion of births prevented by antenatal diagnosis and selective abortion ranged from only 4% (Belfast) to 27% (Paris) [21]. With the application of maternal serum alphafetoprotein and of the "triple test" (serum alphafetoprotein, human chorionic gonadotrophin and oestriol) screening, greater reductions in the number of infants with Down's syndrome are to be anticipated.

Undoubtedly, in the prevention of Down's syndrome, primary prevention will have the greatest impact on birth prevalence rates. However, the cause of non-disjunction, the main aetiological factor remains unknown. The European collaborative pilot study looking for causal factors in non-disjunction was unable to identify any specific responsible factor [32]. Although, at present primary prevention of Down's syndrome is not possible, satisfactory approaches are available to mothers to avoid the birth of an infant with Down's syndrome.

Table 1.8. Impact of antenatal diagnosis and induced abortion on birth prevalence of Down's syndrome, 1980–1986

Centre	Induced abortion		Birth prevalence rate (B)	Proportion of birth prevalence prevented (A/B)
	Number	Rate (A)		
Belfast	12	0.06	1.47	4%
Liverpool	12	0.10	1.15	9%
Glasgow	14	0.16	1.12	14%
Paris	91	0.42	1.54	27%
Marseilles	4	0.17	1.46	12%
Strasbourg	10	0.16	1.12	14%
Firenze	17	0.26	1.56	17%
Groningen	5	0.10	1.23	8%

Conclusions and Recommendations

Continued monitoring of congenital abnormalities in live- and stillborn infants is important. Registries of congenital abnormalities should use multiple sources of ascertainment and also should extend the period of registration beyond seven days after birth. For the effective audit of the impact of antenatal screening and diagnosis, a statutory register of fetal abnormality of induced abortions should be established. Because of the problems inherent in a national register, registers of congenital abnormalities among live births, stillbirths and induced abortions would best be organised at regional level, although for some disorders national registers may be necessary. Regional registers would also be more appropriate because of the geographical variations of congenital abnormalities, the trends of prevalence rate with time and the variation between regions of the utilisation of antenatal screening and diagnosis. In order to improve the accuracy of diagnosis, all terminations of pregnancy for fetal abnormality should have appropriate laboratory investigations, including a complete post-mortem examination by a specialist with experience in fetal pathology, karyotyping of fresh (unfixed) fetal tissue and biochemical and molecular studies. The storage of unfixed fetal tissue is encouraged, particularly, when the diagnosis is uncertain or the fetus has a rare congenital abnormality. Liaison of the staff of the congenital abnormality register with the regional clinical genetic services would be helpful, especially in the recognition of complex syndromes and congenital abnormality associations. The accurate recognition of abnormalities will ensure that reliable and effective genetic counselling can be provided.

References

1. Medical Research Council. Review of clinical genetics. A report of the Council's Cell Biology Board by the MRC sub-committee to review clinical genetics. London: MRC, 1978.
2. Nevin NC. Aetiology of genetic disease. In: Turnbull AC, Woodford FP, eds. Prevention of handicap through antenatal care. North–Holland: Elsevier, Excerpta Medica, 1976; 3–12.
3. UNSCEAR report. Genetic and somatic effects of ionizing radiation. New York: United Nations, 1986.

4. Baird PA, Anderson TW, Newcombe HB, Lowry RB. Genetic disorders in children and young adults: a population study. Am J Hum Genet 1988; 42:677–93.
5. Department of Health report on Health and Social Services. 35 Guidelines for the testing of chemicals in food, consumer products and the environment. London: HMSO, 1989.
6. Elwood JH, Nevin NC. Factors associated with anencephalus and spina bifida in Belfast. Br J Prev Soc Med 1973; 27:73–80.
7. Elwood JH, Nevin NC. Anencephalus and spina bifida in Belfast (1964–1968). Ulster Med J 1973; 42:213–22.
8. Nevin NC, McDonald JR, Walby AL. A comparison of neural tube defects identified by two independent routine recording systems for congenital malformations in Northern Ireland. Int J Epidemiol 1978; 7:319–21.
9. MacMahon B, Yen S. Unrecognized epidemic of anencephaly and spina bifida. Lancet 1971; i:31–3.
10. Janerich DT. Epidemic waves in the prevalence of anencephaly and spina bifida in New York State. Teratology 1973; 8:253–6.
11. Bradshaw J, Weale J, Weatherall J. Congenital malformations of the central nervous system. Popul Trends 1980; 19:13–18.
12. Nevin NC. Neural tube defects. Lancet 1981; ii:1290–1.
13. Radic A, Dolk H, De Wals P. Declining rate of neural tube defects in three eastern counties of Ireland: 1979–1984. Ir Med J 1987; 80:226–8.
14. Czeizel A. Spina bifida and anencephaly. Br Med J 1983; 287:281.
15. Owens JR, Harris F, McAllister P, West L. 19-year incidence of neural tube defects in area under constant surveillance. Lancet 1981; ii:1032–5.
16. Romijn JA, Treffers PE. Anencephaly in the Netherlands: a remarkable decline. Lancet 1983; i:64–5.
17. Mathers CD, Field B. Some international trends in the incidence of neural tube defects. Community Health Stud 1983; 7:60–6.
18. Kallen B, Lofkvist E. Time trends in spina bifida in Sweden 1947–1981. J Epidemiol Community Health 1984; 34:103–7.
19. Cuckle H, Wald N. The impact of screening for open neural tube defects in England and Wales. Prenat Diagn 1987; 7:91–9.
20. Carstairs V, Cole S. Spina bifida and anencephaly in Scotland. Br Med J 1984; 289:1182–4.
21. EUROCAT Report 3. Surveillance of congenital anomalies. Years 1980–1986. Department of Epidemiology, Catholic University of Louvain, Brussels: 1989.
22. Nevin NC. Genetic disorders. Clin Obstet Gynaecol 1982; 9:3–27.
23. Ferguson-Smith MA, Yates JRW. Maternal age specific rates for chromosome aberrations and factors influencing them: report of a collaborative European study on 52 965 amniocenteses. Prenat Diagn 1984; 4:5–44.
24. Hook EB. Chromosome abnormalities and spontaneous fetal death following amniocentesis: further data and associations with maternal age. Am J Hum Genet 1983; 35:110–16.
25. Mikkelsen M, Ayme S. Chromosomal findings in chorionic villi. A collaborative study. In: Vogel F, Sperling K, eds. Human genetics. Proceedings of the Seventh International Congress, Berlin and Heidelberg: Springer-Verlag, 1986; 596–606.
26. Hassold T, Chen N, Funkhouser et al. A cytogenetic study of 1000 spontaneous abortions. Ann Hum Genet 1980; 44:157–78.
27. Mikkelsen M. Incidence of Down's syndrome. Phil Trans R Soc London B 1988; 319:315–24.
28. Mikkelsen M, Fischer G, Stene E, Petersen E. Incidence of Down's syndrome in Copenhagen 1960–1971: with chromosome investigation. Ann Hum Genet 1976; 40:117–82.
29. Lowry RB, Jones DC, Renwick DHG, Trimble BK. Down syndrome in British Columbia, 1952–73: incidence and mean maternal age. Teratology 1976; 14:29–34.
30. Evans HJ, Lyon MF, Czeizel A et al. ICPEMC Meeting Report no. 3. Is the incidence of Down syndrome increasing? Mutat Res 1986; 31:263–6.
31. Iselius L, Lindsten J. Changes in Down's syndrome in Sweden during 1968–1982. Hum Genet 1986; 72:133–9.
32. Aymé S, Baccichetti C, Bricarelli FD et al. Factors involved in chromosomal non-disjunction. In: Vogel F, Sperling K, eds. Human genetics. Proceedings of the Seventh International Congress, Abstracts 1, Berlin and Heidelberg: Springer-Verlag, 1986; 167.

Chapter 2

The Northern Regional Fetal Abnormality Survey

A.F.J. Atkins and E. N. Hey

Introduction

Antenatal care took a quantum leap forward when techniques, including in particular sonar examination, made it possible to assess certain aspects of fetal health and development antenatally in a non-invasive way. Initially such techniques were largely used in the review and assessment of high risk pregnancies, but alphafetoprotein (AFP) and sonar screening are now being widely used to screen all mothers for fetal abnormality in early pregnancy. The reliability of such techniques in expert hands has been well documented, but the reliability of these approaches when used in a routine district health service setting has not so far been monitored. False positive and false negative diagnoses are almost inevitable when any newly developed screening programme is first set up, but the only available information on the accuracy of sonar diagnosis was coming from a small number of referral centres with a concentration of specialist experience serving an ill-defined catchment area.

In our region screening was also bringing to light ill-defined problems that had not previously been recognised before birth, but experience on the handling of such problems was thinly spread across the region and not easily assessable. Difficulty had also been encountered in handling newly born infants with significant treatable anomalies who required help unexpectedly sometimes during the night, despite the fact that in many cases the anomalies had been diagnosed, or at least suspected, before delivery.

With these thoughts in mind, in September 1983 a Newcastle paediatric surgeon, with the approval and support of the Region's specialty subcommittees of obstetrics and gynaecology and of paediatrics, wrote to a small multidisciplinary group of clinicians interested in the prenatal diagnosis of fetal abnormality and in the management of these conditions. He pointed out that, although such management is of concern to obstetricians, geneticists, radiologists, paediatricians and paediatric surgeons, there was no formal mechanism for consultation and advice so that most of these cases were being handled on an ad hoc basis.

This group, with the generous support of the Northern Regional Health Authority, set up a pilot study for one year for the reporting and central registry of all suspected cases of fetal abnormality. The Northern Region already had a perinatal mortality survey with its own central register, and this had shown that more than a quarter of all deaths in the first year of life were due to fetal abnormality [1]. The pilot study aimed to demonstrate that a fetal abnormality register could establish the true community incidence of significant fetal abnormality, identify the false positive and false negative antenatal diagnosis rates for various conditions, compare management, and monitor outcome.

The Northern Region, with a population of 3 million and 40 000 deliveries per year, had geographical advantages for any such study, being bordered by the sea to east and west, with a sparsely populated national border to the north, and an area of low population to the south. There is, therefore, very little cross-boundary flow for maternity care. Furthermore, such cross-boundary migration as there is nearly all flows inwards, into the region. In order to benefit from this geographical advantage, it was necessary to obtain the willing cooperation of all the relevant clinicians in the region, particularly those practising prenatal diagnosis, so that they would not feel inhibited in reporting *all* suspected anomalies for fear of subsequent embarrassment should the final diagnosis not match their prediction. A letter was sent to all consultant obstetricians and paediatricians in the Region, and this was followed by a personal visit by one of the group charged with responsibility for the pilot study. It is gratifying to note the unanimous support that was given at that time, and has continued without reservation ever since.

When the results of the pilot study for 1984 became available, the various specialty subcommittees of the Regional Medical Committee agreed to endorse the launch of a continuing collaborative region-wide study. A steering group with representation from each medical specialty involved was established to oversee the direction and development of the study. Ethics committees in each district in the region also approved the creation of a confidential register of clinical case material stored in non-attributable form on the Region's main-frame computer, and one year later the Regional Health Authority (RHA) found the money to establish a post of "survey coordinator" to cope with the expanding clerical and administrative work being generated by the survey.

Methods

The survey now registers all suspected fetal abnormality, even if the diagnosis is not confirmed after delivery, and any significant abnormality diagnosed for the first time after birth, to all babies of mothers resident in the Northern Region at the time of delivery (even if delivery takes place outside the region). Minor abnormalities (such as skin tags, birth marks, failed testicular descent, glandular hypospadias, isolated talipes and minor syndactyly) are excluded. However, all conditions requiring surgery in the first year of life (except inguinal hernia, patent ductus and congenital dislocation of the hip) qualify for inclusion. Children with congenital hypothyroidism, and familial conditions such as phenylketonuria, haemophilia, Duchenne dystrophy and cystic fibrosis are not at present registered unless the pregnancy is terminated on those grounds, because other specialist registers covering such cases already exist in this region.

The register details how and when each diagnosis was made. Autopsy information is usually available on file for children who died, and for all stillbirths and abortions after more than 19 completed weeks' gestation. For pregnancies terminated because of a suspected fetal abnormality there is no lower gestational age limit to registration. The register contains confidential information on the mother's age and home address at delivery (including the postcode) together with information on management and outcome, including disability at one year, in those who survived.

Validation

The survey has been validated by cross-checking the information available from District Health Authority SD56 returns to the Office of Population Censuses and Surveys (OPCS). This showed an almost complete match in respect of the number of registered births with a neural tube defect, isolated congenital hydrocephalus, exomphalos, rectal atresia, and cleft lip and palate. For all other conditions notifications to the Regional survey consistently outnumbered those submitted voluntarily to OPCS. Not a single condition was identified where the number of cases known to OPCS exceeded the number known to the RHA study. This independent validation of the long-established national SD56 notification procedure confirms that the OPCS reporting system provides a reasonably accurate register of those conditions that are immediately visible to the naked eye at birth, but a very inadequate register of those that are not [2].

Cross checks have also been undertaken against the mandatory termination of pregnancy returns to OPCS [3]. Terminations for a "CNS abnormality" and for "other specified problems" match those known to the Regional Health Authority, but OPCS returns for "suspected chromosome abnormality" were double those known to the RHA. At first it was thought that this represented under-reporting, but a confidential analysis of the OPCS returns for 1985, 1986 and 1987 has since shown that the whole of the discrepancy can be accounted for by pregnancies terminated before 12 weeks of gestation. Before 1988 chorionic villus biopsy was being used on only a few patients in this region, and almost all screening for fetal chromosome abnormality was being done by amniocentesis at 16 weeks' gestation. We suspect, therefore, that the remaining terminations notified to OPCS under this heading between 1985 and 1987 were undertaken without further detailed screening because of the increased risk of chromosomal abnormality in the fetuses of mothers more than 34 years old. Cross checks with the three cytogenetic laboratories in the region confirmed that every pregnancy terminated because of a proven chromosomal abnormality had been reported voluntarily to the survey by the obstetrician concerned.

Input

Whenever an abnormality is suspected antenatally a notification form is sent to the survey coordinator by the person responsible (who may be a radiologist, a radiographer/ultrasonographer, a geneticist or an obstetrician). All chromosomal abnormalities detected either antenatally or postnatally are now also automatically reported to the survey office by the region's three cytogenetic laboratories. It has taken some time to get such a system working, however, and the laboratories themselves are often short of basic background information on much of the clinical material they are asked to examine. Often, for example, they know neither the mother's date of birth nor that of the child. Information on all miscarriages after 20 weeks' gestation or more, on all stillbirths, and on all deaths in the first year of life, is available to the survey because such information comes in automatically to the Region's long-standing Collaborative Survey of Perinatal Mortality. The fetal abnormality survey would probably never get to know about many of the children who die were this second, parallel, regional initiative not also in operation.

Reasonably complete ascertainment is essential, but accuracy is also important, and for this experienced clinical input is necessary. The survey coordinator always has instant access to expert help with the interpretation and coding of survey returns – something that would have been badly lacking had the survey office been set up in an office at the RHA rather than in one of the region's main maternity units. It is now increasingly clear that clinical help at district level is also going to be equally important to the continued viability of the study, which was carried along over the first few years by the dedication and commitment of a few enthusiasts on the survey steering group itself. Children coming to surgery in the first year of life have always been conscientiously reported to the survey by the region's paediatric surgeons: the register now needs to look to an equal commitment from some other groups of clinicians if its early promise is to be fulfilled.

Outcome

Infants with abnormalities carrying an uncertain prognosis are followed for at least one year or to death, and this information is also sent to the survey coordinator. All deaths in the first year of life are made known to the survey anyway because of the register's link with the mortality survey. In practice, of course, much of this information does not reach the survey office automatically but has to be sought. The role of the full-time paid survey coordinator in interpreting and synthesising all the information coming into the survey office over a five-year period without computer support, and in charming essential follow-up information out of clinicians when it did not reach the survey office spontaneously, has been central to any success the survey has achieved to date.

The survey has also depended for a lot of its success on the part played by a number of (often underpaid) clinical secretaries across the region sending copy letters, operation notes, autopsy reports, etc., into the survey office automatically, because the support of clinicians across the region for the survey, while strong in theory, has often been weak in practice!

Output

Teething problems were encountered with the computerisation of the mass of material coming into the survey office as the work of the register expanded between 1987 and 1989, but the backlog of data generated at that time has now been dealt with, and plans are in hand to prepare a report each year identifying the number and type of abnormalities correctly diagnosed, incorrectly diagnosed or missed antenatally. The register can also be consulted at any time (observing strict rules of confidentiality) for the information currently available on a particular specified condition.

Difficulties

The greatest difficulty expected – that of full clinical cooperation – was not a problem, any errors being errors of omission rather than intent. The greatest problem has been a delay in getting expert help with the management of the data coming into the survey. In the recent past this has often almost overwhelmed the single, paid, survey coordinator. Data were initially batch processed and stored on the RHA mainframe computer, but the interrogation of this database proved slow and inflexible. A parallel database on a desk-top "micro" computer has now been installed but many problems and delays have been encountered in getting this operational. Budget limitations resulted in the purchase of hardware that was slower than expected, and an insistence that we use "in house" NHS personnel meant that it took more than 18 months to get the programming finished. In summary: inadequate "hardware"; slow and inefficient "software"; and short of "warm wear".

Results

Certain typical examples have been selected for discussion, to illustrate the issues of population screening, the selective screening of high risk populations, and the interpretation of selected antenatal ultrasound findings.

Population Screening

This can provide an opportunity for the early termination of pregnancy, if the couple so wish, when a lethal abnormality is identified; it can identify conditions amenable to antenatal treatment; it can influence the management of delivery; it can make the immediate management of certain conditions requiring urgent post-delivery treatment easier to organise; it can alert paediatric staff to the existence of conditions benefiting from early postneonatal surgery; and it can provide scope for reducing the amount of disability in surviving children. We thought it might be useful to illustrate each of these considerations.

Lethal Abnormality

Short-limbed dwarfism and renal agenesis and dysgenesis are reviewed. Skeletal dysplasias are rare, but some of the conditions associated with gross limb shortening, and the pulmonary hypoplasia seen as a result of the associated poor rib development, are uniformly lethal. The incidence of such conditions in the Northern Region is only about 0.17 cases per 1000 registrable births, but it seems as though a reasonable proportion of these cases is now being recognised early enough for termination of pregnancy to be an option (Fig. 2.1). Although a precise diagnosis is not always achievable before the fetus has been delivered, no fetus has yet been terminated in this region for suspected short-limbed dwarfism that did not have some lethal abnormality. Clinicians are also increasingly aware of the importance of making a precise radiological diagnosis after delivery to aid subsequent genetic counselling.

Bilateral renal agenesis (or functional agenesis due to severe bilateral renal dysplasia) is rather commoner (0.25 cases per 1000 births). Here, too, an antenatal diagnosis is now being made with reasonable confidence early enough for termination to be an option, more often than it was a few years ago (Fig. 2.2). However, at least two pregnancies associated with oligohydramnios and fetal growth retardation have been terminated in this region in the last five years in the belief that there was renal agenesis when absolutely no structural abnormality could be found on subsequent autopsy. It is not surprising, therefore, that some pregnancies are currently being watched expectantly even when renal agenesis is

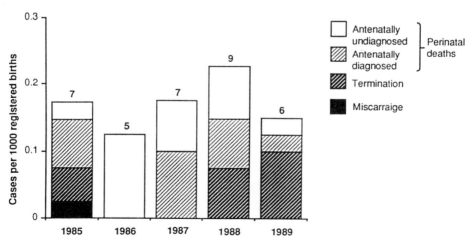

Fig. 2.1. Identification of lethal short limbed dwarfism 1985–1989 (34 cases).

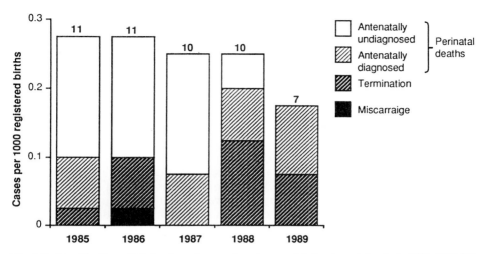

Fig. 2.2. Identification of bilateral renal agenesis and severe bilateral renal dysplasia 1985–1989 (49 cases).

suspected. Families sometimes prefer to "let nature take its course" when warned of problems antenatally, especially if there is some lingering chance of the diagnosis being wrong.

Antenatal Treatment

The list of conditions currently amenable to antenatal treatment is not, as yet, very long. Bladder outlet obstruction and cardiac arrhythmia [4] are discussed here. Our experience with urethral valves and other forms of bladder outlet obstruction including the "Prune Belly" syndrome is not encouraging. The incidence in this region is only about 0.16 cases per thousand births, and only ten of the 33 cases currently known to the survey are alive a year after birth (Table 2.1). Only 12 cases were correctly diagnosed antenatally, and in 14 cases no abnormality of any nature was suspected before delivery (Table 2.4). In the remaining seven cases while some abnormality of the renal tract was suspected it was not appreciated that there was bladder outlet obstruction. There were a further six "false positive" diagnoses (all first suspected in the second half of pregnancy).

Table 2.1. Identification and management of bladder outlet obstruction in the Northern Region 1985–89

Spontaneous miscarriage	3
Pregnancy terminated	7
Stillborn	5
Death in infancy	8
On dialysis at 1 year	2
Well at 1 year	8
Total	33

Five babies correctly identified antenatally were assessed for possible treatment using a fetal vesico-amniotic shunt. Only two were actually shunted: one died antenatally shortly after the shunt was inserted and the other has been on long-term peritoneal dialysis ever since delivery. Seven of the eight babies who died after birth died with respiratory problems within a few days of delivery, and the eighth died with medical problems while on dialysis. Only one of the 12 children correctly diagnosed before birth is alive, and that child is on dialysis (as already noted). It would seem, therefore, that it is only the severe, lethal, cases of bladder outlet obstruction that are currently being diagnosed antenatally.

Neither have we had very much more success with the antenatal treatment of cardiac arrhythmia. Only three of the seven children with partial or complete heart block known to the regional cardiologists were diagnosed before the onset of labour, and only seven of the 24 children with a neonatal tachycardia. Five of the 17 cases of supraventricular tachycardia were identified antenatally, but only two of these children were got back into sinus rhythm before delivery. In one child who survived, flecainide caused its hydropic changes to resolve before birth, but in a second successfully digitilised child hydropic changes did not resolve and the child died two days after delivery when the severe arrhythmia recurred.

Management of Delivery

We have chosen to present our data on abdominal wall defects. It has recently been argued that the known presence of these defects is an indication for delivery by caesarean section to reduce the risk of fetal trauma [5]. Such defects are seen in 0.6 per thousand births in this region, and each case merits careful individual antenatal assessment, because in nearly one-third of cases there are also other lethal abnormalities such as a diaphragmatic defect, with pulmonary hypoplasia, other complex limb:body wall defects with spinal problems, the omphalocele, bladder extrophy, imperforate anus and sacral agenesis (OEIS) syndrome, or a trisomy (Fig. 2.3). The majority of these complex abnormalities are now

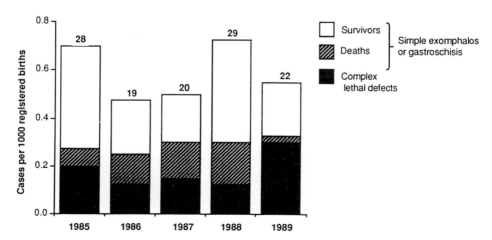

Fig. 2.3. Incidence and outcome of abdominal wall defects 1985–1989 (118 cases).

suspected early enough for termination to be possible if the fetus does not abort spontaneously (Table 2.2).

Table 2.2. Identification of abdominal wall defects (118 cases)

	1985	1986	1987	1988	1989
Simple anomalies found before birth	10/20	9/14	8/15	14/24	6/10
Complex anomalies terminated or miscarrying	4/8	4/5	4/5	5/5	9/12
Total surviving	13/28	9/19	8/20	13/29	9/22

Of the babies with a straightforward uncomplicated exomphalos or gastroschisis delivered in this region in the last five years 57% also had their defects recognised antenatally, but there is relatively little evidence that antenatal diagnosis actually improved the chance of survival although it can certainly reduce morbidity and make the surgeon's task easier (Table 2.3).

Table 2.3. Identification and management of abdominal wall defects in the Northern Region 1985–1989 (excluding 35 cases with chromosome defects, or other complex lethal abnormality)

Cases diagnosed antenatally (83% delivered vaginally)			Cases not diagnosed antenatally (86% delivered vaginally)		
Miscarriage		2	Miscarriage		1
Termination	on social grounds to comply with family views	1 2			
Antepartum SB	rhesus hydrops	1	Antepartum SB	second twin (fetus papyraceous)	1
	no other anomaly	1		with anophthalmos	1
Intrapartum SB	unexplained fetal death in early labour	1		no other anomaly	3
Neonatal death	severe prematurity	1	Neonatal death	bowel gangrenous at birth	1
	bowel gangrenous at birth	1			
1st year death	late surgical complications	2	1st year death	unrelated cot death at 6 months	1
Survivors	(no apparent handicap)	35 (74%)	Survivors	(no apparent handicap)	28 (78%)

A larger review of all the births in the region between 1981 and 1989 with an anterior abdominal wall defect has failed to show a single livebirth or intrapartum death where there was trauma to the fetus, and in particular to the fetal liver or bowel, during birth, although there were two cases where the exomphalos itself probably ruptured during delivery, and only 15% (17/117) of all the viable babies with a diagnosed or undiagnosed anterior abdominal wall defect alive at the onset of labour were delivered by caesarean section in those years. The case for delivery by caesarean section does not seem to have been made as yet. Such an analysis would not have been possible, of course, if the register had not recorded details of case management as well as the antenatal and postdelivery diagnosis.

A more detailed review of the case notes, however, reveals a definite case for arranging delivery in a unit with ready access to surgical care, because the blood supply to the bowel is easily compromised when the gut starts to fill with air after birth (especially if the neck of the abdominal defect is tight), while the liver can become grossly engorged in a matter of minutes if the venous return is obstructed (as it can be soon after delivery in those cases where the liver is exposed and abnormally mobile). Delivery in a unit with surgical facilities also minimises separation of the mother from her child.

Management at Birth

We have chosen to review diaphragmatic hernia. Few conditions are more vulnerable than this to the potential hazard of mismanagement or diagnostic delay at birth. Unfortunately it is not a condition that is very often diagnosed before delivery, and experience shows that the differential diagnosis between diaphragmatic hernia and extensive pulmonary cystic dysplasia can also present problems. Antenatal diagnosis seems to have improved in the last two years, but there is no evidence, as yet, that this has influenced survival even if complex cases with other lethal abnormalities are excluded from consideration (Fig. 2.4). Realistically, however, much larger numbers would be necessary to show that antenatal diagnosis can alter survival rates by improving management immediately after birth.

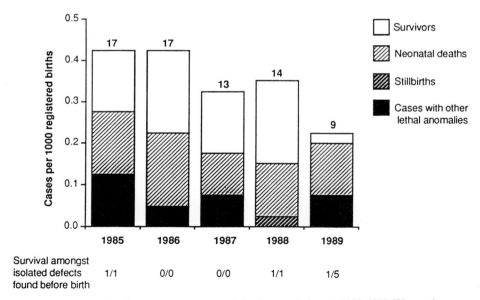

Fig. 2.4. Identification and management of diaphragmatic hernia 1985–1989 (70 cases).

Post-neonatal Care

One area where antenatal ultrasound has had an undoubted impact is in the early identification of urological abnormality such as pelvi-ureteric or ureterovesical obstruction, and the unilateral, non-functioning, dysplastic kidney. Table 2.4 shows the incidence of these conditions (as judged by the number of children dying or coming to surgery within two years of delivery), and the proportion of these cases identified as a result of ultrasound assessment in the antenatal period. Many babies currently identified antenatally as having non-specific hydronephrosis or a dilated renal pelvis appear to have no obstructive lesion or loss of function on further investigation after delivery. Antenatal reporting rates in different units vary widely, and there is clear scope for further research to determine the most appropriate way of responding when a relatively minor abnormality is identified in the second half of pregnancy. Clinicians are keen not to overlook a treatable condition, but large numbers of "false positive" reports generate a lot of postnatal work and can expose families to unnecessary stress and anxiety if the issue is not handled sensitively.

Table 2.4. The incidence and antenatal diagnosis of urological abnormality in the Northern Region 1985–1989

Condition	Number of cases	Incidence per 1000 registered births	Proportion of the cases identified by 2 years diagnosed antenatally (%)
Unilateral renal dysplasia (no significant function)	71	0.35	83%
Pelvi-ureteric junction obstruction	44	0.22	95%
Ureterovesical obstruction	12	0.06	75%
Grade 3+ ureteric reflux	27	0.13	48%
Bladder outlet obstruction	33	0.16	36%

Estimates of the community prevalence of ureteric obstruction in Table 2.4 must, however, be falsely low at present, because most registered cases were diagnosed only in the second half of pregnancy, and many mothers in this region still only get one antenatal scan during pregnancy. There can be little doubt that ureteric obstruction is worth relieving when it is severe, and that antenatal ultrasound is now bringing such conditions to light at a much earlier age. A further routine ultrasound scan early in the third trimester would be necessary to increase antenatal recognition. It has also been traditional for many years to recommend the removal of any functionless dysplastic kidney identified at birth, but there are sceptics who now suggest that such surgery seldom serves any useful purpose.

Reflux, on the other hand, is well worth identifying before infection causes renal scarring: unfortunately relatively few cases of reflux serious enough to merit surgery or even prophylactic antibiotic treatment seem to be brought to light by antenatal ultrasound. Only 27 children have been registered with the survey as having Grade 3 reflux or worse in the last five years, and only 13 (48%) of these children were identified as a result of antenatal screening (Table 2.4). While cases with antenatal abnormalities have probably been fairly consistently registered, many postnatally diagnosed cases have probably gone unreported if they were managed locally and never referred for urological assessment. It seems clear, therefore, that antenatal ultrasound is producing only a modest reduction in the number of children at risk of unrecognised reflux nephropathy at present. The impact of antenatal screening can be assessed properly only if all cases coming to light after birth are also registered with the survey.

Long-term Disability

Neural tube defects and isolated hydrocephalus are reviewed here. Antenatal screening has had a particularly dramatic impact on the identification and management of neural tube defects [6,7], and this is well illustrated by the data now available to the survey. There is clear historic evidence that the natural incidence of these conditions nearly halved between the early 1970s and the mid 1980s, while the number among registrable births was further halved again by the impact of antenatal screening (Table 2.5).

Table 2.5. Incidence of neural tube defects in the Northern Region 1971–75 versus 1983–87

Period	Number		Incidence per 1000 registered births
	Anencephaly	Other	
Registered births 1971–5 known to OPCS[a]	334	482	3.62
Registered Births 1983–7 known to OPCS (and RHA)	12	173	0.92 ⎫
Terminations known to RHA 1983–7	123	76	0.99 ⎭ 1.91

[a]Antenatal screening was not generally available in the Northern Region before 1976.

The motivating factor behind the programme for universal antenatal screening was a recognition that neural tube defects were not only common, but also caused a lot of distressing (and expensive) long-term disability. With this end in view the Northern Regional Survey committed itself from the outset to measuring not only the birth prevalence and the mortality but also the amount of long-term disability in survivors with hydrocephalus, an encephalocele or spina bifida (with or without hydrocephalus). The survey data now available suggest that the region's policy of relying almost exclusively on ultrasound screening (only about 15% of mothers in the region are offered AFP screening) has been more successful than is generally appreciated.

In any analysis of neural tube defects it is important to separate out babies with anencephaly from babies with lesser degrees of the same basic disorder, because anencephaly seems to be better identified by routine antenatal ultrasound screening than by AFP screening. The few cases of anencephaly not terminated in the Northern Region in the last five years (Fig. 2.5) were almost all associated with a failure to present for antenatal care before the third trimester of pregnancy, or associated with twin pregnancy.

The diagnosis of other, lesser, neural tube defects is less consistently achieved by routine ultrasound examination at 17–18 weeks' gestation, but in the years covered by the survey no pregnancy terminated in the belief that there was a neural tube defect did not have an abnormality meriting termination. Furthermore, although the number of cases terminated has remained much the same

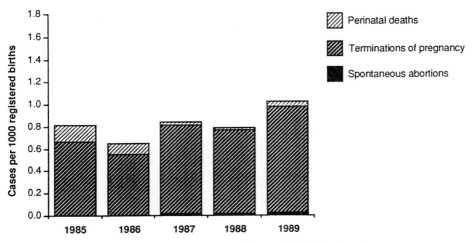

Fig. 2.5. Incidence of anencephaly 1985–1989 (165 cases).

each year (Fig. 2.6), the proportion terminated has doubled. More importantly, follow-up has shown that the majority of the 71 long-term survivors have no serious disability (Fig. 2.7) and these cases should not, therefore, be looked on as "failures" of the screening process. It seems clear that the spinal lesions that are hardest to detect antenatally are the least likely to be associated with disability.

The unrelated condition of isolated hydrocephalus is the other major central nervous system (CNS) defect for which routine population screening is clearly indicated. AFP screening is of no use in this moderately common condition (birth prevalence 0.26 cases per 1000 births). Most cases presenting early are now terminated, but two-thirds of all cases seem to manifest themselves only in the

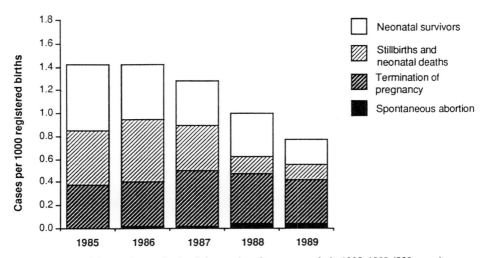

Fig. 2.6. Incidence of neural tube defects other than anencephaly 1985–1989 (239 cases).

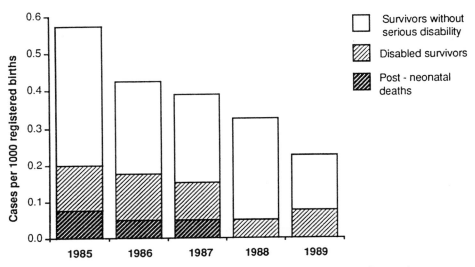

Fig. 2.7. Neonatal survivors with a neural tube defect 1985–1989 (78 cases).

third trimester of pregnancy. Perinatal mortality in these babies is high, but only one of the six long-term survivors born in the last five years with congenital hydrocephalus is currently disabled. As with neural tube defects, birth prevalence and perinatal mortality are misleading indices of the amount of long-term disability in the community.

Selective Screening with a View to Termination

The data on chromosome abnormality are presented here because the birth prevalence of trisomy 21 is a good index of the amount of consequential long-term disability in the community. The diagnosis can only be made antenatally by amniocentesis or chorion villus biopsy which are not without risk, and screening has therefore been limited until now to mothers who are more than 34 years old, because from this age the benefits of screening start to justify the risks.

From data available to the register it can be seen that selective screening has not been as successful as might have been hoped (Table 2.6). Others have had a similar experience [8]. Nevertheless a review that encompasses all trisomy, or all

Table 2.6. Chromosome abnormality and its antenatal recognition in the Northern Region 1985–1989

Defect	Number of cases[a]	Incidence per 1000 registered births	Proportion of all affected pregnancies terminated (%)	
			All mothers	Mothers >34 years
Trisomy 21	308	1.54	16	35
Other trisomy	76	0.38	41	86
Other chromosome defect	131	0.66	34	53

[a] Excluding spontaneous abortion.

structural chromosome abnormality, rather than just trisomy 21, shows that the total yield from the present policy is higher than a simple review of the birth prevalence of trisomy 21 might suggest. What also comes out of the expanded database available to the regional register is some understanding of why the uptake of selective screening for mothers over 34 years is currently incomplete. A few mothers booked too late for screening to be possible, some were not prepared to contemplate termination of pregnancy for a fetal condition that was not in itself lethal, and others were unhappy at the prospect of amniocentesis merely to rule out what had been presented to them as a relatively unlikely outcome anyway. In half the cases, however, there was no evidence that the issue of screening had ever been put to the mother. Such insights may help us to improve selective screening. They will be of even more importance when an attempt is made to expand the present selective screening programme to a programme that covers every pregnancy [9].

Our recent regional analysis also highlights the danger of relying on the present national voluntary reporting scheme for congenital defects run by OPCS and set up by the Government in the wake of the thalidomide disaster. Even with an abnormality as rapidly identifiable after birth as Down's syndrome (trisomy 21), more than a third of all cases never get reported to OPCS at present, and some that do are "false positive". Our own regional register is complete only because all returns are checked against the information files held by the region's three cytogenetic laboratories. In the light of this it is understandable why special funding should have been sought to set up a separate national reporting system for trisomy 21 that links into all the main cytogenetic laboratories in the country. There are, nevertheless, dangers and disadvantages inherent in the development of multiple different "ad hoc" monitoring schemes for gauging the impact of differing antenatal screening programmes. We believe very firmly that a single comprehensive regionally based system with enough funding to be reasonably accurate and complete has the advantage of being able to produce regular, prompt, local feed back, and that without such feed back the whole raison d'être of monitoring is lost.

The Interpretation of Selected Ultrasound Problems

The usual way of looking at the problem of congenital abnormality is to start with the number of children confirmed as having a specific abnormality after delivery, and then to look retrospectively to see if anything had been suspected before delivery, and, if so, what. What the obstetrician or ultrasonologist needs to know, however, is what the true diagnosis (and therefore the ultimate prognosis) is likely to be when a certain ultrasound appearance is detected in the antenatal period, and this is something that the Northern Region's Fetal Abnormality Survey uniquely makes possible.

We have chosen cystic hygroma [10] as an illustration. Almost all the 55 antenatally diagnosed cases (0.27 per thousand registered births) were first suspected on ultrasound examination at 16–20 weeks' gestation. A total of 33 of the pregnancies were terminated, a further nine fetuses died during the second trimester, four aborted spontaneously, and three were stillborn. One mother had Noonan's syndrome (a known risk factor). Two children correctly diagnosed as having cystic hygromatous change limited to the head and neck were liveborn but only one, with modest cystic changes, survived. One child suspected of having a small discrete cystic hygroma of the neck turned out to have a small cervical meningocele. This child is alive and well and has no neurological deficit. The only other neonatal survivor was registered as having definite but limited hygromatous changes at 18 weeks' gestation, but nothing was done because these changes resolved. The child was found to have classical Down's syndrome at birth (with no sign of any hygromatous change) but died from congestive heart failure with an atrioventricular septal defect at 6 weeks.

Of the cases in this series 40% were shown to have a chromosome anomaly (Table 2.7), and this must be a minimum estimate of the true incidence, because some cases were too macerated for chromosome analysis at delivery, and in some cases such an examination was never attempted. However, at least six fetuses with cystic hygromatous change, but no other structural abnormality, had a normal chromosome karyotype. Caution is necessary in recommending termination of pregnancy on the basis of a single ultrasound examination even where the scan appearance is typical, especially if the changes are not extensive, because in three of the 55 cases in this series there was no detectable abnormality on full autopsy examination after delivery. In some cases such changes are only a transient feature. On the other hand, clinicians need to be alert to the fact that transient change seems to be particularly common in trisomic pregnancy.

Table 2.7. Final diagnosis after delivery in 55 fetuses found to have cystic hygromatous changes antenatally

Cystic hygromatous changes only	22
Turner's syndrome with cystic hygroma	13
Trisomy (13, 18 or 21) with cystic hygroma	9
No abnormality at the time of autopsy	3
Generalised hydropic changes only	2
Pulmonary cystadenoma with oedema	2
Cystic hygroma with a limb–body wall defect	1
No abnormality except cleft lip at autopsy	1
Small cervical meningocele only	1
Iniencephaly only	1
Total	55

It is interesting to note that not a single case was identified for the first time after birth. This means that many fetuses miscarrying at 20–26 weeks' gestation are not currently subject to detailed autopsy examination [11], that too few of these cases are subjected to chromosome analysis, and that it is all too easy to confuse hygromatous change for hydropic change or maceration in a mid-trimester fetus.

Discussion

Since the dawn of modern medicine doctors treating patients have had little idea of what they are doing to whom, and with what effect. There has always been the dilemma of using inadequate resources either to treat patients or to analyse practice. The challenge of obstetrics is to sense and anticipate problems early enough to take pre-emptive action when progress is not straightforward so as to avoid being drawn into crisis management.

To this end it has long-been traditional to monitor maternal health throughout pregnancy. All antenatal care is based on a belief that screening for early evidence of abnormality can reduce the hazard of childbirth. Current policies and routines have grown empirically over the years, however, and relatively little has been done, until recently, to assess which of the many aspects of current antenatal practice are truly cost effective [12].

Increasingly we now have the opportunity to monitor certain aspects of fetal health, as well as maternal health, during pregnancy. Such monitoring is expensive and labour-intensive, however, and we urgently need to identify the best way of utilising the limited resources available to us. Audit is, in any case, the current watchword. We cannot afford the luxury of continuing to employ unvalidated screening routines, and we cannot assume that what is shown to work in a specialised research setting, with skilled and motivated research staff, still works reliably when employed in everyday clinical practice. It is possible that some screening programmes currently do more harm than good. Antenatal diagnosis does not always increase a child's chance of survival. Liveborn children with an uncomplicated abdominal wall defect, for example, have an excellent chance of survival and a negligible risk of long-term disability, but antenatal diagnosis can bring with it ill-justified pressure for the pregnancy to be terminated (Table 2.3).

The voluntary collaborative register now in operation in the Northern Region makes it possible to monitor, for the first time, the efficacy of the antenatal and perinatal techniques now being developed to screen for significant fetal abnormality (structural, metabolic or genetic) both during pregnancy and in early infancy. The register also provides clinicians with an invaluable, continually updated, confidential database summarising a whole region's collective experience of dealing with complex fetal abnormality. Access to such a confidential reference file of past experience can be of great help to a clinician faced with an uncommon problem. The information now available on file certainly helps clinicians to know how much weight to give to individual antenatal findings.

Nothing of what has been achieved by the study would have been possible without the corporate goodwill and support of midwives, clerical staff and maternity unit consultants across the region. To them, rather than the steering group, should go the credit for any success achieved.

Appendix

Only with an integrated, clinically managed, register of the type now operational in this region is it possible to piece together case histories such as the following.

A fetal scan at 17 weeks suggested the presence of some posterior cervical hygromatous changes, but a scan 2 weeks later was within normal limits so no further action was taken. The heart appeared normal and mother was 25 years old (retrospective report from the ultrasonologist involved). Trisomy 21 was diagnosed at birth (report from the paediatrician) and the genetic laboratory later reported this to be a translocation trisomy. The child was then sent to a paediatric surgeon at 4 days and eventually shown to have Hirschprung's disease. A paediatric cardiologist later confirmed the child as having an asymptomatic acyanotic atrioventricular septal defect following a fresh referral by the general practitioner at 6 weeks. The child was then unfortunately reported as a cot death at 6 months (notification submitted by a consultant community paediatrician) and finally established as having died with severe bronchopneumonia (coroner's autopsy report made available to the register after it was submitted to the Regional Perinatal Mortality Survey).

Seven different clinical specialities based in four different hospitals in three different health districts knew of part of what happened to that poor unfortunate baby, but only a collaborative joint register that had earned the trust and support of local clinicians was in a position to piece the jigsaw together.

Acknowledgments. This study would not have been possible without the pivotal role played by Mrs Marjorie Renwick, the survey coordinator, in collecting, collating and analysing the wealth of information acquired from the 24 maternity units serving patients in the Northern Region; and the work done by Miss Katherine Denham in programming all the data into the Region's mainframe computer.

References

1. Northern Regional Health Authority Coordinating Group. Perinatal mortality: a continuing collaborative regional survey. Br Med J 1984; 288:1717–20.
2. Knox EG, Armstrong EH, Lancashire R. The quality of notification of congenital malformations. J Epidemiol Community Health 1984; 38:296–305.
3. Office of Population Censuses and Surveys. Abortion Statistics 1985–1987. Reports AB12, 13 and 14. London: HMSO, 1986–8.
4. Maxwell DJ, Crawford DC, Curry PVM et al. Obstetric importance, diagnosis and management of fetal tachycardias. Br Med J 1988; 297:107–10.
5. Lenke RR, Hatch EI. Fetal gastroschisis: a preliminary report advocating the use of caesarean section. Obstet Gynecol 1986; 67:395–8.
6. Owens JR, Harris F, McAllister E et al. 19-year incidence of neural tube defect in an area under constant surveillance. Lancet 1981; ii:1032–5.
7. Carstairs V, Cole SE. Spina bifida and anencephaly in Scotland. Br Med J 1984; 289:1182–4.
8. Walker S, Howard PJ. Cytogenetic prenatal diagnosis and its relative effectiveness in the Mersey Region and North Wales. Prenat Diagn 1986; 6:13–23.
9. Wald NJ, Cuckle HS, Densen JW et al. Maternal serum screening for Down's syndrome in early pregnancy. Br Med J 1988; 297:883–7.
10. Edwards MJ, Graham JM. Posterior nuchal cystic hygroma. Clin Perinatol 1990; 17:611–40.
11. Clayton-Smith J, Farndon PA, McKeown C et al. Examination of fetuses after induced abortion for fetal abnormality. Br Med J 1990; 300:295–7.
12. Hall MH (ed.) Antenatal care. Baillieres Clin Obstet Gynecol 1990; 4:1–232.

Discussion

Drife: Does the uniqueness of the Northern Region mean that nobody else could do a similar kind of study?

Atkins: I wish I knew the answer to that but I should like to open the challenge to people to try. Certainly because of some of those unique factors we have always tended to work as a cohesive whole.

Rodeck: Recently North-West Thames Region has started a similar survey. We have not seen any of the results, but the news I hear is very encouraging.

Does Mr Atkins think the OPCS system should be abandoned and that the funding that has been used for the congenital malformations part of it should be diverted to a multiple series of regional surveys?

Atkins: That is a very interesting suggestion. Certainly our survey was disappointed in the efficiency of the OPCS system and if other systems are working they clearly would be better. The great difficulty we had was to find sufficient resources.

Lilford: Here we are quite properly concerned exclusively with secondary prevention. But I think we should also establish whether there is anything firm that we can say about primary prevention, and whether for instance there is any new knowledge pertinent to neural tube defects (NTDs) or anything about occupational hazards which can be used to give definitive advice.

Wald: Professor Nevin described a number of congenital abnormalities the birth prevalence of which either remained more or less the same or declined. Are there any congenital abnormalities that have shown increases over the past 15 or 20 years in birth prevalence? For example, what about testicular abnormalities or hypospadias?

Nevin: We have no evidence in our own data that there has been an increase. One or two communities, Hungary for example, are now suggesting that there might be an increasing prevalence of NTDs, but we have had no evidence of an increase other than what could be attributable to an improved diagnosis or improved autopsy rates, e.g. congenital heart defects. I know that in the Scandinavian countries there has been some suggestion of increasing rates of hypospadias, but looking at the European data and our own data we have no evidence of any increased rate of those anomalies.

One of the problems, and it comes back to the question about OPCS, is that they have no mechanism for validating what is put into them, and this is particularly important in areas of genitourinary (GU) anomalies.

Wald: I believe there is some concern that there is an increased prevalence of undescended testicle that has also been linked to increases in the rate of testicular cancer later in life. There is speculation as to whether this is real and whether there might be some causal factor.

Nevin: Many of the systems for monitoring defects may not include what they call minor anomalies, such as undescended testicle, inguinal hernia, or epispadias. It all boils down to resources to cope with the volume of data.

Campbell: I am still not 100% convinced that there is a true fall in prevalence in terms of NTDs, and that it is not principally due to prenatal diagnosis. The principal reasons are: first that NTDs are the only abnormality where there is a widespread screening test. None of the others seems to have fallen – only the one with the widespread screening test.

Second, anencephaly showed the most profound fall and the earliest fall. Anencephaly is quite obvious on ultrasound screening and is more easily diagnosed with α-fetoprotein (AFP) screening, and as there is no register of terminations I suspect perhaps some of the patients from Northern Ireland will have come over to the mainland for their terminations of pregnancy.

With these points in mind, can Professor Nevin reassure me that there is a true fall in prevalence, not reflecting prenatal diagnosis?

Nevin: I am absolutely convinced it is a true fall. First of all, we have no routine serum AFP screening. There is ultrasonography. There is no situation where anencephaly or NTD is diagnosed on ultrasonography where the mother would have to go to the mainland. Terminations for NTDs would be carried out in any obstetric unit.

Second, in terms of prenatal diagnosis, since 1969 we have been running a central genetic obstetric clinic where most of the antenatal screening would be done – ultrasonography, amniocentesis, etc.

I am absolutely convinced that this is a true fall and that only a small proportion of that fall is attributable to antenatal screening.

Campbell: How much of the fall in prevalence is due to this natural fall and how much to prenatal diagnosis?

Nevin: In round figures I would recknon that no more than 15%–20% of that fall in Northern Ireland is attributable to antenatal screening.

R. Harris: If with one possible exception there has been no increase in prevalence of any birth defect or chromosomal defect, can one take reassurance that the worries of the Green Party are perhaps not reflected in birth defects, or could we review the other possibility, that spontaneous early abortion rates are obscuring true frequencies?

Whittle: If I could come back to the spina bifida/anencephaly story. The data from the West of Scotland AFP screening programme showed that there are two trends, both of which result in a downward incidence of delivery rate for anencephaly and spina bifida. There is clearly an impact from screening by whatever method (there, it is AFP) that there is a downward trend in the incidence of the condition generally. I could not put figures on which was the bigger influence, but glancing at the figures it would look as if in that particular population screening was having a bigger impact than any secular downward trend in NTD.

Wald: In England and Wales in a survey that Howard Cuckle and I performed, supported by the Department of Health, the majority of the decline was due to screening and antenatal diagnosis by a mixture of ultrasound and AFP. Giving an exact proportion is extremely difficult. One thing that did come to light, and I believe it was also the Scottish experience, is that notification of induced abortions is far from complete even though there is a statutory requirement, and the extent of that under-reporting is unknown and could be much larger than people suspect. So, any study that looks at births plus the notified induced abortions would not be expected to reach the expected birth prevalence in the absence of intervention.

Atkins: Our figures include, as far as we know, all the NTDs and showed a marked fall in those other than anencephaly. With a termination for NTD someone has to make the diagnosis and the ultrasonographer will notify the case automatically. So even if the termination is not notified, it gets picked up.

The surprising thing was anencephaly, which seems to be running quite level for the last five years. But that is over 200 000 births, perhaps not enough.

J. Harris: Are any steps taken to tell patients of the false positive and false negative rates affecting centres where testing is done? Are mothers or families told that the diagnosis in the test has been confirmed or otherwise on post-mortem examination?

Atkins: I can answer that for us but it would have to be done specifically for each of the problems.

Take, for example, NTDs. We have an "opt in" system; the mother is asked if she wants such a test. We do AFPs. If she gets an alert factor on an AFP or if she gets a suspicious ultrasound, she is very carefully counselled. She is told, for example, that there is no such thing as a 100% diagnosis. She knows the risks both ways and she and her family have talked through it. If she elects for termination she is usually shown the baby, and most mothers want to see it. Usually a counselling nurse will go round to her house a few days later and there is a formal review at 4 to 6 weeks where the details are discussed. But each district will have its own answer to that question.

Trudinger: Dr Atkins presented a very full account of the problems of prenatal diagnosis and indicated the lack of evaluation of a programme. Apart from education, what changes would he like to see in his region on the basis of the information he presented? I had the feeling that perhaps prenatal diagnosis was being dispensed in the wrong way.

Atkins: I would like to see the survey continue and I think I would see a response in how people actually work in the district general hospitals, if we can give them regular alerts. For example, such things as "Beware of the transient hygromatous change", or "You do not have to do a C-section for this or that condition, but it is worth (or not worth) sending that patient to a centre for delivery".

I am happy with it as it is developing, but I hope it stays on the rails, otherwise initial enthusiasm will run out of steam.

Williamson: I would like to comment on Professor J. Harris's questions as far as they are relevant to DNA analysis of thalassaemia, cystic fibrosis, Duchenne muscular dystrophy and Friedreich's ataxia. Certainly in these cases the question of false positives does not arise particularly but the question of false negatives does arise and the patients in my experience are suitably counselled. Second, the DNA analysis is normally confirmed on products of conception wherever possible although we do not always get material to do so. But where it is possible it is done, and if there is a question from the family we tell them.

But there is an implication in that question that there is a major problem here and Mr Atkins identified a very much greater problem. In my experience at least half, if not more than half of parents at risk are not appropriately counselled as to the availability of these facilities. That, it seems to me, is a much greater problem than a very marginal problem which occurs occasionally in relation to those who are counselled.

Drife: We shall be discussing counselling later.

Chapter 3
Routine Fetal Anomaly Screening

M. J. Whittle

Introduction

Probably at least 90% of women in the UK undergo an ultrasound examination at some time during pregnancy. The vast majority of these examinations are in the early weeks to establish dates and viability, or in later pregnancy to determine fetal size or placental site. As the standard of equipment available has improved and the technical skill of the operators has advanced, the ability to visualise fetal structures has increased to the extent that routine screening for anomalies has become feasible. Indeed it is the view of some that the stage has been reached where a routine anomaly scan may be considered almost mandatory.

There can be little doubt that some form of prenatal screening is necessary. In most studies of perinatal mortality, between 20% and 25% of the losses are due to congenital anomaly. In addition, congenital anomaly is responsible for the death of about 25% of the babies lost in the first year of life and approximately 15% of those dying between the ages of 1 and 4 years [1]. Morbidity is also important and serious mental and physical handicap may coexist with congenital defect.

One of the best examples of the impact of a screening programme on congenital anomalies is found in the maternal serum alphafetoprotein assessment, which, in the West of Scotland, has led to a steady fall in the incidence of babies born with anencephaly and open spina bifida. The effect of ultrasound screening for structural anomalies is less clear but this may be for a number of reasons. First, the condition being sought may be difficult to detect and therefore missed by the screening technique. For example, in screening the fetal heart the identification rate for anomalies is probably around 50%–60% and only the severest conditions or those associated with other anomalies may be found. Second, the condition may not always be detectable, or even present, at the screening period but may appear later. Examples of this include heart problems and some renal conditions and also certain forms of microcephaly.

The feasibility of prenatal screening depends on the ability of the technique to identify lesions, a clear idea of the objectives of screening and the difficulties of organisation and implementation.

Identification

Before it is possible to establish whether ultrasound can identify anomalies it is necessary to determine the reliability of the technique to confirm the normality of structures, and the stage of pregnancy at which this is best achieved.

A list of the structures that should be visible by about 20 weeks is shown in Table 3.1, but it is important to realise that not all will be seen in every case every time. Indeed the ability of ultrasound to identify normal structures was assessed [2] by determining how often it was possible to gain clear views of certain organs in over 2000 pregnancies screened between 14 and 27 weeks (Table 3.2). It is apparent that some structures were more easily seen than others, the bladder and stomach being almost always noted whereas a four-chamber heart was clearly

Table 3.1. Structures that should be visible on ultrasound scanning by 20 weeks' gestation

Head	– Cerebral ventricles
	– Ventricular/hemispheric ratio
	– Cerebellum
	– Choroid plexus
Face and neck	– Lips, palate, nose, chin
	– Orbits
	– Muscular insertion on eyeball
	– Nuchal thickening
	– Nuchal cysts
Spine	– Longitudinal/transverse section
Thorax	– Heart size, shape
	– Four chamber view, outflow tracts
	– Pleural cavities, chest shape
	– Diaphragm
Abdomen	– Stomach
	– Renal outlines, bladder, genitalia
	– Anterior abdominal wall, cord
Limbs	– Arms, legs
	– Hands, feet

Table 3.2. Frequency of visualisation by ultrasound of individual structures

Organ	% Visualised
Head	93–94
Heart	30–33
Stomach	92–97
Kidneys	80–87
Bladder	80–100
Diaphragm	70–80
Spine	70–83
Extremities	80–93

Adapted from Zodor et al. [2].

observed in only about 30%–40% of cases. Interestingly, the spine was seen satisfactorily in only about 70%–80% of cases. Other factors may reduce the ability of ultrasound still further and, for example, in a later paper, the same group [3] showed that maternal obesity had a significant effect on the visualisation of nearly all structures.

Although it is vital that normal structures can be reliably identified, the problem of observing the derangement of normal anatomy is different. For example, although a normal fetal kidney may be difficult to see at 20 weeks' gestation there should be no problem in identifying one which is multicystic. Clearly, some anomalies will be recognised more easily than others so that conditions such as anencephaly, large cystic hygromata and massive bladder distension arising from urethral agenesis are unlikely to be missed even in early gestation.

It would seem possible that the vast bulk of anomalies will be best seen at about 18–20 weeks (Table 3.3). Rizzo (personal communication) found that about 40% of anomalies would be identified at 20 weeks compared with about 15% at 16 weeks; this same group found that about 30% of anomalies were not discovered until delivery. The optimum time at which to perform a scan is unclear, although anomalies will be identified with greater reliability as gestation advances. In a study looking for markers of Down's syndrome only 25% were seen at 14–24 weeks whereas 75% were seen after this time [4]. The optimal time for screening must take into account not only the stage at which a lesion is most likely to be seen but also the limitations imposed by the law relating to abortion.

The type of anomaly has an important influence on when it is likely to be identified, and it is difficult to quantify the accuracy for all individual anomalies. The pathogenesis of some lesions is not clear and indeed ideas about the evolution of several conditions are only just emerging. For example, it seems possible that in a proportion of fetuses, cystic hygromata which are detected early will resolve leaving the baby apparently normal [5]. Conversely there is evidence that certain types of renal pathology may not appear until the third trimester. Thus, in a Swedish study only 9% of renal anomalies were detected at 17 weeks compared with 91% at 33 weeks [6]. In a more recent study the sensitivity of a scan at 28 weeks to identify renal problems was at least 88% and may be found to be higher with more prolonged follow-up [7].

Table 3.3. Time of diagnosis for congenital anomaly

Antenatal		
Early	Up to 13–14 weeks	Anencephaly
		Cystic hygromata
		Bladder obstruction
		Body stalk
Mid-pregnancy	18–22 weeks	Neural tube defect
		Hydrocephalus
		Bladder obstruction
		Anterior abdominal wall defects
		Diaphragmatic hernia
		Duodenal atresia
		Cardiac lesions
		Dwarfism
		Clefting
		Multiple abnormalities
Late pregnancy	30 weeks	Mild/moderate urinary obstruction
		Reflux
		Hydrocephalus
		Cardiac lesions
		Hydroceph/microcephaly
Postnatal		
Neonatal		Cardiac lesion
		Urethral valves
		Ureteric reflux
		Hydrocephalus
		Microcephaly

The Objectives of Ultrasound Screening

Reduction of Perinatal Mortality and Morbidity

As mentioned above, congenital anomalies contribute significantly not only to perinatal loss but also to morbidity, and anything that will reduce this should be welcomed. The majority of anomalies occur in apparently low-risk groups and so universal screening is the only method by which they can be identified. The implication of screening is that if a serious anomaly is found termination of the pregnancy will be offered and this particular aspect should be carefully explained to the mother at the time of the scan.

Termination of the pregnancy is an important issue in screening and although the decision is an easy one when the lesion is clearly incompatible with life, as for example with anencephaly or renal agenesis, things may be less certain with some cardiac lesions or, to go to an extreme, with facial clefting. Here the purpose of screening becomes less clear since all it might achieve is the generation of worry and anxiety in the parents, and the advantages of prenatal warning may seem of dubious value.

Probably the greatest value of screening is the identification of lesions which lead to long-term morbidity, and probably the two best examples of this are spina bifida and serious cardiac anomalies. The efficacy of screening in the detection of spina bifida has been assessed and although specificity and sensitivity can be around 100% in units dedicated to scanning [8], things are less clear for departments in which, perhaps, equipment and staffing levels are less favourable. In fact the data are not available on a national basis so it is impossible to provide figures, although a prospective study by the Medical Research Council is currently in progress to assess the efficacy of routine screening in the detection of neural tube defect.

The value of screening for cardiac lesions is even harder to assess but such data as are available suggest that about half of the most severe conditions will be identified. In fact severe cardiac disease will in general be associated with other lesions in about 40% of cases [9]. The value of prewarning parents and attendants about the presence of a cardiac anomaly is debatable, though, since heart problems may not become apparent until some time after birth, it seems desirable. The issue of termination of pregnancy for heart conditions is also controversial, especially as many are potentially capable of correction. However, there is less debate when the heart condition is associated with other lesions and also when it is a marker for more serious abnormality such as trisomy.

Severe renal abnormality will also be useful to diagnose in the fetus, and non-survivable conditions such as renal agenesis and renal polycystic disease will be indications for termination. A failure to diagnose these conditions will often lead to difficulties in late pregnancy, when the mother may present with a small-for-dates uterus, a fetus with oligohydramnios and a presumptive diagnosis of severe growth retardation. If labour ensues, fetal distress is almost inevitable and a pointless caesarean section will be performed for a baby with a fatal condition.

Evidence that ultrasound screening would lead to a reduction in perinatal mortality is not extensive. In a study of over 9000 patients, half of whom were screened between 16 and 20 weeks, perinatal mortality in the screened half was 4.2/1000 compared with 8/1000 in the controls. Some of this reduction arose from a more accurate estimation of gestational age in the screened group and a precise diagnosis of multiple pregnancy. However, about half the reduction was accounted for by fewer anomalies [10], even though half the anomalies were missed, there being no terminations for fetal abnormality in the control group.

The Potential for In Utero Treatment

One of the potential benefits of screening for anomalies is that it may allow the identification of a group of babies for whom treatment in utero, usually in the form of shunting procedures, may be appropriate. A review of cases in which shunts had been used both in hydrocephalus and in obstructive uropathy suggested that shunting in the former was not helpful whereas in the latter some improvement was observed [11]. The matter is controversial but one of the problems with the available literature is that clearly defined diagnostic groups have not been established, which makes comparisons impossible. Considering posterior urethral valve problems alone, one study [12] has suggested a 92% survival with conservative management compared with a 76% survival in cases undergoing in utero shunting [11].

More aggressive in utero treatment in the form of extrauterine surgery is still experimental but the successful repair of a diaphragmatic hernia has been reported [13]. Screening would bring some such cases to light although whether the risk of the surgery to the mother is justifiable remains to be seen.

The Identification of Conditions Amenable to Neonatal Surgery

When a structural anomaly occurs as an isolated lesion and is deemed survivable the option of continuing the pregnancy should be considered. Prior knowledge that the baby is abnormal but has an operable condition may confer some advantage, but there is no real evidence to confirm this. Experience with anterior abdominal wall defects has failed to show any benefit from antental diagnosis in a small personal series. The idea that diagnosis would allow the selection of those cases best delivered by caesarean section has not been borne out.

Forewarning is probably useful in some conditions, but whether this significantly influences outcome is unclear. However, when the fetus has been found to be abnormal there must be an advantage in the transfer of the mother for delivery to the hospital with the necessary intensive care nursery and nearby surgical expertise. Diaphragmatic hernia provides an example since it is a condition which can be difficult to diagnose soon after birth but one in which early intubation and ventilation may produce the best results. Other chest conditions such as effusions and cysts may also be helpful for the paediatrician to know about ahead of time.

More recently it seems that evidence of mild hydronephrosis in the fetus may indicate a group of babies at risk of subsequently developing reflux and infective damage as children. In one study [7] of 92 babies with abnormal fetal renal scans, 42 had anomalies confirmed postnatally. These investigators considered that about 55% of the babies had benefited from antenatal diagnosis because it allowed early postnatal surgical intervention or chemoprophylaxis. This was confirmed in a similar study in which it was estimated that prenatal warning of a renal problem had been of either definite or probable value in the subsequent management of 75% of the babies born with renal disease but without clinical manifestations [14]. It should be noted, however, that the false positive rate seems high so the unnecessary anxiety generated in the parents whose baby ultimately proves to be normal needs to be taken into account.

Organisation of a Screening Programme

Before a screening programme is adopted it is essential to establish that sufficient numbers of trained staff are available and that there is adequate equipment. It seems likely that 15–20 minutes would be required for each scan and a reasonable number of the mothers would need to be recalled because of inadequate or suspicious views. The Royal College of Obstetricians and Gynaecologists and the Royal College of Radiologists recommended that a medically trained, experienced ultrasonographer should be always available and preferably working in the department at the same time as the radiographers. To this end the two Colleges have established advanced training programmes in many departments throughout the country in an attempt to raise the standards of ultrasound departments. One unpublished series shows the steady improvement over a number of years in the rate of identification of anomalies, with a final pickup rate of about 75% (Table 3.4). However, the efficacy of scanning is strongly operator-dependent

Table 3.4. The improved rate of identification of anomalies with increasing experience; Hillingdon Hospital, London

	1985	1986	1987	1988	1989
Total number of anomalies	4	12	18	19	27
Terminations	1	8	12	7	15
Births	2860	2843	3137	3341	3076

With grateful acknowledgements to Dr D. Shirley.

and success rates may vary widely so that in one study the identification rate for fetal anomaly ranged from 36% in a city hospital to 76% in a university department [10]. Although the reasons for these differences may be multifactorial they must be considered important in the evaluation of a universal policy of screening by ultrasound.

The timing of the screening scan is uncertain and to some extent depends on the objectives. Early scanning, before 16 weeks, will identify many of the most serious anomalies, most of which will be fatal and suitable for termination. In many ways the earlier these diagnoses are made the better, and ideally they should be before 13 weeks' gestation, when suction termination would be feasible; gross anomalies such as anencephaly can certainly be confirmed by this time.

However, if the aim is to obtain a general view of the fetal anatomy then a scan between 18 and 22 weeks is probably ideal, although the severe fetal anomalies, capable of earlier diagnosis, will then require late termination. Further, if a single scan is performed at a stage selected on menstrual history alone the dates are likely to be out by two weeks in about 15%. This will usually mean the mother is scanned too early but occasionally it will be too late. In practical terms, it seems to be the general experience that about 60%–70% of women will have a scan at booking in any case, because of uncertainty of dates or because of complications such as bleeding.

Thus it seems likely that in the organisation of a screening scan, account must be taken of the need for booking scans as well as those examinations which have to be repeated to obtain more adequate or confirmatory views. In addition the possibility that a later scan may be necessary to identify renal tract anomalies means that screening programmes may produce a considerable increase in the workload of a department.

The cost effectiveness of a screening programme has yet to be assessed. It seems likely that about half the 2% of babies born with a structural anomaly will be found by a screening scan so that a unit delivering 4000 babies a year will expect to identify about 40 anomalies. Based on other studies, about 50% of these anomalies will involve either the central nervous system or cardiovascular system and a good proportion will probably be fatal. Others, such as cystic hygromata or duodenal atresia will provide markers of karyotypic abnormalities which may in themselves have serious morbid potential.

The degree of anxiety produced in those couples told that their baby has a potentially non-fatal condition is uncertain, but there may be positive benefits to advance knowledge which if managed sensitively can prepare the parents for what will happen following the birth of their baby.

Conclusions

The evidence that second trimester screening for fetal anomalies is a cost effective exercise which should be generally adopted is not available. There can be little question that a routine scan in the first half of pregnancy is essential in modern-day obstetrics since not only might it identify a serious fetal anomaly but it also enables the viability of the fetus to be checked, dates established and multiple pregnancy confirmed.

Before a policy of universal screening by ultrasound is adopted it is necessary to examine other possible regimes which may, in fact, be more cost effective because they place a less severe demand on manpower and equipment. Such programmes include maternal serum alphafetoprotein (MSAFP) estimation, ideally preceded by a "booking" scan. Detailed ultrasonography could then be restricted to those pregnancies with a raised MSAFP result, but this approach will usually not allow the diagnosis of conditions other than those involving neural tube lesions or anterior abdominal wall defects.

The purpose of ultrasound screening should be clearly defined. Its greatest value is undoubtedly the potential to reduce perinatal mortality and morbidity but the evidence that it does so is scanty. The number of babies with anomalies which may be suitable for intrauterine therapy is small and the efficacy of this treatment is debatable. Whether antenatal diagnosis may help in providing suitable postnatal treatment is uncertain although there are good reasons to expect that, in some circumstances, it might. The antenatal diagnosis of renal tract anomalies may select out a group of babies with no other markers of renal disease who may be at risk of developing serious problems later in life.

There is a need for a large randomised study to assess the efficacy of anomaly scanning in pregnancy. If this demonstrates that there are clear advantages its implementation should be strongly recommended even though the cost implications may be very large.

References

1. Registrar General (Scotland). Death Registration Tapes.
2. Zador IE, Bottoms SF, Tse GM, Brindlay BA, Sokol RJ. Nomograms for ultrasound visualization of fetal organs. J Ultrasound Med 1988; 7:197–201.
3. Wolfe HM, Sokol RJ, Martier SM, Zador IE. Maternal obesity – a potential source of error in sonographic prenatal diagnosis. Obstet Gynecol 1990; 76:339–42.
4. Nyberg DA, Resta RG, Lathy DA, Hickok DE, Mahoney S, Hirsch JH. Prenatal sonographic findings of Down's syndrome – Review of 94 cases. Obstet Gynecol 1990; 76:370–7.
5. Abramowicz JS, Warsof SL, Doyle DL, Smith D, Levy DL. Congenital cystic hygroma of the neck diagnosed prenatally: outcome with normal and abnormal karyotype. Prenat Diagn 1989; 9:321–7.
6. Helin I, Persson PH. Prenatal diagnosis of urinary tract abnormalities by ultrasound. Pediatrics 1986; 78:879–83.
7. Livera LN, Bookfield DSK, Egginton JA, Hawnaur JM. Antenatal ultrasonography to detect fetal renal abnormalities: a prospective screening programme. Br Med J 1989; 298:1421–3.
8. Van de Hof MC, Nicolaides KH, Campbell J, Campbell S. Evaluation of the lemon and banana signs in one hundred and thirty fetuses with open spina bifida. Am J Obstet Gynecol 1990; 162:322–7.
9. Allan LD, Crawford DC, Chita SK, Tynan MJ. Prenatal screening for congenital heart disease. Br Med J 1986; 292:1717–19.

10. Saari-Kemppainen A, Karjalainen O, Ylostalo P, Heinoneu OP. Ultrasound screening and perinatal mortality: controlled trial of systematic one-stage screening in pregnancy. Lancet 1990; 336:387–91.
11. Manning FA, Harrison MR, Rodeck CH. Catheter shunts for fetal hydronephrosis and hydrocephalus. N Eng J Med 1986; 315:336–40.
12. Arthur RJ, Irving HC, Thomas DFM, Watters JK. Bilateral fetal uropathy: what is the outlook? Br Med J 1989; 298:1419–20.
13. Harrison MR, Adzick NS, Longaker MT et al. Successful repair in utero of a fetal diaphragmatic hernia after removal of herniated viscera from the left thorax. N Engl J Med 1990; 322:1582–4.
14. Greig JD, Raine PAM, Young DG et al. Value of antenatal diagnosis of abnormalities of the urinary tract. Br Med J 1989; 298:1417–19.

Chapter 4

Some Practical Issues in the Antenatal Detection of Neural Tube Defects and Down's Syndrome

N. Wald and H. Cuckle

Introduction

Maternal serum alphafetoprotein (AFP) and ultrasound screening for neural tube defects is now established in Britain and recently serum screening for Down's syndrome has been introduced in a number of centres. Our purpose in this chapter is not to review the subject as a whole but to identify some practical issues that perhaps need wider recognition.

Monitoring the Screening Programmes

In Britain there is no designated authority responsible for implementing screening for neural tube defects or Down's syndrome. Programmes are developed locally and access to screening facilities can be patchy. There is a need for a central authority that would ensure that worthwhile screening programmes were available to all, but more important, there is a need to monitor the impact of screening programmes with continuous reliable information on the birth prevalence of the disorder in question and the proportion of affected pregnancies identified through screening.

Fig. 4.1, based on routinely collected official data from the Office of Population Censuses and Surveys (OPCS), shows that between 1964–72 and 1988 there was a more than 90% reduction in the estimated birth prevalence of neural tube defects in England and Wales (from 37 to 3 per 10 000 births; 18 to 1 per 10 000 for anencephaly and 19 to 2 per 10 000 for spina bifida). The timing of this decline is what one would have expected from the introduction of screening for neural tube defects in Britain but the OPCS data cannot distinguish between a decline in the natural birth prevalence of neural tube defects and a decline due to antenatal screening and selective abortion. In a special survey commissioned by the Department of Health we collected information on births and therapeutic abortions occurring in 1985 from AFP laboratories and ultrasound departments throughout England and Wales [3]. In our survey we estimated that there were 979 therapeutic abortions carried out on account of an antenatal diagnosis of a neural tube defect (anencephaly in 534 and spina bifida in 445), a rate of 15 per 10 000 births (eight for anencephaly and seven for spina bifida). In 1985, the estimated birth prevalence was 7 per 10 000 births; the proportion of the reduction since 1964–72 attributable to antenatal diagnosis and selective abortion was, therefore, 50% (15/(37−7)). The proportion attributable was similar for anencephaly (47%) and spina bifida (54%). Even though our survey was specially

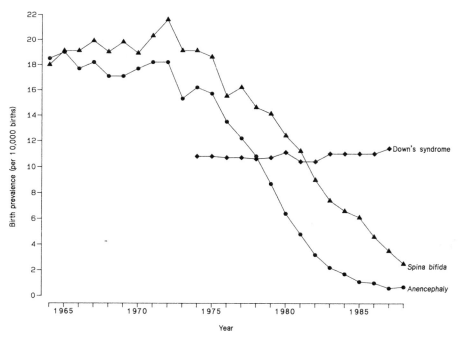

Fig. 4.1. Estimated annual birth prevalence in England and Wales of neural tube defects from 1964 to 1988 (from ref. [1] after adjusting the numbers for under-reporting to OPCS (19% for anencephaly and 13% for spina bifida) and extending the data to 1986–88), and Down's syndrome from 1974 to 1987 (from ref [2]).

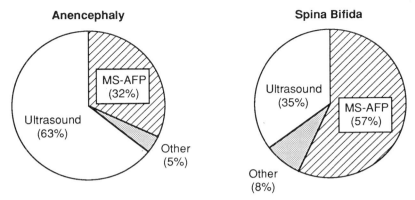

Fig. 4.2. Therapeutic abortions performed on account of a diagnosis of fetal anencephaly or spina bifida in England and Wales in 1985 according to the indication for performing the diagnostic procedure [3].

concerned with assessing the impact of screening for neural tube defects ascertainment of therapeutic abortions performed because of a neural tube defect diagnosis may have been incomplete. The estimates of the proportion attributable to antenatal diagnosis and selective abortion are, therefore, conservative.

The under-ascertainment is even greater if OPCS figures on the number of legal abortions carried out because of a suspected central nervous system malformation are used. In 1985, 522 were reported to OPCS compared to our own direct estimate of 979. A national neural tube defect register is urgently needed based on births and terminations. This could be achieved by building on the existing scheme of notification to OPCS, but to ensure complete notification a named individual in each obstetric unit should be appointed with the task of notifying to the register all terminations of pregnancies and all births in respect of a neural tube defect. Confidentiality would be ensured by using registration numbers identifiable only to the unit concerned.

Data collected in our 1985 survey show that nearly two-thirds of cases of anencephaly that were detected were first identified by means of an ultrasound examination and nearly 60% of cases of spina bifida that were detected were first identified by serum AFP screening (Fig. 4.2). If the designated person responsible for registering neural tube defects also recorded whether the defect was detected antenatally and if so how, together with the method of diagnosis, this would represent an important example of medical audit.

Fig. 4.1 also shows that between 1974 and 1987 the estimated birth prevalence of Down's syndrome has not materially changed. On the basis of OPCS figures, we have estimated that the proportion of affected pregnancies that were terminated was on average 14% (range 11%–19%) [2]. With the introduction of maternal serum screening for Down's syndrome it is expected that this proportion will increase. It is fortunate that a national Down's syndrome register has recently been set up, which could be exploited to monitor the change in the birth prevalence associated with the general introduction of serum screening.

Screening for Neural Tube Defects

Ultrasound Dating Scan to Interpret AFP values

Serum AFP values are normally expressed in multiples of the median (MoM) for unaffected pregnancies of the same gestation. By ensuring that this is always calculated using an estimate of gestational age based on a biparietal diameter (BPD) measurement rather than 'dates' will mean that the screening detection rate is increased from about 75% to 90% for a cut-off level of 2.5 MoM [4]. This increase in detection arises from the fact that spina bifida fetuses on average have small heads in utero so that the use of a BPD measurement to estimate gestational age leads to such fetuses being, on average credited with a less mature gestational age than is in fact the case. A given concentration of AFP (in iu/ml) is, as a result, converted into a higher MoM value than when gestational age is based on "dates". Since in many centres an ultrasound examination is carried out on at least one occasion during pregnancy it is only necessary to arrange for the ultrasound scan result to be available prior to the interpretation of the AFP test to obtain the advantage of increased detection. This simple administrative arrangement needs to be explicitly implemented in antenatal clinics or else cases of spina bifida will be missed that would otherwise be detected.

Some centres, although aware of the advantages of a simple dating scan performed immediately prior to the serum AFP test, prefer to delay the ultrasound scan examination until 18–20 weeks' gestation when it is easier to detect structural fetal birth defects. Regardless of the value of such an anomaly scan in the detection of neural tube defects it is, we believe, unwise to abandon the earlier simple scan in favour of the later detailed scan. The early one should be offered as a matter of routine and centres that wish to perform an anomaly scan should do so as a separate exercise. An early dating scan will also have the advantage of increasing the detection rate in serum screening for Down's syndrome (see below).

Complementary Use of Amniocentesis and Ultrasound in the Diagnosis of Open Spina Bifida

Some centres have abandoned amniocentesis in favour of ultrasound in the diagnosis of neural tube defects following a positive serum screening test. Anencephaly can be confidently diagnosed by ultrasound alone but none of the available diagnostic tests for open spina bifida are free from error and this is as true for ultrasound as it is for amniotic fluid AFP and acetylcholinesterase (AChE) determination. Fig. 4.3 examines the effect of error in the performance of ultrasound as a diagnostic test for open spina bifida among women with a high serum AFP level, assuming that the ultrasound detection rate for spina bifida is 90% and the false positive rate 1%. It can be seen that if the ultrasound diagnosis were positive, the odds of being affected given a positive result would be about 10:1, that is for every 10 affected fetuses with a positive result there would be one unaffected fetus with a positive result. In general, it would be unwise to avoid offering a woman with a positive diagnosis a confirmatory amniocentesis. Most of

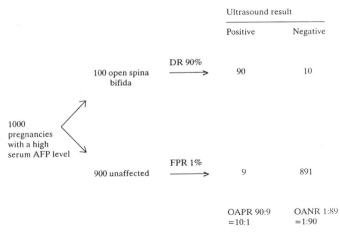

Fig. 4.3. An illustration of the performance of ultrasound in the diagnosis of open spina bifida among women with a high serum AFP level. It is assumed that the spina bifida detection rate (DR) is 90% and the false-positive rate (FPR) is 1%. (OAPR = odds of being affected with a positive result; OANR = odds of being affected given a negative result.)

the cases that would be "ultrasound false-positives" are likely to be corrected using the amniotic fluid tests. Among women with a negative ultrasound diagnosis of spina bifida the odds of having an affected fetus would be about 1:90, so even if the ultrasound were to provide a reassuring result it would be important to advise women that for every 90 unaffected fetuses with a negative result, there would be one affected fetus with a negative result. Fig. 4.3, therefore, shows that,

Table 4.1. The performance of ultrasound in the diagnosis of anencephaly and spina bifida in women known to be at high risk of neural tube defects

Centre[a]	Detection rate[b]				False positive rate[c]	
	Anencephaly		Spina bifida			
	%	(no.)	%	(no.)	%	(no.)
Glasgow	100	(39/39)	79	(15/19)	0.0	(0/302)
New Haven	—		100	(6/6)	0.0	(0/22)
South Wales	100	(57/57)	58	(22/38)	2.2	(53/2414)
London	100	(48/48)	94	(112/119)	0.2	(3/1306)
Oxford	100	(23/23)	100	(19/19)	3.0	(6/201)
Chicago	100	(5/5)	75	(6/8)	2.2	(2/92)
San Francisco	100	(12/12)	100	(32/32)	0.0	(0/94)
North Carolina	100	(12/12)	100	(8/10)	0.0	(0/587)
Virginia	100	(11/11)	100	(10/10)	0.0	(0/167)
Combined	100	(207/207)	88	(230/261)	1.2	(64/5185)

From Wald et al. [5].
[a]In one further study from Michigan, there were nine neural tube defect pregnancies diagnosed out of 12, but the number that were anencephaly and spina bifida were not specified separately; there were no false positives out of 245 pregnancies.
[b]Proportion of affected pregnancies with ultrasonographic diagnosis of neural tube defects.
[c]Proportion of unaffected pregnancies with ultrasonographic diagnosis of neural tube defects.

assuming reasonable estimates for the detection and false-positive rates of ultrasound, there would be an unaffected fetus in about 10% of positives and an affected fetus in about 1% of negatives.

These results may seem to be in conflict with clinical experience, because the vast majority of diagnostic procedures will be without error, and a false positive or a false negative diagnosis might arise in an average district general hospital only about once in every two years. Table 4.1 summarises the result of the published reports of diagnostic ultrasound in women with a raised maternal serum AFP level in which both the detection and false positive rates were estimated [5]. The overall detection rate in spina bifida is 88% and the false positive rate 1.2% – similar to the assumed rates used in Fig. 4.3. Table 4.2 shows the relative performance of three diagnostic policies, the use of ultrasound alone, amniocentesis alone and both together. The best policy is the one in which amniocentesis and ultrasonography are used as complementary diagnostic investigations. We do not believe that it is justifiable to suggest a general policy that amniocentesis be abandoned in the antenatal diagnosis of neural tube defects in women with positive serum screening tests.

Table 4.2. Comparison of three policies in the diagnosis of open spina bifida among women with raised maternal serum AFP levels

Diagnostic policy	Detection rate (%)	False-positive rate (%)	Fetal loss rate per 1000 women with with unaffected pregnancies due to:		
			Diagnostic error	Amniocentesis	Both
Ultrasonography only	88	1.2	12	0	12
Amniocentesis only	97	0.4	4	8[a]	12
Ultrasonography and amniocentesis[b]	95[c]	<0.1[d]	<1	8	<9

From Wald et al. [5].
[a]From the Copenhagen randomised study, subtracting the fetal loss rate in women allocated to the control group from the rate in women allocated to the amniocentesis group.
[b]Ultrasonographic examination is repeated if it is negative but the amniotic fluid results are positive; a final positive result is one in which both the ultrasonographic examination and amniotic fluid results are positive.
[c]Assumes four out of five cases missed by ultrasonographic examination are detected on reexamination after a positive amniotic fluid result is known (cited in references in [5]).
[d]Assumes that at least three out of four amniotic fluid false positives will be corrected by ultrasonographic examination; in fact the proportion is likely to be greater, but data on this are lacking.

Screening for Down's Syndrome

The poor performance of antenatal screening for Down's syndrome in the past has arisen because the only methods of identifying women who were sufficiently at risk of a Down's syndrome pregnancy to justify offering them an amniocentesis, was advanced maternal age and having had a previously affected pregnancy. Only about 35% of affected pregnancies can be detected by identifying the 7% or so of women aged 35 years or more and less than 1% of affected pregnancies occur in women who have already had one. In addition, advanced maternal age is often not regarded with sufficient importance to alert the clinician or the woman to the risk, so the offer of an amniocentesis may not be made or, if it is, it may be declined.

With the discovery of second trimester maternal serum markers of Down's syndrome the position has changed. The principle markers are AFP, unconjugated oestriol (uE_3) and human chorionic gonadotrophin (hCG) [6]. The most efficient method of screening is to use a woman's age in conjunction with the marker levels to estimate the risk of her having a Down's syndrome pregnancy and offering her an amniocentesis if this risk were above a specified level. Table 4.3 shows how the detection rate varies for different combinations of the screening variables if the false positive rate is held constant and how the false positive rate varies if the detection rate is held constant. (Maternal age is included in all the combinations and AFP in all except the first because AFP is usually measured as a part of neural tube defect screening programmes.) For example, using all three serum markers and the woman's age, 60% of affected pregnancies are detected, whereas, for the same false positive rate, only 30% of affected pregnancies are detected using maternal age alone.

Table 4.3. Detection and false positive rates using different screening variables

	Screening variables				
	Age alone	Age and AFP	Age, AFP and uE_3	Age, AFP and hCG	Age, AFP uE_3, and hCG
False positive rate (%)	*Detection rate (rounded to nearest 5%)*				
1	15	20	25	30	40
3	25	30	35	45	55
5	30	35	45	55	60
7	35	40	50	60	65
9	40	45	55	65	70
Detection rate (%)	*False positive rate (rounded to nearest 1%)*				
30	5	4	2	1	1
40	11	7	4	2	1
50	19	12	7	4	2
60	31	20	12	7	5
70	45	30	18	12	9

Derived from Wald et al. [6].

Screening Policy

Since screening using maternal age together with the three principal serum markers can lead to a 60% detection rate with a false positive rate that is no higher than the rate that had been previously used with maternal age screening alone (thereby doubling the detection rate without increasing the amniocentesis rate) and since tests can be performed on the same sample of blood collected for neural tube defects screening, there is no reason not to introduce serum screening for

Down's syndrome immediately. If cost is regarded as an obstacle, it is possible to adjust the screening risk cut-off level so that the extra cost of the uE_3 and hCG determinations are off-set by the savings through performing less amniocentesis and karyotypes. The scale of this saving can be seen by the fact that to detect 30% of all Down's syndrome cases using serum markers together with maternal age would be achieved by offering only 0.5% of women an amniocentesis instead of about 5% if maternal age were used alone. Serum screening for Down's syndrome is, therefore, a cost-effective development, permitting the identification of a given number of affected pregnancies at lower cost than would be possible using age alone.

Although new tests are likely to arise with a performance better than those already described (for example, urea resistant neutrophil alkaline phosphatase determination [7]) none are at a stage of development where they can be used in routine antenatal care. It is likely to be some time before serum screening using the three principal markers will be replaced by alternative methods.

The Role of Maternal Age

Serum screening for Down's syndrome should be offered to all pregnant women. It is irrational to offer all women above a certain age an amniocentesis and restrict serum screening to younger women. At any given rate of amniocentesis this practice would yield a lower detection rate than if serum screening were to be applied to all women. Also it means that a large proportion of women of advanced maternal age would be offered an amniocentesis when, with the knowledge obtained from the serum screening results that their risk of having a pregnancy associated with Down's syndrome was small, they would decide not to have an amniocentesis. The avoidance of an amniocentesis in older women is as much a benefit of serum screening as is the possibility of detection in younger women. Table 4.4 exemplifies this by giving the risk of having a Down's syndrome term pregnancy for women of different ages:

1. In the absence of serum screening results
2. If the levels of AFP and uE_3 are low and hCG high
3. If the levels of AFP and uE_3 are high and hCG low

Table 4.4. Risk of Down's syndrome according to maternal age and maternal serum levels

Serum level (MoM)			Maternal age at EDD (years)				
AFP	uE_3	hCG	20	25	30	35	40
ND	ND	ND	1:1530	1:1350	1:910	1:385	1:110
0.5	0.5	2.0	1:120	1:100	1:70	1:30	1:10
2.0	2.0	0.5	1:140 000	1:120 000	1:84 000	1:35 000	1:10 000

ND = not done.

It can be seen that in neither of the last two examples does the woman's age alone determine whether she merits an offer of an amniocentesis on the basis of her risk of Down's syndrome.

Refinements to Screening

It is likely that, in practice, the performance of serum screening for Down's syndrome will be somewhat better than shown in Table 4.3. For example, the routine use of a biparietal diameter measurement to estimate gestational age will reduce gestational errors and since the concentration of all three serum markers is dependent on gestational age, this will improve the precision of risk estimation leading to an increase in the detection rate for a given false positive rate. A person's racial background is a factor in determining serum AFP levels (blacks have AFP levels of about 15% higher than whites), and there may be racial differences with respect to uE_3 and hCG [8,9]. If these differences are large enough, it may be useful to allow for race in the interpretation of screening results. AFP, uE_3 and hCG levels are raised in twin pregnancies and, although this is well quantified with respect to AFP, more work needs to be done for uE_3 and hCG. When the magnitude of the effect is known, screening will be able to take into account whether or not the pregnancy is a singleton or a twin. Some factors such as smoking affect the screening markers but by too small an extent to justify modifications of screening protocol [10].

Conclusions

Screening for neural tube defects is widely adopted in Britain but there are still centres that do not offer it in a way which will maximise detection. It is important that maternal serum AFP screening be available generally and that it is performed in conjunction with dating ultrasound scan examination so that all AFP tests can be interpreted with an estimation of gestational age based on a biparietal diameter measurement. The complementary use of ultrasound and AFP screening should be supported by a similar complementary use of ultrasound and amniotic fluid AFP and AChE determination in the diagnosis of spina bifida. In neither screening nor diagnosis should an ultrasound examination or biochemical testing be regarded as mutually exclusive alternatives.

The cost-effectiveness of screening for Down's syndrome has been improved by the discovery of the mid-trimester serum markers, so that for a given detection rate the proportion of women requiring amniocentesis is now much smaller than was the case when maternal age was used alone, or for a given false positive rate (i.e. without increasing the number of women requiring an amniocentesis) the detection rate is now much greater. If the same false positive rate is used as the one adopted in the past with maternal age screening, as is likely, there will be a substantial reduction in the birth prevalence of Down's syndrome in the coming years.

With the potential of making a major impact on the birth prevalence of neural tube defect and Down's syndrome through serum screening there is an urgent need to establish reliable national monitoring schemes for these two disorders.

References

1. Cuckle HS, Wald NJ. The impact of screening for open neural tube defects in England and Wales. Prenat Diagn 1987; 7:91–9.
2. Cuckle HS, Nanchahal K, Wald NJ. Birth prevalence of Down's syndrome in England and Wales. Prenat Diagn, in press.
3. Cuckle HS, Wald NJ, Cuckle PM. Prenatal screening and diagnosis of neural tube defects in England and Wales in 1985. Prenat Diagn 1989; 9:393–400.
4. Wald NJ, Cuckle HS, Boreham J, Stirrat G. Small biparietal diameter of fetuses with spina bifida: implications for antenatal screening. Br J Obstet Gynaecol 1980; 87:219–21.
5. Wald NJ, Cuckle HS, Haddow JE, Doherty RA, Knight GJ, Palomaki GE. The ultrasonographic diagnosis of open spina bifida. N Engl J Med, in press.
6. Wald NJ, Cuckle HS, Densem JW et al. Maternal serum screening for Down's syndrome in early pregnancy. Br Med J 1988; 297:883–7.
7. Cuckle HS, Wald NJ, Goodburn SF, Sneddon J, Amess JAL, Dunn SC. Measurement of activity of urea resistant neutrophil alkaline phosphatase as an antenatal screening test for Down's syndrome. Br Med J 1990; 301:1024–6.
8. Simpson JL, Elias S, Morgan CD, Shulman L, Umstot E, Anderson RN. Second trimester maternal serum human chorionic gonadotropin and unconjugated oestriol levels in blacks and whites. Lancet 1990; 335:1459–60.
9. Muller F, Boué A. A single chorionic gonadotrophin assay for maternal serum screening for Down's syndrome. Prenat Diagn 1990; 10:389–98.
10. Cuckle HS, Wald NJ, Densem JW et al. The effect of smoking in pregnancy on maternal serum alpha-fetoprotein, unconjugated oestriol, human chorionic gonadotrophin, progesterone and dehydroepiandrosterone sulphate levels. Br J Obstet Gynaecol 1990; 97:272–6.

Discussion

Lilford: Could I disagree on the need for a randomised trial to establish the value of ultrasound as a useful routine screening test? My own view would be that it is too late to do a randomised trial of routine ultrasound screening at 18 weeks. The issue that it can reduce congenital anomalies is solved; it certainly can reduce congenital anomalies. The attempt to prove that deaths due to intrauterine growth retardation could be cut would require a study – I originally estimated with Tim Chard – of 50 000, and a recent study by De Bono [1] estimated >100 000 pregnancies.

Two kinds of study are required. The first is cross-utility studies, and I entirely agree on that. As to the second, if we are to do randomised trials, it may be better to restrict those to randomised trials of different ultrasound regimes, such as having early ultrasound in the first trimester or adding an ultrasound in the third trimester, e.g. to test the hypothesis that late ultrasound in the third trimester cuts down morbidity and mortality in childhood from chronic renal disease.

Whittle: I would like to make it clear that I do not feel that any randomised study would be of screening against not screening. There is no question at all that there is a need for screening. I was trying to make the point that there may be different methods that could be employed, some of which might be more cost effective than others. That is perhaps the question that needs to be addressed.

Galjaard: Prenatal diagnosis in most countries has until now been restricted to a certain group and the answers have been nearly 100% sure. In other words, for the last 20 years people have been educated that if we do prenatal diagnosis in terms of chromosomes, biochemistry or DNA, despite a few false negatives and positives one could rely on that answer. Now for the first time a test is offered that does not give certainty. Testing is offered to more people, and on a population basis it is all very cost effective and more defects come to light. But at the individual level for the first time one is no longer talking about certainties, but about risks and about odds.

In several Western European and Scandinavian countries I foresee a trend against genetics. One of the strong points of genetics so far has been the certainty of our answers. Once we go to large populations we will be prone to more criticism.

In Sweden, for example, they started serum AFP and have given up after a few years.

Wald: The first point is easier to deal with. I would suggest that in the question there is confusion between diagnostic tests and screening tests. Amniocentesis and DNA analyses are diagnostic tests with a very low probability of error.

As far as Down's syndrome is concerned, the screening test has always been a screening test. It has been 'How old are you?'. On that basis the performance of the test has been extremely poor, with a detection rate of only about 35% and a false positive rate of about 5%. Using age as the test instead of oestriol, or AFP or hCG, only about 35% of affected pregnancies in the community will be picked up and the amniocentesis will have to be offered to 5% of unaffected pregnancies. That is far from certainty, and I would suggest that asking a woman her age and using that as a test is not intellectually any different from using the serum AFP, especially if the AFP is to be done in any event. And so all we really have is a better method of screening than we had before. Patients in general, in antenatal care and in medicine, are aware that there are questions of probability and I think we underestimate their intelligence if we think otherwise. What is the probability that they have got cancer? What is the probability that they have got this or that? And that probability will become surer as we go through a process of testing. Hopefully, at the last stage there will be almost no error, but certainly at the first stage it is a question of risk estimation.

Donnai: The key to the success and the acceptability of most of these tests is education – education of the population being screened but also education of the professional people at the interface with the patient. Not the people in the laboratory doing the calculation but the people actually talking to the patient. I am constantly surprised with a test like serum AFP applied for NTDs, how wrong the professional clinical staff often get it in terms of thinking it is "THE spina bifida test" and not being aware that at best only one in 10 of the population identified by serum screening would be carrying a spina bifida fetus, even now when the test has been in place for a long time.

It might be useful for us to hear more about education, although it will be featured in a later discussion. For example, Dr Whittle mentioned the Joint

Recommendations that Professor Campbell had helped draw together. Did they address the educational issues with regard to counselling aspects? How do radiographers doing the test cope with identifying an anomaly, and should any training in counselling be recommended for them? Many women who have had a fetus terminated refer back to the nightmare of somebody not communicating and yet knowing there was something wrong.

Campbell: This is a doctor training programme. There is a radiographer training programme. Clearly counselling is an important aspect of this training, but also performing the ultrasound screening test.

What has come through to me strongly from Dr Whittle's talk is that there is still great variability. But the general levels are rising. The fact that most departments now are detecting 70% of malformations detected in the neonate effectively – we do not know what the long-term detection rate is for abnormalities that are detected later on – is a fantastic improvement over say 10 years ago.

Dr Whittle was saying he was anxious about resources being poured into routine screening, but in fact the last National Audit Office survey shows that 100% of women in England and Wales have a routine scan. So what we really now have to decide is when do we do it, and who does it, and the priority. This has just grown up. The DHSS has not said that it recommends a routine scan and will put resources into this, it has just grown up. If they had put as much effort into this as they have put into the rest of genetics then ultrasound would be much further ahead at this moment in time.

Donnai: And it is not just the technical provision. It is the actual delivery of the service to the woman.

Campbell: Absolutely. There is much too much unnecessary anxiety engendered and this has got to be part of the training.

Donnai: What about the educational aspects of serum AFP and the triple screening for Down's syndrome? How has that developed and are there areas that need attention?

Wald: One point that needs to be taken on board by the profession probably more than by the public is that in screening the aim is to identify a high-risk group: it is not to exclude the disease. In the screen negatives, unless the detection rate is very high the risk is not substantially different from background; it might be a third of background or a quarter of background. So the notion that a screen negative is reassurance is misleading, and I do not think the profession has got that straight.

What one can say is that in the high-risk group it can be a very high risk, ranging from one in four to one in 20. So we need to get our perspectives on screening right, and it will then work in the community. With proper information, leaflets, counselling and opportunity for discussion, I think that will work.

Having said that, at meetings people often say "All our patients get detailed counselling about the tests". The reality is that that is not the case. The cost of detailed counselling when there are thousands of patients coming through an antenatal clinic would be so great that one could not do other things that were more essential. What is important is to get the programme right, with good written information that does not focus just on one disease but looks at the main ones, so that people get the right impression about what is happening. That is the kind of philosophy we have got to put forward.

Donnai: That aspect will be covered in a later chapter.

Read: When he discussed NTD screening, and the merits of doing amniotic AFP and acetylcholinesterase, Professor Wald indicated that if the serum AFP and ultrasound suggested a spina bifida, what was to be lost in doing the amniotic AFP. It is always very nice to do lots of tests provided the tests all agree with one another and it leaves one feeling extremely comfortable and secure, but there is no point in doing extra tests unless one is prepared to do something different if the extra test shows something unexpected.

Supposing a woman has high serum AFP, the ultrasonographers say they can see a spina bifida, and then the amnio comes out "AFP and acetylcholinesterase negative", then my guess is – and others may have different views – that what will be said is that it must be a closed lesion and the woman would probably opt for a termination in any case. But does it actually alter the management of the patient?

Wald: I think it would but in an unpredictable way. What one would certainly do is to make the patient aware that there is doubt, at the outset, so that if an unsatisfactory outcome emerges, she is fully appraised of that. What is a real catastrophe is when one confidently says that it is spina bifida, and the baby does not have it. If there are tests – and they are both good tests – and one says that the tests seem inconsistent and there is a problem here, at least one or other or both of the tests can be repeated, the patient can be sent for another opinion, but most of all she can be made aware that there is uncertainty.

References

1. De Bono M, Fawdry RD, Lilford RJ. Size of trials for evaluation of antenatal tests of fetal wellbeing in high risk pregnancy. J Perinat Med 1990; 18:77–87.

Chapter 5

Heterozygote Screening for Cystic Fibrosis

D. J. H. Brock, M. E. Mennie, I. McIntosh, C. Jones and
A. E. Shrimpton

Introduction

Cystic fibrosis (CF) is usually regarded as the most common life-threatening genetic disorder in populations of north-European ancestry. It is inherited as an autosomal recessive trait and there is now abundant evidence that all cases of CF are the result of either homozygosity or compound heterozygosity or mutant alleles at a single genetic locus. The birth prevalence in the UK is estimated to lie between 1 in 2000 and 1 in 2500, and the heterozygote frequency is thus between 1 in 22 and 1 in 25. About 300 children with CF are born in the UK each year, and more than 2 million of the British population are symptomless carriers of the CF gene.
    A national survey of CF in the UK covering the years 1977 to 1985 was published by a British Paediatric Association Working Party in 1988 [1]. It estimated the total prevalence at mid-1985 to be near 5000, and increasing by about 100 cases a year. The median survival for a subset of the 1977 to 1985 cohort who attended single clinics was 19 years, with males (21 years) doing better than females (17 years). Statistics from the Cystic Fibrosis Foundation in the USA [2] point to a median survival for patients under care in sponsored centres in 1986 to 26 years. Analysis of survival curves in both the UK and the USA has suggested a steadily improving prognosis. However, at least part of this may be accounted for by improved diagnosis of milder forms of the disease. There is thus no clear evidence that the life expectation of a CF patient will continue to improve.

CF is an expensive disorder to manage and treat. The Royal College of Physicians Report [3] "Prenatal Diagnosis and Genetic Screening" estimated the minimum annual cost to the National Health Service (NHS) as £4000–£6000 per patient giving a total cost of £100 000–£150 000 for a life-expectancy of 25 years. This is a necessarily crude estimate and takes no account of the costs of the newer range of expensive antibiotics or of heart-lung transplants. There are justified fears that in a cash-limited NHS budget a continued high prevalence of CF will necessitate suboptimal treatment. Although this cannot be seen as a primary reason for attempting to reduce the birth incidence of CF by heterozygote screening and prenatal diagnosis, it is a consideration which cannot be ignored.

The CF Gene

The cloning and identification of the CF gene by a combined team from Toronto and Ann Arbor, reported in September 1989 [4–6], was a triumph for proponents of the reverse genetic approach to otherwise intractable Mendelian disorders. The first publication showed that about 70% of CF chromosomes carried a 3 base-pair deletion in exon 10 about the gene, the ΔF_{508} mutation. Hopes that a limited set of further mutations would comprise most of the remaining CF alleles were rapidly dashed, and it is now apparent that the disorder is quite heterogeneous at the molecular level. More than 60 different mutant CF alleles have been privately notified to an international consortium and details of 24 published [7–11]. Many of these are rare mutations, found to date in members of a single family. More common alleles, identified in the Scottish population, are shown in Table 5.1.

Table 5.1. CF alleles in the Scottish population

CF allele	Exon	Frequency	Reference
ΔF_{508}	10	0.728	6
G551D	11	0.063	9
G542X	11	0.051	8
R117H	4	0.018	7
1717-1G→A	intron 10	0.015	8
R560T	11	0.007	8
W1282X	20	0.007	11
621+1G→T	intron 4	0.004	8
A455E	9	0.004	8
Unpublished (5)		0.026	
Total		0.923	
Note detected			
D110H	4		
R347P	7		
ΔI_{507}	10		
S549N	11		
S549I	11		
R553X	11		
2566insAT	13		

Detection of CF Alleles

The product of the CF gene locus, the cystic fibrosis transmembrane conductance regulator (CFTR), is a 168 kD integral membrane protein. Its tissue distribution is limited and it seems unlikely that it will be found to be expressed in cells (such as blood components) which are easily accessible to large-scale observations. There is thus no easy solution to the problem of the molecular heterogeneity of this disorder, and diagnosis of either CF affecteds or symptomless heterozygotes will rest on the ability to detect specific mutant CF alleles.

Measurement of the predominant ΔF_{508} allele is technically simple. Exon 10 of the CF gene locus can be amplified to high copy number by means of the polymerase chain reaction (PCR) and the resulting products inspected after electrophoresis and ethidium bromide staining of acrylamide or agarose gels. Because the ΔF_{508} allele has three base pairs of DNA less than the normal allele, it migrates slightly faster (Fig. 5.1). Furthermore, if an individual has both ΔF_{508} and normal alleles, a highly characteristic heteroduplex band is seen at a slower migration point. Thus the heterozygote status of about 73% of carriers of the CF gene in Scotland and 75% in England can be readily detected by a single test. Carriers of the rarer ΔI_{507} allele will be simultaneously revealed on these gels [8].

Detection of other mutant alleles in exon 10 of the CF locus can also be made on the same PCR-amplified material by additional manipulations. These do not add a great deal to the cost of heterozygote detection, though in labour-time they are only justified if the other alleles have a reasonable frequency. However, if mutant alleles in other exons are to be included in the analysis, additional PCR amplifications are required. We estimate that each separate amplification increases the cost of detection by about 60%. DNA-based heterozygote detection thus involves a compromise between sensitivity and cost. Detecting all the known mutant alleles will be impossibly expensive and hard decisions must be taken about the practical sensitivity limits.

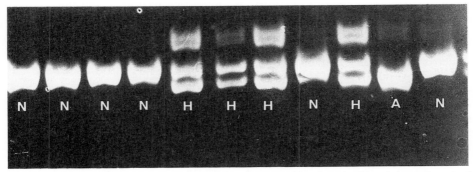

Fig. 5.1. Detection of ΔF_{508} alleles after amplification of exon 10 and acrylamide gel elecrophoresis of products. N, normal homozygotes; A, affected ΔF_{508} homozygotes; H, heterozygote for normal and ΔF_{508} alleles.

Heterozygote or Prenatal Screening

The ability to detect three-quarters of CF heterozygotes by a simple test made on either a blood or saliva specimen raises the question of instituting programmes of population screening in countries like the UK or USA where the birth incidence of the disorder is high. Mass heterozygote screening might be targeted at a number of potential age or situation groups. Neonatal heterozygote screening could be comparatively easily grafted into existing programmes for the detection of phenylketonuria, but it would be many years before the information (if retained) would be useful to the heterozygous individual. High-school screening for carriers of the recessively inherited Tay–Sachs disease has been carried out with some success in Quebec province in Canada [12], but in the UK the standards of biology teaching are low and there are obvious problems in getting informed consent for legal minors. A consensus has begun to emerge that any form of mass heterozygote screening for CF in the UK should aim at individuals or couples whose plans for reproduction are in the not too distant future.

The rationale for targeting heterozygote screening at young people near reproductive age is that it allows them to make informed decisions about avoiding or accepting the risks of conceiving an affected child. In theory two known heterozygotes could change partners, seek artificial insemination of the woman by a non-heterozygous donor or avoid reproduction altogether. In practice evidence from other recessively inherited disorders like Tay–Sachs disease and β-thalassaemia [13] suggests that most heterozygotes do not use knowledge of their genotype in selecting a mating partner. Few abstain from reproduction because they are at high risk of conceiving an affected child. Instead a more normal pattern is to seek a monitored pregnancy using prenatal diagnosis of the condition to ensure unaffected offspring.

If these conclusions are also true for CF, it profoundly affects the strategy of heterozygote screening. When the objective of screening is to permit informed reproductive decisions, it must be carried out before conception and ideally before marriage or partner bonding. If, on the other hand, the objective of screening is to allow high-risk couples the chance of avoiding the birth of an affected child, it can be delivered during pregnancy. This has obvious advantages, since virtually all pregnant women attend antenatal clinics and are thus an easily targetable group. They are likely to be susceptible to advice and information about the future health of their offspring, and most (though not all) will be able to identify the biological father of their child. In contrast heterozygote detection before pregnancy is likely to be a diffuse and difficult operation and biased towards the socially and financially advantaged. It is very difficult to see it operating without the backup of antenatal clinic screening.

It therefore seems likely that pilot trials of heterozygote screening for CF, designed to be incorporated into general health care, will need an antenatal clinic as an essential turnstile. Since the objective of such screening will be to allow high-risk couples to avoid the birth of an affected child, it is as much a programme of prenatal screening as it is of heterozygote screening. However, as one of the important side effects is the identification of heterozygotes in the wider families of carrier women, the term heterozygote screening should probably be retained.

Table 5.2. Risk of a CF child when a woman has a positive test and partner a negative test compared to proprotion of CF alleles that are detectable

Proportion of CF alleles detectable	Risk of CF child
0.75	1 in 400
0.80	1 in 500
0.85	1 in 667
0.90	1 in 1000
0.95	1 in 2000
0.96	1 in 2500
0.97	1 in 3333
0.98	1 in 5000

Heterozygote Screening in the Antenatal Clinic

A programme for heterozygote screening for CF in the antenatal clinic envisages testing being offered on a voluntary basis at as early a stage in pregnancy as is practical. If a woman tests positive her partner would also be offered a heterozygote test. If both members of the couple were found to be heterozygotes for CF alleles, they would be offered prenatal diagnosis and the chance of termination of pregnancy if the fetus were shown to be affected. However, within this simple scheme there are a host of intertwined technical, social and ethical problems.

A major difficulty stems from the incomplete nature of present and probable future heterozygote testing. It is instructive to consider this problem on the assumption that only the ΔF_{508} allele is being screened for, and that it is found on 75% of CF chromosomes. If a pregnant woman lacks the ΔF_{508} allele, her risk of being a carrier reduces from 1 in 25 to about 1 in 100. Her chance of bearing a CF child reduces from 1 in 2500 to 1 in 10 000. Although this is a low risk it is not a zero one, and counselling information will have to make this point clear. However, a more serious problem confronts women who test positive but whose male partners test negative. The risk of having a CF child is now about 1 in 400. Thus the net effect of screening for such couples is to increase the probability over prior odds of their having an affected child. This problem has been referred to [14] as putting a couple into "genetic limbo", an unfortunate and emotive phrase which is better expressed as "moderate risk". If only the ΔF_{508} mutation is screened for, nearly 3% of couples will find themselves in this group.

As additional CF alleles are included in the screening process, the residual odds for the moderate-risk group decrease (Table 5.2). Only when 96% of CF alleles are detectable do the odds in the moderate-risk group approximate the prior odds of unscreened couples. It has been suggested [15] that only when this 96% figure is reached should mass heterozygote screening be contemplated. In view of current understanding of the large number of different CF alleles, and the high cost of detecting all of them, this is not a practical proposition.

A number of suggestions have been made to attempt to circumvent the moderate risk problem. Brock [16] proposed that such couples might be offered amniocentesis and microvillar enzyme testing. If amniotic fluid cells were rapidly screened for the ΔF_{508} mutation (and other detectable CF alleles), the possibility of a homozygous CF affected could be eliminated in 50% of samples. A negative microvillar enzyme test on the amniotic fluid supernatant would reduce the chances of a CF fetus to very low levels (Table 5.3). There would still remain a small group of couples, where the amniotic fluid had a positive microvillar test, whose possibility of an affected fetus was substantial. Many of these couples would probably opt for termination of pregnancy. However the fact that between 3% and 4% of screened couples might need amniocentesis makes this proposal impractical, since obstetrical resources could probably not cope with the increased load.

Table 5.3. The use of amniocentesis and microvillar enzyme testing (MVT)[a] on women who carry a CF allele but whose partners appear negative
Assumptions: CF allele detection rate 88%
Risk of CF in pregnancy 1 in 800
Proportion of screened women 3.4%

Finding	Risk	Proportion
No CF allele in fetal cells	0	1.7%
CF allele present, MVT negative	1 in 7000	1.5%
CF allele present, MVT positive	1 in 34	0.22%

[a]Sensitivity of MVT 95%, specificity 92%.

An ingenious proposal has recently been made by Professor N. J. Wald and is referred to as "couple screening". In this model no heterozygote testing for CF would be carried out unless samples were available from both partners in a pregnancy. The samples would be coded and presented to the laboratory as a unit of two. Only if both samples had detectable CF alleles would the couple be informed that they were at high risk of bearing a CF child. A unit with two negatives or a unit with one negative and one positive would be treated in the same way, and not viewed as an indication for further action. Any couple where only one partner was positive would not be informed of their increased risk. This model has the advantage of economy and efficiency. Provided that it were made clear in advance that screening was solely directed at detecting couples with a 1 in 4 risk of bearing a CF child, it should be a feasible option. However, it does not allow for any extended family screening in which the finding of a heterozygous individual is used as a starting point for carrier testing in the immediate relatives.

In the Edinburgh trial we intend to adopt a slightly different approach to antenatal clinic screening. Volunteer women will be screened for the three most common CF alleles, ΔF_{508}, G551D and G542X, representing 84% of mutations. Those who are negative for these three alleles have a residual risk of bearing a CF child of 1 in 15 000, which we believe can be ignored. For women who are positive there are two immediate consequences. The first is that there is a CF allele segregating in her branch of the family and that she might have sibs or more distant relatives who would wish to avail themselves of genetic counselling and testing. The second consequence is that the risk to her pregnancy is 1 in 100, because she is a CF carrier. We will endeavour to refine this risk by offering her partner a more extended screening investigation, perhaps covering 90% of CF alleles. If the partner is positive the couple's risk is 1 in 4 and prenatal diagnosis can be suggested. If the partner is negative the residual risk is of the order of 1 in

950. It will be emphasized that a risk of 1 in 950 is really very low for a couple where one partner is known to be a CF carrier.

One advantage of this approach is its relative economy. It is not practical to search for minor CF alleles in all women who volunteer for screening. It makes more sense to concentrate detailed investigations on the partners of the 3.5% of positive women, and to try and reduce residual risks in these couples to an acceptable figure. It has to be emphasized that we have as yet little experience in delivering molecular genetic tests to large numbers of individuals, and that any over-ambitious primary screen may encounter unsuspected difficulties.

It is perhaps useful to compare this situation with a more familiar one – maternal serum alphafetoprotein (MSAFP) screening for open neural tube defects. In an area with a spina bifida birth prevalence of 1 in 1000 the prior odds of an unselected, unscreened woman having such a child is just that figure. MSAFP screening changes these odds, and any value above 1 multiple of the normal median increases the risk. There is obviously no possibility of carrying out detailed further investigations on the 50% of women in this category; instead a robust commonsense approach is adopted and an MSAFP cut-off is selected for further action. Women with values below the action cut-off are usually told that their MSAFP value is "normal", even though many of them have a substantially increased risk of having a child with a neural tube defect. It is not customary to talk about such women as being in any type of risk limbo.

Social and Organisational Considerations

A distinction must be made between mass heterozygote screening programmes and pilot trials of different models of screening. One represents a blind plunge into an unknown dark lake, the other a delicate probing at whether to dive at all, and if so, just where. There is now a consensus, both in the UK [17], and the USA [18], that pilot trials are urgently needed, which would address a number of clearly defined questions. These include the central role of counselling and its implications for the staffing of clinical genetic centres, laboratory aspects, cost–benefit considerations and the acceptability of screening to those at whom it is targeted. In many of the discussions of screening it has been assumed that the sensitivity of CF heterozygote detection will rapidly approach the 96% figure at which any given couple will either have a 1 in 4 risk of conceiving or bearing an affected child or a risk at least as low as that pertaining before screening. This may not be the case. It is entirely possible that the cost and the technical difficulties of amplifying all the regions of the CF gene in which mutations lie (even in multiplex PCR) for several thousands of samples a year may lead to an acceptance of the lower detection figure. Thus pilot trials must be prepared to address the practicality of screening at detection levels which would not satisfy the purist.

This will undoubtedly complicate the process of information transfer to the prospective targeted group. Modell [17] has emphasized the importance of the screening infrastructure and pointed out that CF screening might bring about a re-allocation of health service resources into community genetics. Although this is a desirable ideal, any suggestion that CF screening is only possible with a massive increase of funding is not likely to be well-received by the current generation of UK health managers. One of the principal reasons that we have chosen the antenatal clinic approach to screening is that it does offer the prospect of delivering a programme with minimal increase in staff numbers. It is our contention that any pilot trial must be planned as a module which is capable of being scaled into a full programme without major reorganisation of the essential framework. We further believe that evidence will have to be presented that amplification will not be linear, and that larger programmes will show economies of scale. In Scotland this should be possible, since molecular genetic services have been planned on a national basis and the four major genetic centres participate fully in a consortium approach to the delivery of services [19].

Conclusions

It is now possible to detect between 75% and 85% of mutant CF alleles by comparatively simple laboratory tests on blood or mouth-wash samples. Early expectations that virtually all CF alleles would be detectable in this way have faded, and it now seems likely that mass heterozygote testing for CF may have to accept a cost-determined practical sensitivity limit of around 90%. Several pilot trials of CF heterozygote screening have been instituted in the UK, aimed at evaluating the social and technical problems of testing large numbers of reproductive-age individuals in clearly-defined target groups. It is obviously premature to draw any conclusions about whether any of these models will prove both acceptable and practical, and how they might be scaled up to a full-sized programme. This chapter presents a model for CF heterozygosity screening in the antenatal clinic which will be the subject of an extended trial over the next three years.

Acknowledgements. Work described in this paper was supported by the Cystic Fibrosis Research Trust, the Scottish Home and Health Department and the Ludovici Bequest to the University of Edinburgh.

References

1. British Paediatric Association Working Party on Cystic Fibrosis. Cystic fibrosis in the United Kingdom 1977–85: an improving picture. Br Med J 1988; 297:1599–602.
2. Boat TF, Welsh MJ, Beaudet AL. Cystic fibrosis. In: Scriver CR, Beaudet AL, Sly WS, Valle D, eds. The metabolic basis of inherited disease, 6th edn. New York: McGraw-Hill, 1989; 2649–80.
3. Royal College of Physicians. Prenatal diagnosis and genetic screening: community and service implications. London: Royal College of Physicians, 1989.
4. Rommens JM, Iannuzzi MC, Kerem B et al. Identification of the cystic fibrosis gene: chromosome walking and jumping. Science 1989; 245:1059–65.
5. Riordan JR, Rommens JM, Kerem B et al. Identification of the cystic fibrosis gene: cloning and characterisation of complementary DNA. Science 1989; 245:1066–73.
6. Kerem B, Rommens JM, Buchanan JA et al. Identification of the cystic fibrosis gene: genetic analysis. Science 1989; 245:1073–80.
7. Dean M, White MB, Amos J et al. Multiple mutations in highly conserved residues are found in mildly affected cystic fibrosis patients. Cell 1990; 61:863–70.

8. Kerem B, Zielenski J, Markiewicz D et al. Identification of mutations in regions corresponding to the 2 putative nucleotide (ATP)-binding folds of the cystic fibrosis gene. Proc Natl Acad Sci. (in press)
9. Cutting GR, Kasch LM, Rosenstein BJ et al. A cluster of cystic fibrosis mutations in the first nucleotide binding fold of the cystic fibrosis conductance regulator protein. Nature 1990; 346:366–9.
10. White MB, Amos J, Hsu JMC, Gerrard B, Finn P, Dean M. A frame-shift mutation in the cystic fibrosis gene. Nature 1990; 344: 665–7.
11. Vidaud M, Fanen P, Martin J et al. Three point mutations in the CFTR gene in French cystic fibrosis patients. Hum Genet 1990; 85:446–9.
12. Clow CL, Scriver CR. Knowledge about and attitudes towards genetic screening among high-school students; the Tay–Sachs experience. Pediatrics 1977; 59:86–91.
13. Modell B, Berdoukas V. The clinical approach to thalassaemia. London; Grune and Stratton, 1984.
14. Gilbert F. Is population screening for cystic fibrosis appropriate now? Am J Hum Genet 1990; 46:394–5.
15. Ten Kate LP. Carrier screening in CF. Nature 1989; 342:131.
16. Brock DJH. Population screening for cystic fibrosis. Am J Hum Genet 1990; 47:164–5.
17. Modell B. Cystic fibrosis screening and community genetics. J Med Genet 1990; 27:475–9.
18. Workshop on Population Screening for the Cystic Fibrosis Gene. Special Report. N Engl J Med 1990; 323:70–1.
19. Brock DJH. A consortium approach to molecular genetic services. J Med Genet 1990; 27:8–13.

Discussion

Williamson: We have also been carrying out a pilot study and I completely agree that we have a lot to learn about this. It is a very different situation from most of what we have been talking about. We are identifying carriers who are at risk of having a CF child in the next generation and we are not for the most part identifying affected individuals. It is very important to make that clear.

We have chosen to go forward through approved family planning centres and general practices on the grounds that it increases the range of reproductive choice available to couples at high risk. Once someone is pregnant they have the choice of continuing or terminating the pregnancy, whereas if they can be identified as a couple at risk before the woman is pregnant then they have a larger range of options; the fact that many people do not make use of them is not relevant. They also have time to think about those choices.

So far we have screened in rather similar numbers. We can identify 86% of CF mutations with four or five tests. Like Edinburgh, we are hitting the partners hard. Another reason for going through general practices is that we feel there will be a tendency if women are screened first to project some of the genetic blame, if someone is a carrier, on to women, whereas if the whole population can be screened equally that problem does not arise.

We have now screened nearly 1000 people through general practices and family planning clinics in our own studies. To the extent that one can assess anxiety – and in our experience there are no good measures – by going to the doctors, the practice nurses, and family planning counsellors, we have no evidence of anxiety among those who test negative and no evidence of anxiety after roughly an hour's counselling among those who score positive, and a group in Denmark has had exactly the same experience.

Brock: The thalassaemia and the Tay–Sachs stories suggest that people do not take the options, and if they do not we have to bring other factors into account, such as equal access and no social class bias. This would tend to favour the well educated and the affluent rather than the poor. One tends to work with GPs who are enthusiastic rather than with those who have difficult practices and numerous patients.

Another point is that if we screen couples, we shall double the number who will be in the positive/negative category, and we will need to throw the book at twice as many people.

Rodeck: One of the crucial things about these various schemes for screening would be the acceptance rate. What evidence have we got of that so far?

Brock: We have evidence that it will be extremely high.

Williamson: Population surveys, for what they are worth, indicate that 100% of CF families feel this should be offered and will take advantage of it, and in the general population something between 70% and 90% express an interest in being screened with minimal counselling. In practice, in GPs' surgeries and family planning clinics where people are already there and sitting around (and I do not think there is as much bias in Watford in the GPs' surgeries as Professor Brock implied) 68%–70% make use of it. When we sent out a letter to a random sample on a GP register, roughly 12% came in to make use of it. So a very high proportion will have the test after seeing a leaflet and a counsellor for a few minutes if they are sitting there and it is offered, but only a small proportion will leave their home and come somewhere to be tested.

Lilford: Whilst I am not disagreeing with the direction of Professor Brock's cost/benefit analysis, the way it has been done does exaggerate slightly. The assumption is that £1 spent now is the same as £1 spent in 10 or 15 years' time. The economists would want us to discount inflation and to discount compound interest on the money now.

Brock: Yes, but there is another thing bouncing it the other way. I have taken an absolute minimal cost for CF. Everybody thinks that CF costs £5000 a year. Yes – if they are not given any of the new range of expensive drugs, and if no account is taken of heart–lung transplant and similar procedures. And our trial is costing about three times as much as it would cost, because we are looking at all kinds of details. So things will balance out.

Pembrey: One thing that does not seem to tie with what we had always thought of in general genetic counselling, is this obsession to have a test which is as good as the population risk of 1 in 2500. The truth of the matter is that people do not live with a risk of 1 in 2500: couples live with a risk of one in four or zero. This idea that there is something magical about trying to achieve that is ridiculous. All the evidence from ordinary genetic counselling suggests that when the figures get down to absolute risks of less than 1%, and certainly when they are down to about 1 in 500, people are very reassured if it is put in perspective. And really one should go for what is easier and perhaps not worry too much about throwing the book to get all below a figure of about 1 in 500.

Brock: I would agree. We have been talking about serum AFP and about screening for Down's syndrome. A woman whose serum AFP is greater than 1 multiple of median, has an increased risk of having a child with a neural tube defect, and yet we do not say to her, "What are we going to do for you?" We in fact take a very practical, robust, commonsense attitude and say we can only deal with the top 5% or whatever.

Wald: It is important to identify the objectives of screening. If the objective is to identify the affected individual, in general in screening programmes one should go to where the trouble is greatest, be as simple as possible, and provide information as close as possible to the time when a decision has to be taken. Whatever else we do, the first point to go would be the antenatal clinic, and if the couple is regarded as the unit and both have to be positive, the detection rate would be $>60\%$, the false positive rate would be under 1 in 1000, and if the couple was positive the odds of the fetus being affected are 1 to 3. It would be one of the most powerful antenatal screening tests we have available, and the only problem in my view is the medical profession's mind set into thinking that we have got to identify carriers, heterozygotes. That is what generates the problem. If the objective is getting affected pregnancies, then the profession has in its hands a highly cost-effective screening test that will outperform serum AFP and Down's syndrome screening.

Donnai: But there is a practical problem for inner city areas. Many people booking in hospitals are no longer in contact with the father of the pregnancy or in some cases are not aware who that is. Some sort of system has to be built in to any programme to take account of that.

Wald: What one would do is to give the mother at the first antenatal visit two tubes and tell her to go away, take a mouthwash, and give one tube to the father, to do it too and post it in. It would be the best evidence of consent if they send it in, and the lab would measure one; it does not matter which is the mother's or which is the father's. If that is negative there is no purpose in testing the second. If it is positive, then the second is tested. What is required is a very clear information process to explain to the woman. But it is the two together that count.

I know it is different from the practice of antenatal care so far, but we are constantly having to face new steps. We have to be bold enough to try it. I agree that there will be some obstetricians that will be uncomfortable about it, but one or two at St Bartholomew's are willing to give it a try.

Donnai: But if a couple is reported as negative when in fact the woman is positive or the man is positive, they may have reassembled into a different partnership by the time another pregnancy comes and yet have been reassured by the negative result first time.

Wald: In what I am proposing, each pregnancy is a fresh screen. In what I am proposing there is no memory. We start from scratch every time, as with AFP or Down's. It is true that occasionally we shall waste information, but that way we cut through many of the problems that have been referred to.

R. Harris: I agree very strongly indeed with Professor Wald, because these are screening tests, and the payoff is in what one is actually looking for, not all the potential benefits. For example, if somebody is identified as being a carrier of a very common disorder, then that could lead to the benefit of counselling close relatives that they were carrying the disorder. But one cannot do everything. The analogy is between cervical screening by a doctor and by a nurse. It is always said that if a nurse does it she will miss the ovarian tumours because she does not do a pelvic examination. If one is screening for cervical carcinoma one is not looking for all the other things.

Section II
Special Techniques: 1

Chapter 6

Chorion Villus Sampling: The MRC European Trial

T. W. Meade

Introduction

In 1984, the Medical Research Council (MRC) and the Royal College of Obstetricians and Gynaecologists (RCOG) held a joint meeting to consider the implications of the growing interest in chorion villus sampling (CVS) in first trimester prenatal diagnosis. As a result, the MRC established a working party whose main concern from the outset was the feasibility of a randomised comparison of CVS with standard practice, i.e. amniocentesis during the second trimester. The clinical rationale was the possibility of earlier and easier termination of abnormal pregnancies after CVS, the benefits of which would have to be set against potentially higher fetal loss rates.

The possibility of a trial of this kind immediately raised the question that arises from the intended evaluation of any new technique – the "window of opportunity". On the one hand, a trial carried out too soon, while the technique is being developed and potential operators are obtaining experience with it, may not give a true picture of its eventual value under routine conditions and widespread familiarity with its requirements. On the other hand, by the time the technique has been fully developed and the "learning curve" completed, the method may have become routine to such an extent that its practitioners are not prepared to take part in a trial at all. In addition, further technical developments while a trial is in progress may diminish the practical value of its results, or even render them completely obsolete. This point is considered again later. The working party considered that a trial would be timely, particularly if it could be initiated quickly.

In 1985, within a few months of the working party's first meeting, the MRC's Systems' Board approved a proposal for a randomised comparison of first trimester CVS with second trimester amniocentesis. The main objectives of the trial were:

1. To compare the short-term effects of the two fetal diagnostic policies, CVS and amniocentesis, on the progress and outcome of the index pregnancies in terms of fetal and neonatal morbidity and mortality
2. To assess the extent to which CVS leads to accurate diagnosis
3. To establish two randomised groups of children among whom hypotheses about any possible long-term adverse effects of CVS and amniocentesis could be studied.

Based on the information then available from observational studies of complications following CVS, it was estimated that the trial would need to include between 2000 and 4000 pregnancies.

The detailed planning and coordination of the trial were carried out by the National Perinatal Epidemiology Unit in Oxford. A pregnant woman was considered eligible for the trial if:

1. Ultrasonography suggested that the pregnancy was progressing normally
2. The condition for which her fetus was to be examined could be diagnosed both by chorion villus sampling and by amniocentesis
3. She and those responsible for her care were uncertain which of the two methods was more likely to be in her interests.

For a participating centre to be eligible, the same criteria as those established in the Canadian trial [1] were agreed:

1. An obstetrician was eligible to participate if he or she had performed 30 or more practice procedures and obtained at least 10 mg of villous tissue in 23 of the most recent 25 cases examined
2. A laboratory was eligible to participate if karyotyping had been achieved successfully in 19 out of 20 consecutive chorion villus samples of at least 10 mg weight.

Progress

At the time of the RCOG Study Group's meeting, the main analyses of the MRC European trial were nearing completion and the working party had agreed that for a number of reasons the findings should remain confidential until publication. However, two developments during the trial's planning and fieldwork stages were of particular interest and of potential relevance not only to any further trials of methods for prenatal diagnosis, but also more generally.

The first was the active involvement of several special interest groups in the planning of the trial, particularly the preparation of information for women being approached about entry to it. These were:

Association for Improvements in the Maternity Services
Association for Spina Bifida and Hydrocephalus
Down's Children's Association
Maternity Alliance
Sickle Cell Society
Spastics Society
UK Thalassaemia Society
National Childbirth Trust

Their support for the trial, set out on the first page of an information leaflet, undoubtedly played a major part in the recruitment of adequate numbers – so much so that it would be unwise to contemplate any further trials of prenatal or similar techniques without their advice.

Many points about the trial were raised during a BBC television "Horizon" programme broadcast while recruitment was still in progress. If anything, the difficulties to be overcome in a trial of this kind, including ethical aspects, were given greater prominence than the medical justification – though they were explained very fairly. Contrary to what might have been expected, the programme was followed by a marked increase in the rate of recruitment.

When the trial had been in progress for some months, the working party began to receive reports – not very many, but obviously needing careful thought – of patients who, having not been allocated to CVS in a participating centre, arranged referral to another centre not taking part in the trial, which would undertake CVS. Any substantial tendency of this kind might have seriously affected the conduct and results of the trial. The Canadian trial had dealt with this question by a virtually countrywide agreement that CVS would be available in Canada only within the trial. Thus, a patient with a preference for CVS would have a 50% chance of being investigated by CVS, but, equally, a 50% chance of having amniocentesis, and she would not easily be able to have CVS elsewhere. Although a similar policy appealed to the MRC working party, it was by no means sure that it could or should aim for a similar arrangement in the United Kingdom. Its doubts were, however, resolved when the lay organisations themselves approved the policy of trying to confine the availability of CVS in participating centres to the trial. The effects of this decision were not as complete in the United Kingdom as they were in Canada but nevertheless it helped and it could not have been achieved without the support of the lay organisations.

Taken together, aspects of the trial involving lay organisations and lay comment in its planning and during recruitment provide considerable encouragement for future trials of a similar nature and should allay any remaining anxieties that lay involvement will automatically have the effect of making trials more difficult rather than less difficult.

The second feature was the recruitment of other European centres. These included a few entering the trial de novo after discussion with the working party. Most of this further recruitment, however, resulted from the agreement of centres in other countries – Denmark, Italy and Finland – that had independently started their own trials. These latter centres agreed to provide their data for what then became an MRC European trial, while at the same time retaining their freedom to publish their own results centre by centre at their discretion though, it is to be hoped, after the publication of the results of the MRC European trial itself.

Recruitment began during the second half of 1985 and ended in July 1989, and resulted in just over 3200 entrants.

Canadian Trial

Early in 1989, i.e. towards the end of recruitment to the MRC European trial, the results of the Canadian trial were published [1]. These are summarised in Table 6.1. As with the MRC European trial; maternal age and its association with an increased risk of Down's syndrome was by far the commonest indication for prenatal diagnosis. There was a small but non-significant excess of fetal loss in those investigated by CVS. Laboratory failures were significantly more frequent for CVS samples than for those obtained by amniocentesis. The proportion of abnormal findings was also significantly higher in CVS samples, most of this excess being explained by false positive results.

Table 6.1. Summary of the Canadian Trial

	Investigation Allocated	
	CVS	Amniocentesis
Number entered	1392	1396
Mean age (years)	38.1	38.1
Mean gestation at randomisation (days)	67.4	67.8
% Fetal losses[a]		
All entrants	16.8	15.1
Eligible entrants	7.6	7.0
% Laboratory failures	1.5	0.1 $P<0.001$
% Abnormal results	4.6	2.4 $P<0.01$
% False positive results	2.0	0.3 $P<0.001$

[a]Randomisation preceded ultrasonagraphy in some cases. "All entrants" refers to all patients randomised, some of whom were later found ineligible because of a non-viable fetus or multiple pregnancy, for example. This group experienced a higher loss rate than "eligible entrants" which excludes those with non-viable pregnancies, etc.

It is with these results that the MRC European trial's findings will naturally be compared.

Conclusion

It is likely that the results of the MRC European trial will be published during 1991. While the trial has been in progress, there have of course been further developments in invasive techniques including, in particular, growing interest in the possibility of first trimester transabdominal CVS and in early amniocentesis. The implications of these developments, however, are by no means sufficiently clear to pre-empt the results of a comparison of first trimester CVS (mainly though not exclusively transvaginal in the MRC European trial) and second trimester amniocentesis. Provided the results of the MRC European trial are judged to be sound, they will in themselves be valuable in decisions about clinical practice and will form a dependable basis for proposing, carrying out and eventually interpreting the results of other trials of which (as already indicated) a comparison of trans-abdominal CVS and amniocentesis in the first trimester is perhaps the most likely. Therefore apart from its specific contribution to methods for prenatal diagnosis, the trial will also very probably be seen to have supported the principle of using the "window of opportunity" when the need for evaluation first becomes apparent.

Acknowledgements. I am grateful to Dr Adrian Grant and Sarah Ayers for comments on this chapter. The views expressed are my own and do not necessarily represent those of other members of the working party.

Reference

1. Canadian Collaborative CVS-Amniocentesis Clinical Trial Group. Multicentre randomised clinical trial of chorion villus sampling and amniocentesis. Lancet 1989; i:1–6.

Chapter 7

Invasive Diagnostic Procedures in the First Trimester

R. J. Lilford

Introduction

Prenatal diagnostic tests can be divided into those involving measurement of chemicals in maternal blood; imaging the fetus; and invasive tests to remove tissues of fetal origin. The last may be divided into those carried out beyond about 14 weeks' gestational age (GA), before 14 weeks but after implantation, and those in the preimplantation period. The first group (beyond 14 weeks) includes fetal blood sampling, fetal tissue biopsy, "good old-fashioned" mid-pregnancy amniocentesis, and transabdominal chorion biopsy. The last group (preimplantation diagnosis) includes embryo biopsy and polar body analysis. This chapter considers the middle group; early amniocentesis, transabdominal chorion villus biopsy or sampling (CVS) and transcervical CVS as the main contenders, with rival sideshows such as aspiration of chorionic villi, amniotic or blastocyst fluid through the vaginal fornices and cytological sampling of the lower uterine segment. Although consumer choice is regarded as a good thing, you can have too much of all good things and excessive choice is bewildering [1]. It is time for one or two "market leaders" to emerge: each test will be considered in turn.

Techniques

Transcervical CVS

This is the traditional method of chorionic sampling and the first attempts at this method were made in 1974 [2]. It is now a widely used technique and, as medical tests go, it is also extremely well evaluated. The proper appreciation of this technique, and especially of its inaccuracies, requires an understanding of the anatomy of the first trimester pregnancy.

The chorion consists of an outer layer of trophoblast and an inner mesodermal layer containing blood vessels. At five weeks of pregnancy (three weeks after fertilisation) the entire chorion participates equally in villus formation, producing 200 tree-like colonies of villi which completely surround the embryo. Over the following weeks the villi facing towards the uterine cavity gradually degenerate to form the chorion laevae while those adjacent to the decidua basalis proliferate to form the placenta. Further growth takes place in this area due to fresh villus formation from the villus stems and progressive arborisation of previously formed villi. At six weeks of pregnancy (four weeks after fertilisation) the chorionic vesicle begins to intrude into the cavity of the uterus with the decidua capsularis covering the vesicle the decidua parietalis on the maternal aspect of the vesicle. The decidua parietalis lines the remainder of the body of the uterus. The cervical mucous membrane does not undergo decidual change but secretes a thick mucous plug in the cervical canal.

Chorionic villus sampling is seldom carried out before eight weeks of pregnancy (six weeks of embryonic life). The embryo is now 15 mm in length and the chorion frondosum is 3–6 mm thick, while the decidua capsularis measures less than 2 mm. At ten weeks' GA (eight weeks of embryonic life), probably the optimal time for CVS, the embryo measures 25 mm and further thinning out of the chorion laevae has taken place. The chorionic vesicle almost fills the uterine cavity but this will not be completely obliterated until the decidua capsularis fuses with the parietalis at 16–20 weeks of pregnancy [3]. By the end of the first trimester (10 weeks after fertilisation) the villi adjacent to the decidua capsularis are reduced to microscopical stumps and the extra-embryonic coelom has disappeared by fusion of the amnion with the chorion. Trophoblast is derived from the shell of the blastocyst. The embryo and extra-embryonic mesenchyme derive from the inner cell mass. The mesodermal cores of the chorionic villi are therefore embryologically closer to the fetus than is the trophoblast.

The current techniques for transcervical chorion biopsy may be grouped into those dependent on biopsy forceps [4,5] and more popular methods based on cannulae. Gustavii and coworkers [6–8] describe a method of biopsy under direct vision through a fibreoptic scope which is introduced among the villus fronds (villoscopy or chorionoscopy). The great majority of operators use metal or plastic cannulae [9–11]. The placental site is identified by ultrasound, with the operator's fingers in the vagina so that the exact relationship to the cervix may be ascertained. The malleable cannula is then bent to the appropriate shape and inserted through the cervix and guided into the chorion frondosum. The aspiration is normally taken about half way between the edge of the forming placenta and the umbilical insertion. The aspirate is immediately examined in theatre. "Bush-like" villi with many sprouts will contain more mitotic figures than smooth "root-like" villi [12]. Chorionic villi have a characteristic fluffy, white appearance and float to the surface of the culture medium. There are, however, many traps for the unwary operator with this deceptively simple technique and we will see later that low miscarriage rates can be achieved only by experienced operators whose technique causes minimal tissue damage.

Transabdominal CVS

Blind transabdominal placental biopsy was carried out for the diagnosis of hydatidiform mole in 1965 [13]. Ultrasound directed biopsy was developed independently by my team and a Danish group [14–17].

The fundus of the pregnant uterus usually comes to lie against the abdominal wall after about eight weeks' GA. The uterus can therefore be reached through the abdominal wall at this point without traversing the bladder or bowel. This route provides access to both the anterior and the posterior walls of the uterus but, in order to reach the latter without penetrating the amniotic cavity, an empty bladder is required. In very rare cases, the uterus may still be acutely retroverted at this stage of pregnancy but even this can be corrected by vaginal manipulation.

At St James's Hospital, Leeds, an 18 gauge stilette pointed needle is inserted through the abdominal and uterine walls to reach the placental edge. This is used as a conduit for a thinner 20 gauge needle which is inserted to a point midway between the placental edge and the umbilical cord insertion. It is from here that villi are aspirated and, if the first attempt is unsuccessful, the outer needle can again serve to guide the aspirating needle to the correct position. In this way, the need for multiple insertions through the uterine wall is avoided.

The technique may be carried out "free-hand" [18] or by means of a biopsy attachment on the ultrasound transducer. The latter is easier for beginners and modern equipment allows the exact depth of required penetration to be measured. We are now working, in association with Huddersfield Polytechnic, on the use of robotics to provide a stable platform for the technique and, in association with our Bioengineering Department, on the use of a mechanical "gun" to insert the biopsy needle. Progress on the former has been impeded by the difficulty in obtaining indemnity for this work.

Amniocentesis

In most cases, culture of amniotic fluid cells is required for prenatal diagnosis and this takes a further ten days to four weeks, depending on the number of cells required and individual variations in the speed of cell growth.

The correct line for insertion of the needle is chosen by ultrasound. There is some difference of practice between those who select a point of entry and then carry out the technique with no further reference to ultrasound, and those who guide the needle into the correct position under continuous ultrasound control. The latter technique is greatly to be preferred, not because it is necessarily safer but because it is more reassuring for the patient. Patients are aware that the fetus is highly mobile within the amniotic sac and are therefore concerned about the possibility of a direct fetal hit if the operator cannot watch the needle tip at all times. In any event, ultrasound control is essential for amniocentesis before fourteen weeks of gestational age.

A number of measures have been recommended to decrease the chance of miscarriage and, although none of these has achieved universal acceptance, I recommend that they should all be followed where possible. A small needle should be used, and in practice this usually involves gauge 23 or 22. As little fluid as possible should be removed. The placenta should be avoided and if the procedure is guided by ultrasound in real time, it is almost always possible to thread the needle into the amniotic cavity, avoiding both the placenta and the fetus. The fetal surface of the placenta contains a large number of thin-walled veins, which do not have the retractile properties of the main umbilical vein, and fatal haemorrhage following puncture of these chorial vessels has been documented. Full aseptic skin preparation is also recommended since both acute and chronic infections are possible following amniocentesis. Transabdominal amniocentesis before 14 weeks of gestation is a somewhat deceptive technique, since the membranes are easily indented at this gestation and the needle tip may appear to lie within the amniotic cavity when it has not, in reality, penetrated the amnion.

Accuracy

Single Gene Defects

Enzyme assays and gene probe diagnosis may be inaccurate, because of the problems inherent in a laboratory technique (e.g. meiotic cross-over between marker and genetic locus in the case of linkage studies using random variations in base-pair sequences adjacent to the gene of interest) or because of sampling errors. The latter can result from maternal cell overgrowth of cultured tissue or, in the case of direct analysis of CVS samples, from inadvertent sampling of the placental remnant of a resorbing second sac. Chorionic sampling would seem to have the advantage over amniocentesis for diagnosis of single gene defects, since time is of the essence in these cases, and this technique usually provides sufficient tissue for direct analysis without culture. Transcervical CVS has been consistently shown to produce larger samples than those obtained by transabdominal methods [19–21], but we and others seldom, if ever, fail to obtain adequate samples for enzyme or DNA diagnosis. Moreover, the availability of the polymerase chain reaction (PCR) has decreased the demand for large samples for the purposes of gene probe diagnosis. Apparently all metal contains trace elements and transcervical CVS using a plastic cannula is the preferred method for diagnosis of Menke's syndrome, a rare X-linked recessive disease involving copper metabolism.

Chromosomal Abnormalities

The situation is far more complex for cytogenetic diagnosis. First, in these cases, the fetus is usually at much lower genetic risk, and therefore the relative importance of early diagnosis is not as great as in cases of higher risk [22]. Second, diagnostic inaccuracy, inherent in sampling technique, is greater than in the case of single gene defects. This is the result of cytogenetic disparity between different cell lines within the placenta and between placenta and fetus. Chorionic sampling seldom provides a complete false negative or false positive diagnosis provided that short (e.g. overnight) and long-term cultures are used. However, cytogenetic disparity resulting from postzygotic mitotic errors (mosaicism), between long- and short-term culture, or within either of these methods, is a formidable problem, occurring in 0.5%–4% of cases. The anxiety caused to parents and the expense to the Health Service are extremely important factors and probably constitute the single strongest argument against CVS as the standard method for cytogenetic diagnosis, especially in women of relatively low genetic risk.

Transabdominal CVS has not been shown to be any more accurate than transcervical methods, but laboratories report that the samples are much "cleaner" in the case of the former and one could hypothesise that maternal contamination will be less of a problem. However, it is not difficult to dissect decidual fragments away from chorionic villi when contaminating material is present. This is necessary, not only as a prelude to long-term culture, but also before short-term cultures (direct preparations), as maternal metaphases can be found in decidua in the first trimester [23].

It would appear that first trimester amniocentesis is superior to CVS (carried out by any route) with respect to ambiguous karyotype results. However, a note of caution must be sounded, since the proportion of chorion-derived cells in amniotic fluid is higher in earlier gestation. It is, therefore, likely that chromosome mosaicism or perhaps even false negative results will be more common in amniotic fluid cells sampled at an earlier gestational age than in conventional amniotic samples. However, trophoblast, the predominant tissue in short-term CVS cultures and that having the greatest potential cytogenetic disparity with the fetus, is not represented in long-term amniotic fluid cell cultures. It is, therefore, almost certain that the rate of mosaicism will be lower in early amniocentesis than in chorion samples. Initial experience would seem to bear this out [24–31]. The advantage that amniotic fluid may be analysed for alphafetoprotein and acetylcholinesterase is not as great as might at first be suspected, since false positive results may be more common at earlier gestations [32].

Multiple pregnancy always presents a special challenge in prenatal diagnosis. Amniocentesis is feasible because ultrasonic identification of the septum separating all dizygotic and many monozygotic fetuses allows liquor to be drawn from each sac independently [33]. Selective fetocide can then be carried out if one twin is found to be affected. Separate sampling from the relatively amorphous placental mass is more difficult to ensure. Again, the transabdominal method is advantageous as it affords precise localisation of the needle tip. Confirmation of the separate origin of each sample is virtually assured if they are of separate sex, if the results are different or if different chromosomal banding patterns are observed. If these differences do not exist, identification must rely on HLA antigens which have been demonstrated on cultured mesenchymal cells [34] or on genetic finger-printing. The latter has been found very useful and it is used routinely when cytogenetic and other laboratory results are the same. Following diagnosis it may be more difficult to ensure selective termination of the appropriate fetus, although this has been accomplished in the first trimester for a haemophiliac fetus [35]. We obtain a tissue sample at the time of fetocide for retrospective confirmation that the correct fetus has been killed and this is particularly important if fetuses are of the same sex.

Safety

Evaluating Safety of Invasive Prenatal Diagnostic Test

The technique which is safest in general is not always the method which is safest in one's own hands. Nevertheless established centres will wish to monitor pregnancy outcomes and audit these against reported results and doctors planning early pregnancy diagnostic services will wish to compare the safety of various procedures. Comparisons may be drawn from different centres doing different tests, or from particular centres offering alternative procedures (i.e. direct comparative studies). Non-randomised studies of both types are subject to numerous potential biases (some examples are given in table 7.1) and practitioners should also be aware of these when comparing their results with those reported in the literature or published by other centres. The principal maternal risk of prenatal tests is septicaemia. The principal fetal risk is miscarriage. The other documented risk is a degree of pulmonary under-development with amniocentesis. However, as with drug effects in pregnancy, any rare complications could not be distinguished from the natural incidence of these outcomes [36]. For example, if transabdominal CVS were associated with, say, a 1 in 300 risk of massive placental abruption or if amniocentesis caused brain damage, through preterm birth, in one baby in 400, we would never know about these risks.

Transabdominal versus Transcervical CVS

A good overview of the results of different types of CVS is presented in Professor Laird Jackson's Newsletters from the Division of Medical Genetics, Jefferson Medical College, Philadelphia. For example, one letter contains a report on a non-randomised study in the State of Victoria giving fetal loss rates of 5.9% and 2.4% for transcervical and transabdominal CVS respectively, though operator experience and gestational age at sampling may have differed between techniques, thereby biasing the comparison (Jackson Newsletter – May 1990). Amalgamating all data in the Newsletter, the transabdominal technique is associated with a consistently lower fetal loss rate. Take, for instance, the report from March 1987 when the total register had 25 000 patients. In those days transabdominal sampling was very new and therefore results may have been biased away from this technique. Nevertheless, this method was associated with a total loss rate of 2.1%, in contrast to over 4% for all transcervical methods. However, the very best centres, with the greatest experience with transcervical CVS, had fetal loss rates close to 2%. My tentative conclusion is that with great skill and experience, transcervical CVS can be as safe as transabdominal CVS but the learning curves are different. If we refer back to 10 December 1984, the time when the Danish group and Maxwell and the author were starting out on

Table 7.1. Sources of bias of particular relevance in non-randomised studies comparing safety of pre-natal diagnostic procedures

Factor	Applicable to	Mechanism of bias
Maternal age	1. Any comparison 2. Possibly any comparison	Miscarriage rate rises with age and a younger age distribution in one group could bias results away from the other
Removal of aneuploid fetuses from follow-up	Calculation of procedure related losses from background and total miscarriage rates	Many fetal losses have aneuploidy and these are removed from follow-up by testing, thereby biasing results in favour of the diagnostic test. This factor may counterbalance the maternal age effect, but to an unspecified degree
Gestational age at testing	Any comparison	Ultrasonically intact pregnancies may be lost at any time. First trimester procedures may be carried out at different mean gestational ages and this could bias results away from transcervical versus transabdominal CVS on the one hand or CVS versus amniocentesis on the other
Operator experience	Any comparison	Bias results against test where operators are on earlier part of learning-curve. Tests may have differently shaped learning curves
Unequal follow-up	Any comparison	If good news travels better than bad, results biased against test with better follow-up and vice versa

transabdominal CVS, Golbus (San Francisco) had done 400 transcervical samplings with a 6% fetal loss rate. Gilmore in Glasgow had an 18% loss rate. Rodeck in London, after doing 90 cases, had a 7% loss rate and Ward, also in London, had a 19% loss rate. It seems to me that the learning curves for the two techniques are different.

One of the worries with transcervical CVS has always been the fear of maternal septicaemia and septic abortion. Although it cannot be claimed that these complications are impossible following transabdominal procedures, nevertheless they would appear to occur much more commonly following transcervical aspiration and this would agree with our knowledge of the bacteriology of the cervix. It also fits with the experience of doctors who are well aware of the hazards of transcervical manipulations in countries where abortion is illegal. What about comparative studies within institutions? Finikiotis and Gower [37], on direct comparison of the two techniques, found two septic abortions in 84 transcervical CVS compared to no such cases amongst 126 transabdominal procedures.

Non-randomised studies are subject to selection bias, especially that which would arise if transabdominal sampling were done at a greater mean gestational age. A large randomised comparison of the two techniques was carried out by Brambati and et al. [19,38] who found that the transabdominal method was easier to learn. Analysable samples were produced more frequently with transabdominal CVS whereas uterine infection was more common after the transcervical method. It is fair to say, however, that total fetal loss rates were very similar with both techniques; 3.2% transabdominal (29/850) versus 3.9% transcervical (53/1350). Nevertheless, they have come to prefer the transabdominal method, having originally been pioneers of transcervical sampling. Wapner et al. [39] have also done a randomised trial as part of the, as yet unpublished, NIH trial, and compare abortion rates among pregnancies which were not terminated following the transabdominal and transcervical methods. They report 16 miscarriages among 659 pregnancies with transabdominal CVS (2.4%) versus 11 with miscarriages among 698 unterminated pregnancies following transcervical CVS (1.6%). This difference is not statistically significant and these authors, when amalgamating data from non-randomised and randomised patients also report a few cases of clinical infection among 1636 transcervical procedures versus no such cases among 964 transabdominal procedures. Bovicelli et al. [20] in a small (120 cases) randomised study report identical miscarriage risks with each method. Results from the transabdominal versus transcervical arms of the British trial and the Danish trial (co-ordinated by Dr Steen Smidt-Jensen) are awaited with interest.

CVS versus Traditional Amniocentesis

Abortion rates for pregnancies destined to continue following either transabdominal or transcervical CVS in centres with the greatest experience are of the order of 2%. Background total abortion rates at various stages of pregnancy have been documented [7,40] but these are of little relevance to the risk of miscarriage where the fetus is seen to be intact on ultrasound. A number of studies have now addressed specifically the probability of fetal loss in pregnancies, shown by ultrasound to be intact and viable at around ten weeks' gestation [41]. Exclusion of patients with missed abortions shows that the background abortion rate for ultrasonically viable pregnancies at ten weeks' gestation is about 2% [42]. This could be an overestimate for women of unselected maternal age, as patients having ultrasound in the first trimester may be a selected group with high risk of miscarriage, e.g. because of previous or threatened miscarriage. Furthermore, simple subtraction of background abortion risk from the total abortion rate following the procedure may not give a completely accurate measurement of the procedure-related risks. Patients requiring CVS are often older and may, therefore, have a higher background risk of abortion than unselected populations. For example Gilmore and McHay [43] estimate a 4% background fetal loss rate before 28 weeks for women of 35–39 years of age. On the other hand, it could be argued that this higher rate of fetal loss among older mothers is due to a higher incidence of chromosomally abnormal conceptions and, as affected fetuses are eliminated from the quoted abortion rates, this factor might be less important. Direct comparison of miscarriage risks following CVS and amniocentesis may also be confounded by gestational age at testing since the possibility of fetal loss in the interval between these procedures would bias results in favour of the later test. Despite this a prospective trial found no increased risk with CVS [44]. The only way to resolve this issue is by randomised trials using amniocentesis patients as controls. The Canadian trial suggests an excess of fetal loss of at least 1% with transcervical CVS although losses were extraordinarily high in both groups and the relatively small sample size (2600) of this trial makes measurement imprecise. Results of the large British trial are awaited with interest. Experienced centres can offer total fetal loss rates of under 2% with CVS. The background loss rate of euploid fetuses of similar aged mothers may be 1%. Therefore the procedure-related loss rate is likely to be little higher than the 0.3%–1% rate associated with amniocentesis [45,46].

Abortions after invasive tests follow two patterns. A proportion occur within the first week of chorion villus sampling and manifest with bleeding followed by abortion. A second, and equally common, form of abortion occurs between one and five weeks after the procedure. In many cases the fetus grows initially but this is followed by severe oligohydramnios, loss of the fetal heart beat and then abortion. The first pattern can be ascribed to mechanical disruption of the placenta and this is seldom seen in units with considerable experience. The second pattern of abortion, however, is presumed to be due to chronic infection [47], with an organism such as *Listeria*, *Chlamydia* or *Mycoplasma* since these are commonly found in the pregnant cervix [48]. A number of factors are thought to increase the risk of abortion following transcervical chorion biopsy. These include the need for repeated aspiration, a gestational age of greater than 11 weeks and immediate bleeding or sac puncture. In addition, ultrasonic demonstration of a subchorial haematoma or tract after the procedure may be associated with a higher chance of subsequent abortion [47]. Unfortunately, it is not possible to predict which pregnancies are at greatest risk of spontaneous abortion; gestational age-related human chorionic gonadotrophin or gestation sac volume measurements correlate very poorly with pregnancy outcome [49].

Safety of "Early" Amniocentesis

The safety of early amniocentesis is less well explored. Nevin et al. [50] report an abortion risk of less than 1%. Miller et al. [27] find that this technique is safer than transcervical chorionic sampling, but do not give fetal loss rates of either. Garrison et al. [51] do not give their overall miscarriage risks. Elejalde et al. [52] claim the remarkable result of no miscarriage in some 300 cases. Arnovitz et al. [26] quote a miscarriage rate of 1%. Godmilow et al. [25] quote a miscarriage rate of 3%. These results should all be taken as very preliminary, but it would seem that early amniocentesis is going to be as safe, or perhaps even safer than CVS as far as miscarriage is concerned. However, early amniocentesis may be more likely than later sampling to lead to pulmonary hypoplasia. Animal experiments show that this occurs to a greater degree when amniotic fluid is removed very early in pregnancy [53]. However, Nevin et al. [50] do not report any increase in this complication. Nevertheless, individual small series cannot detect even large increases in relatively rare events and I think it is essential that carefully documented perinatal follow-up should be carried out whenever early amniocentesis is performed. Pending these studies I think it is quite possible that this technique will largely supplant CVS for chromosome diagnosis, especially in low-risk women. The technique is likely to be at least as safe as transabdominal sampling and the problem of mosaicism will almost certainly be less, for the reasons given earlier. Although the cytogenetic result is likely to be made available somewhat later in pregnancy than is the case of CVS, it should nevertheless be possible to make this available by 14 weeks' GA and most practitioners are prepared to do transcervical termination up to this stage of pregnancy. However, this prediction is made with great caution since only 10% of published early amniocenteses have been carried out before 12 weeks' gestational age. Chorionic sampling will remain the test of first choice, for patients with single gene defects and those of very high cytogenetic risk, such as translocation carriers.

Other Issues

Chorionic villus sampling is more expensive than amniocentesis for the diagnosis of cytogenetic disorders. The clinical costs of early techniques (amniocentesis or CVS) are greater than those of conventional amniocentesis because the operator (usually highly trained and therefore expensive) is required for larger periods of time [54]. Similarly the laboratory costs of CVS are more than double those of amniocentesis because:

1. Both direct (or short-term cultures) and long-term preparations are necessary.
2. Ancillary investigations (to elucidate ambiguous results) are required more often.

In the case of most cytogenetic indications, where genetic risk is low, these costs greatly outweigh the financial advantages of early versus late termination of pregnancy. Foregoing long-term culture for cost reasons would increase the risk of a false negative result for both aneuploidy (a mosaic cell line may be missed in trophoblast or a trisomic chromosome might have been lost in the progenitors of this tissue) and chromosome rearrangements (banding is less clear-cut in short-term preparations). Foregoing the short-term culture increases the risk of maternal cell overgrowth and removes an attractive feature of CVS versus early amniocentesis – namely a broadly reassuring result within 24 hours of sampling.

Transcervical and transabdominal techniques have no significant relative cost advantages over each other except those contingent on any difference in fetal loss rates. However, transabdominal sampling may be more acceptable and certainly less embarrassing to patients. Monni et al. [55] asked 72 Italian women with experience of both techniques to give their preference for sampling method and all but one (who had a long history of needle phobia) voted for the transabdominal route, on the grounds of privacy and comfort.

If the probable greater safety of transabdominal CVS (especially regarding infection and miscarriage risk in the early phases of the learning curve) is added to the other advantages of the transabdominal technique, then I personally think that this emerges as the somewhat superior technique for general use. Its other advantage is that it can be carried out later in pregnancy and therefore staff can learn one technique suitable for application over a wide range of gestational ages. The need for multiple cannula insertions is less and, having observed both methods at firsthand, I think it is a more pleasant procedure for patients to undergo.

References

1. Lilford RJ. "In my day we just had babies". J Reprod Infant Psychol 1989; 7:187–91.
2. Hahnemann N. Early prenatal diagnosis; a study of biopsy techniques and cell culturing from extraembryonic membranes. Clin Genet 1974; 6:294–306..
3. Hamilton WJ, Boyd JD. Development of the human placenta. In: Philipp E, Barnus J, Newton M, eds. Scientific foundations of obstetrics and gynaecology. London: Heinemann Medical Books, 1970; 185–253.
4. Gosden JR, Gosden CM, Mitchell AR, Rodeck CH, Morsman JM. Direct vision chorion biopsy and chromosome-specific DNA probes for determination of fetal sex in first trimester prenatal diagnosis. Lancet 1982; ii:1416–19.
5. Rodeck CH, Morsman JM, Gosden CM, Gosden JR. Development of an improved technique for first-trimester microsampling of chorion. Br J Obstet Gynaecol 1983; 90:1113–18.
6. Gustavii B. First-trimester chromosomal analysis of chorionic villi obtained by direct vision technique. Lancet 1983; i:507–8.
7. Gustavii B, Chester MA, Edvall H et al. First-trimester diagnosis on chorionic villi obtained by direct vision technique. Hum Genet 1984; 65:373–6.
8. Nordenskjold F, Gustavii B. Direct vision chorionic villi biopsy for prenatal diagnosis in the first trimester. J Reprod Med 1984; 29:572–4.
9. Elias S, Simpson JL, Martin AO, Sabbagha RE, Gerbie AB, Keith LG. Chorionic villus sampling for first-trimester prenatal diagnosis: Northwestern University Program. Am J Obstet Gynecol 1985; 152:204–13.
10. Maxwell D, Czepulkowski BH, Heaton DE, Coleman DV, Lilford R. A practical assessment of ultrasound-guided transcervical aspiration of chorionic villi and subsequent chromosomal analysis. Br J Obstet Gynaecol 1985; 92:660–5.
11. Perry TB, Vekemans MJJ, Lippman A, Hamilton EF, Fournier PJR. Chorionic villi sampling: clinical experience, immediate complications, and patient attitudes. Am J Obstet Gynecol 1985; 151:161–6.
12. Verlinsky Y, DeChristopher PJ, Pergament E, Ginsberg NA. Histomorphological aspects of chorionic villi in first trimester fetal diagnosis. In: Fraccaro M, Simoni G, Brambati B, eds. First trimester fetal diagnosis. Heidelberg: Springer-Verlag, 1985; 178–88.
13. Alvarez H. Diagnosis of hydatidiform mole by transabdominal placental biopsy. Fetus Newborn 1965; 95:538–41.
14. Smidt-Jensen S, Hahnemann N. Transabdominal fine needle biopsy from chorionic villi in the first trimester. Prenat Diagn 1984; 4:163–9.

15. Lilford RJ, Maxwell D. The development of a transcutaneous technique for chorion biopsy. Proceedings of prenatal diagnosis group meeting "Progress in first trimester diagnosis", Queen Charlotte's Hospital, London, 1984.
16. Maxwell DJ, Lilford RJ, Czepulkowski B, Heaton D, Coleman D. Transabdominal chorionic villus sampling. Lancet 1986; i:123–6.
17. Lilford RJ, Irving HC, Linton G, Mason MK. Transabdominal chorion villus biopsy: 100 consecutive cases. Lancet 1987; i:1415–16.
18. Brambati B, Oldrini A, Lanzani A. Transabdominal chorionic villus sampling: a freehand ultrasound-guided technique. Am J Obstet Gynecol 1987; 157:134–7.
19. Brambati B, Oldrini A, Lanzani A, Terzian E, Tognoni G. Transabdominal versus transcervical chorionic villus sampling: A randomized trial. Hum Reprod 1988; 3:811–13.
20. Bovicelli L, Rizzo N, Montacuti V, Morandi R. Transabdominal versus transcervical routes for chorionic villus sampling. Lancet 1986; ii:290.
21. Mackenzie WE, Holmes DS, Newton JR. Study comparing transcervical with transabdominal sampling (CVS). Br J Obstet Gynaecol 1988; 95:75–8.
22. Lilford RJ. Does the tradeoff between the gestational age and miscarriage risk of a prenatal diagnostic test vary according to genetic risk? Lancet 1990; 336:1303–5.
23. Blakemore KJ, Samuelson J, Breg WR, Mahoney MJ. Maternal metaphases on direct chromosome preparation of first trimester decidua. Hum Genet 1985; 69:380.
24. MacLachlan NA, Rooney DE, Coleman D, Rodeck CH. Prenatal diagnosis: early amniocentesis or chorionic villus sampling. Contemp Rev Obstet Gynaecol 1989; 173–80.
25. Godmilow L, Weiner S, Dunn LK. Early genetic amniocentesis: experience with 600 consecutive procedures and comparison with chorionic villi sampling (Abstract). Am J Hum Genet 1988; 43 (Suppl 3):A234.
26. Arnovitz KS, Priest JH, Elsas LJ, Strumlauf E. Amniocentesis prior to 14.5 weeks gestation: experience in 142 cases (Abstract). Am J Hum Genet 1988; 43(Suppl 3):A225.
27. Miller WA, Davies RM, Thayer BA, Peakman D, Harding K, Henry G. Success, safety, and accuracy of early amniocentesis. Am J Hum Genet 1987; 41(Suppl):A281.
28. Hanson FW, Zorn EM, Tennant FR, Marianos S, Samuals S. Amniocentesis before 15 weeks gestation: outcome, risks and technical problems. Am J Obstet Gynecol 1987; 156:1524–31.
29. Johnson A, Godmilow L. Genetic amniocentesis at 14 weeks or less. Clin Obstet Gynecol 1988; 31:345–52.
30. Stripparo L, Bruscaglia M, Longatti et al. Genetic amniocentesis: 505 cases performed before the sixteenth week of gestation. Prenat Diagn 1990; 10:359–64.
31. Hanson FW, Happ RL, Tennant FR, Hune S, Peterson AG. Ultrasonography-guided early amniocentesis in singleton pregnancies. Am J Obstet Gynecol 1990; 162:1376–83.
32. Burton BK, Pettenati MJ. False positive acetylcholinesterase with amniocentesis (Abstract). Am J Hum Genet 1988; 43 (Suppl 3):A903.
33. Rodeck CH, Ivan D. Sampling pure fetal blood in twin pregnancies by fetoscopy using a single uterine puncture. Prenat Diagn 1981; 7:43–9.
34. Niazi M, Coleman DV, Mowbray JF, Blunt S. Tissue typing amniotic fluid cells: potential use for detection of contaminating maternal cells. J Med Genet 1979; 16:21–3.
35. Mulcahy MT, Roberman B, Reid SE. Chorion biopsy, cytogenetic diagnosis and selective termination in a twin pregnancy at risk of haemophilia. Lancet 1984; ii:866–7.
36. Orrell RW, Lilford RJ. Chorionic villus sampling and rare side-effects. Will a randomised controlled trial detect them? Int J Gynecol Obstet 1990 (in press).
37. Finikiotis G, Gower L. Chorion villus sampling – transcervical or transabdominal? Aust NZ J Obstet Gynaecol 1990; 30:63–5.
38. Brambati B, Lanzani A, Tului L. Transabdominal and transcervical chorionic villus sampling: efficiency and risk evaluation of 2,411 cases. Am J Med Genet 1990; 35:160–4.
39. Wapner RJ, Davis G, Jackson LG, Johnson A, Ronkin SL, Fischer RL. A randomised prospective comparison of transcervical and transabdominal chorionic villus sampling. Jefferson Medical College – report in preparation.
40. Hobbins JC. Consequences of chorionic biopsy. N Engl J Med 1984; 310:1121.
41. Wilson RD, Kendrick V, Wittman BK, McGillivray BC. Risk of spontaneous abortion in ultrasonically normal pregnancies. Lancet 1984; ii:290.
42. Liu et al. In: Jackson Newsletter: Early prenatal diagnosis. Jefferson Medical College, Philadelphia, 1987.
43. Gilmore DH, McHay MB. Spontaneous fetal loss rate in early pregnancy. Lancet 1985; i:107.
44. Crane JP, Beaver HA, Cheung SW. First trimester chorionic villus sampling versus mid-

trimester genetic amniocentesis. Preliminary results of a controlled prospective trial. Prenat Diagn 1988; 8:355–661.
45. Tabor A, Masden M, O'Bell EB et al. Randomised controlled trial of genetic amniocentesis in 4,606 low risk women. Lancet 1986; i:1287–92.
46. National Institute of Child Health and Human Development. Mid-trimester amniocentesis for prenatal diagnosis. Safety and accuracy. JAMA 1976; 236:1471–6.
47. Maxwell DJ, Lilford RJ. An interesting ultrasonic observation following chorionic villus sampling. J Clin Ultrasound 1985; 13:343–4.
48. Scialli AR, Neugebauer DL, Fabros S. Microbiology of the endocervix in patients undergoing chorionic villi sampling. In: Fraccaro M, Simoni G, Brambati B, eds. First trimester fetal diagnosis. Heidelberg: Springer-Verlag, 1985; 69–73.
49. Jouppila P, Huhtaniemi I, Herva R, Piiroinen O. Correlation of human chorionic gonadotrophin secretion in early pregnancy failure with site of gestational sac and placental histology. Obstet Gynecol 1984; 63:537–42.
50. Nevin J, Nevin NC, Dornan JC, Sim D, Armstrong JC. Early amniocentesis: experience of 222 consecutive patients, 1987–1988. Prenat Diagn 1990; 10:79–83.
51. Garrison CP, Berry PK, Mitter NS, Snoey DM, Johnston JS. Early amniocentesis and rapid prenatal diagnosis. Am J Hum Genet 1988; 43(Suppl 3):A234.
52. Elejalde BR, de Elejalde MM. Early genetic amniocentesis, safety, complications, time to obtain results and contradictions (Abstract). Am J Hum Genet 1988; 43(Suppl 3):A232.
53. Hislop A, Howard S, Fairweather DVI. Morphometric studies on the structural development of the lung in *Macaca fascicularis* during fetal and postnatal life. J Anat 1984; 138:95–112.
54. Lilford RJ, Irving H, Gupta JK, O'Donovan P, Linton G. Transabdominal chorion villus biopsy versus amniocentesis for diagnosis of aneuploidy: safety is not enough. In: Chapman M, Grudzinskas JG, Chard T, Maxwell D, eds. The embryo: normal and abnormal growth. Heidelberg: Springer-Verlag, 1991:91–100.
55. Monni G, Olla G, Cao A. Patient's choice between transcervical and transabdominal chorionic villus sampling. Lancet 1988; ii:1057.

Discussion

Campbell: In terms of CVS versus amniocentesis, the two things we could probably debate are the risks and the diagnostic accuracy. There are no results yet from the MRC study but there has been a fair amount written about the risks and we have some expert practitioners here.

Rodeck: Professor Lilford has shown very well that the data for and against transabdominal or transcervical are not strong. The crucial thing with transcervical CVS is the number of passages that have to be performed in going through the cervix in order to get an adequate sample. There is no doubt in my mind that if one can become good enough with a method that enables the sample to be obtained with one passage in nearly all cases, then one's results are very good indeed. The big transcervical series have produced losses that are as good if not lower than some of the transabdominal ones. That all depends on the technique that is used, the operator and so forth. Otherwise the operator experience is relatively unimportant and the other factors that Professor Lilford mentioned, maternal age, gestational age, become relatively more important. It depends where someone is on the learning curve as an operator.

Regarding women's choice of procedure, it is very striking that in all the series and surveys that have been done, the women tend to agree with the preference of the operator. So the bias has not been eliminated.

Nicolaides: The theoretical advantages of the transabdominal approach that Professor Lilford has raised have not been substantiated by all the studies that he has presented. It is not surprising that Wapner, who made his name by developing a tremendous experience with the transcervical approach, had a much higher complication rate when, after 4000 procedures, he decided to allocate the patients randomly into two groups; with one group he was extending his experience, and in the other he was learning how to develop a new technique.

The main concerns for me are not whether CVS should be performed transabdominally or transcervically, but whether the results of the Canadian study are relevant at all. If they are, we should abandon doing CVS. Unless within the next year the various scientists involved in biochemical screening come up with something that will distinguish chromosomally abnormal from normal babies in the first trimester of pregnancy, we must abandon first-trimester techniques.

Lilford: I would not disagree that it is not yet proven that CVS is better than amniocentesis at low genetic risk. But I would take quite strong issue at high genetic risk. I think CVS is the preferred technique. I have a paper in *The Lancet* [1] which proves mathematically that the higher the genetic risk, the greater the permissible fetal loss rates. So even if the Canadian study is correct, when we are looking at a higher genetic risk rate of 50% or 25% it is still not unreasonable for a woman to choose a test with a 1% higher miscarriage risk.

Ward: That may be due to the previous history of the patient. I imagine they have had previous induced abortions. Has that been taken into account?

Lilford: No. It takes the values into account on losing a pregnancy early, losing a pregnancy late and having a termination of pregnancy. As the chance rises that the woman will have to have a termination of pregnancy, so for a given set of values the expressed utility of an early test increases even if the miscarriage rate is higher.

Ward: I agree with everything Professor Lilford says in his final paragraph, but he started his presentation by saying that transcervical CVS was an unpleasant technique. I was not sure if it was unpleasant to him, or to the patient. And I thought the figures at Charlotte's where they lost three out of the six transcervicals they did, which is 50%, would naturally put one very much against that technique to begin with. So he is quite right in saying that he is likely to be biased. Having said that, I think he has a remarkable loss rate of 1.3%.

Two things he did not mention which Brambati et al. [2] mention about transabdominal. They consider it quicker and cheaper.

As far as Smidt-Jensen is concerned we have heard that Wapner was very expert at the transcervical technique; Smidt-Jensen conversely was a great expert at the transabdominal technique, and the results that will come out of that study for transcervical CVS are disastrous because he has forced himself to do a technique in which he has no interest and probably little expertise. They are two entirely different techniques. A transcervical technique is a technique which a trained gynaecologist can do and a transabdominal technique is a technique that an expert with ultrasound can do. They are two entirely different techniques and they must be complementary; Professor Lilford has not said that. And for twins I think even more complementary. That is a very important occasion when both tests should be available.

Lilford: The thing that struck me was that Smidt-Jensen's study will show a disastrously high miscarriage rate for transcervical. Although Wapner has a slightly higher miscarriage rate with transabdominal, it is only very slight. That would tend to support the notion that a learning curve for transcervical CVS is a more dangerous learning curve than for transabdominal.

As to the two being complementary, my senior registrar keeps telling me that they are complementary and that I should get back to doing transcervical, and I can see what he means. There are some cases where I delay or prevaricate for a week which would be quite easy transcervically. The trouble is that I do not want to get back on that dangerous learning curve. The one case where transcervical CVS has a clear advantage is Menkes' disease, where there is copper metabolism and a metal needle cannot be used.

Nicolaides: On the question of CVS for first trimester diagnosis in multiple pregnancies, we have to await the results of a possible study comparing the advantages and disadvantages of fetocide in the second trimester versus the first trimester. In the absence of any convincing evidence in favour of the one or the other, I personally believe that there should perhaps be no place in the first trimester for CVS or amniocentesis when we are dealing with a very low genetic risk. It is extremely unlikely that there will be any benefit from the disclosure of the results in the first trimester. If the genetic risk is 1 in 100, the chances that both fetuses would be affected must be low, and the danger is that however careful one is in distinguishing between the two fetuses, it will be much more difficult to decide which is fetus one and fetus two when it comes to considering the option of selective fetocide, for example.

Meade: It is not giving away any secrets to re-emphasise something in Chapter 6. The extent to which the MRC trial will be able to shed any light on transabdominal versus transcervical is almost negligible. It was allowed as a subrandomisation option, but the numbers involved are very small in the randomised comparison.

Campbell: Clearly this will be disappointing, for those proponents of transabdominal will claim that the results are biased because those involved in the study have had a big learning curve and the learning curve affects transcervical biopsy. But we shall wait and see what the results are.

Could we get on to accuracy, and here our genetic colleagues will come on the scene. We have heard of placental mosaicism affecting results of CVS and less so amniocentesis, and certainly the false positive rate would suggest it is something which is a concern.

Lilford: I think that it is likely that early amniocentesis will have a lower mosaic rate than CVS. I know Professor Gosden will tell us that there is a much higher proportion of chorion cells in the amnion in early pregnancy, but for all that I do not think there are many trophoblast cells in there, and it is trophoblasts that are a real problem when it comes to mosaicism.

Gosden: There are trophoblast cells in early amnios.

Lilford: But they do not divide, so the result would only be the mesenchyme result. So I am sure earlier amniocentesis will show a lower mosaic rate.

Gosden: The same trial that needs to be done with regard to safety – that is early amniocentesis versus CVS, if a first trimester procedure is wanted – will have to address diagnostic accuracy as well. And it will take a large series to answer these questions effectively.

The components of diagnostic accuracy are false positives, false negatives and maternal cell contamination, and the total of the three seems to be quite large. And in any equation the risks of sampling plus the accuracy risks will have to be added in, and that is what will add up to a total figure.

Williamson: I wanted to comment on DNA analysis in CVS as against early amniocentesis. We have had very few problems with DNA analysis and maternal contamination, which is the major problem for DNA analysis with CVS. We have had rather more limited experience, obviously, with early amniocentesis, but we also have had rather more problems with maternal contamination, partly because everything has to be checked more carefully.

Galjaard: A view from the Continent. When I come to England I hear from laboratory people that they are unfortunate; they consider they are very good but the obstetricians are not so good, and I hear from the obstetricians that they are perfect but the labs are not good.

When the group comes to discuss their recommendations, they might consider limiting expertise in those active things that need not be done everywhere and centralising the laboratories in the same place where the good obstetricians are to be found. It will then not be necessary to speak about differences between 1.5 and 1.6 but about 10 versus five or two. The percentages and the mistakes would be quite different depending on how the service is organised.

Secondly, could I hear a comment on a recent paper by Cohen-Overbeek [3], a pupil of Professor Campbell now in Rotterdam, about timing of transabdominal in which she decided that at the age group it might be better to postpone.

One further comment. Obstetricians always talk of so and so being very good at one procedure, someone else being good at something else. This one is an expert at transcervical, that one has a good record in transabdominal. All obstetricians agree that the operator should have some quality, and then they accept that the labs are all the same. But this is not the truth. The same concepts apply to the labs. There are some very good labs that have had no mistakes so far in 20 000 cases. We can continue to talk about how they see themselves, but they have not made mistakes.

Lilford: The idea that came up in the Rotterdam paper was that a disadvantage of a technique which is done too early is that the miscarriage rate afterwards will be higher in total; not procedure-related, but because the background rate is higher earlier on. The implication there is that people will feel worse having had a miscarriage after the invasive test which may have potentiated the miscarriage.

Like so much of the discussion I suspect that that must hinge on genetic risk. If genetic risk is low, the chance of it happening is high versus genetic risk. But if the genetic risk is very high, my experience with patients would suggest that they want to get on with the diagnosis as early as they possibly can.

Allan: It is not true that the take-up rate by patients is greater the earlier the test, which actually has quite a lot of bearing on the "CVS versus amniocentesis argument". If patients find it more acceptable to have early diagnosis, and therefore be able to make any decision earlier, and they are taking it up, then surely that has to come into the equation.

References

1. Lilford RJ. Trade-off between gestational age and miscarriage risk of prenatal testing: does it vary according to genetic risk? Lancet 1990; 336:1303–5.
2. Brambati B, Lanzari A, Tului L. Transabdominal and transcervical chorionic villus sampling: efficiency and risk evaluation of 2,411 cases. Am J Med Genet 1990; 35:160–4.
3. Cohen-Overbeek TE, Hop WCJ, den Ouden M, Pijpers L, Jahoda MGJ, Wladimiroff JW. Spontaneous abortion rate and advanced maternal age: consequences for prenatal diagnosis. Lancet 1990; 336:27–9.

Chapter 8

Cardiac Ultrasound Scanning

L. D. Allan

Introduction

Many of the structures forming the fetal heart can be identified by cross-sectional ultrasound screening from as early as 14 weeks' gestation. The vaginal probe is most useful at this stage but the abdominal approach with a high resolution transducer will also display the cardiac connections in the majority of patients. More complete fetal heart examination usually takes place at 18 weeks' gestation, when the connections can be seen in every patient. The addition of Doppler examination of the blood flow in the heart, both pulsed and colour flow mapping, provides details of cardiac function as well as confirmation of the structure seen by cross-sectional scanning.

There are two "levels" of fetal cardiac scanning in operation at the present time in the UK. During routine antenatal scanning, the four chamber view of the heart can be examined by the ultrasonographer or obstetrician. This will detect about two major forms of structural heart disease per 1000 fetuses scanned. A complete fetal echocardiogram is more time-consuming and skilled but identifies all the cardiac connections and should detect about four to five forms of heart disease per 1000 patients scanned. The rate of congenital heart disease in live births is about eight per 1000. Thus, the four-chamber scan is suitable for population screening whereas the complete fetal echocardiogram is best concentrated in specialised units run in conjunction with paediatric echocardiographers. Both levels of study will miss some forms of heart disease, some major forms of heart disease being missed by a four chamber view alone, although only minor defects should be overlooked by a more extensive study.

Four-Chamber Screening

The four-chamber view of the fetal heart is visualised by imaging a transverse section of the thorax just above the diaphragm. The heart lies mainly in the left chest with the apex pointing out toward the left anterior chest wall. The normal position of the heart within the thorax and in relation to the spine must become absolutely familiar, whatever the fetal position.

The same method of orientation is always used in order to identify the cardiac chambers. In Fig. 8.1, the first step is to locate the spine. Opposite the spine is the anterior chest wall or sternum. Below the sternum is the right ventricle. Returning to the spine, the descending aorta is seen as a circle in the mediastinum lying anterior to the spine. Related to the aorta anteriorly is the left atrium. The remaining intracardiac chambers, the right atrium and left ventricle, can then be identified.

The appearance of the four-chamber view will vary according to the orientation of the fetus to the ultrasound beam. Fig 8.1 illustrates the image obtained when the apex of the heart is closest to the transducer. (When colour flow mapping is used the forward flow into the ventricles is coloured red.) If the view is correctly orientated the flows into each side should appear equal. When the fetus is in this position, the beam is parallel to the interventricular septum. If the fetus lies with

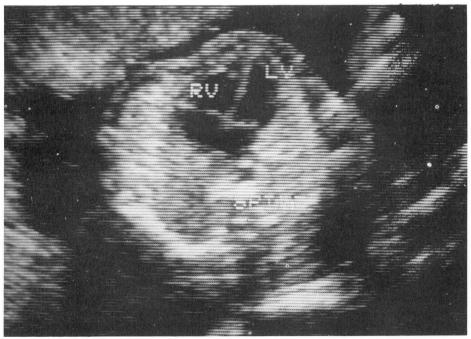

Fig. 8.1. The four-chamber view is seen with the fetus in an ideal position with the apex of the heart closest to the transducer. The ultrasound beam is parallel to the interventricular septum in this orientation.

Cardiac Ultrasound Scanning

the right chest anterior or the left or right back anterior, the appearance will be a little different but the important points can always be seen in the normal heart. They are:

1. The heart occupies about a third of the thorax.
2. There are two atria of approximately equal size.
3. There are two ventricles of approximately equal size and thickness. Both show equal contraction in the moving image.
4. The atrial and ventricular septa meet the two atrioventricular valves at the crux of the heart in an offset cross.
5. Two opening atrioventricular valves are seen in the moving image.

After 32 weeks' gestation, however, the right ventricle may look slightly larger than the left in the normal fetus. The reason the cross at the centre of the heart is not "straight", is that the septal leaflet of the tricuspid valve inserts slightly lower in the ventricular septum than the mitral valve.

Since 1986, we have taught these rules of four-chamber view scanning to those obstetric units expressing an interest and in the last two years there has been a more organised programme running in the South East Thames Region. This has led to the surge in number of abnormalities detected illustrated in Fig. 8.2. However, several problems have been encountered during the screening programme and have limited its success. These are:

1. Gestational age
2. Lack of organised policy
3. Maternal obesity
4. Quality of equipment
5. Fetal position
6. Time and motivation

The most important problem is the gestational age at the routine scan. The ideal for the maximum identification of cardiac and extracardiac anomalies is 18–20 weeks. Many scans are still scheduled for 16 weeks, when the four-chamber

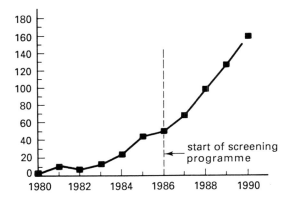

Fig. 8.2. Number of cases of congenital heart disease diagnosed prenatally each year in Guy's Hospital.

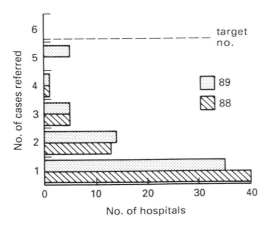

Fig. 8.3. The rate of referral of detected cases of CHD in different hospitals in 1988 and 1989. Most units should expect an annual pick-up of 5 to 6 cases. In 1989, five hospitals achieved this target.

view is only identifiable in about 50% of cases. Some units do not have an organised programme of fetal scanning for abnormality. A standard form or "checklist" should be used to try to establish standards for each operator and each unit. Maternal obesity is also a handicap in fetal imaging. This could be solved by giving routine appointments to mothers at 20 weeks' gestation if the maternal weight is over a given value at booking. The quality of ultrasound equipment in some obstetric units is not always adequate for the standards of fetal scanning which can now be achieved and which are expected by the public. The fetal position at the routine scan can be unfavourable but if there is adequate time and staffing levels the patient can be brought back later in the day or on a subsequent occasion. If the four-chamber view is being correctly examined during every obstetric scan, the following conditions would be recognised prenatally:

1. Aortic, mitral, pulmonary, or tricuspid atresia
2. Atrioventricular septal defect
3. Severe Ebstein's anomaly or tricuspid dysplasia
4. Severe coarctation or interruption of the aorta
5. Double inlet ventricle.

The number of major cardiac anomalies that could be detected in the UK in one year would be of the order of 1300. At present about 10%–12% of those detectable are found but this continues to improve steadily (Fig. 8.3).

The Complete Fetal Echocardiogram

Certain groups of pregnancies are at increased risk of congenital heart disease (CHD), and should be selected for specialised fetal echocardiography. These include:

1. A family history of CHD. If one previous child has had CHD the recurrence risk is one in 50. Where there have been two affected children the risk increases to one in 10. When a parent is affected the risk to the next generation is of the order of one in 10.
2. Maternal diabetes is associated with a statistical risk of cardiac malformation of about one in 50. Good diabetic control in early pregnancy probably diminishes this risk.
3. Exposure in early pregnancy to teratogens such as lithium, phenytoin or steroids is reported to be associated with a one in 50 risk of heart malformation.
4. The detection of an extracardiac fetal anomaly on ultrasound should lead to a complete examination of the fetal heart, as many types of abnormality, for example exomphalos, are often linked with heart disease. Abnormalities in more than one system in the fetus should arouse the suspicion of a chromosome defect and a fetal blood sample or amniocentesis should be performed in order to give an accurate prognosis to the patient. If the chromosomes are normal, a syndrome of co-existing defects may be present.
5. Some fetal arrhythmias are associated with structural heart disease, especially complete heart block which produces a sustained bradycardia of less than 100 beats per minute.
6. Non-immune fetal hydrops can be due to congenital heart disease, and a fetal echocardiogram should be an essential part of the work-up of these patients. Fetal hydrops will have a cardiac cause in up to 25% of cases.
7. By far the most important high-risk group seen in the last two years are those "normal" pregnancies where the obstetrician notices an abnormality of the four-chamber view on a routine scan.

Major cardiac malformations affect the connections of the heart. The heart is made up of six connections, three on each side of the heart. These are the atria, which receive the venous drainage, the ventricles and the great arteries.

On the right side of the heart, the inferior and superior vena cavae drain to the right atrium, which connects through the tricuspid valve to the right ventricle. The pulmonary artery arises from the right ventricle. On the left side of the heart, the left atrium receives the pulmonary veins. The left atrium connects via the mitral valve to the left ventricle which gives rise to the aorta.

Once the method to image these connections is learnt, major forms of heart disease can be excluded. Heart defects which do not involve a connection abnormality are usually relatively minor defects with a good chance of successful correction, although many can be recognised prenatally. However, it is mainly the more severe forms of heart disease which are important to diagnose in fetal life.

Examination of the four chamber view will demonstrate the pulmonary veins and atria, the atrioventricular connections and the two ventricles. Thus, in this one view alone, three of the six connections are seen and clues to the normality of the great arteries can be identified. This, therefore, is an important view to master completely and thoroughly. This view should be identifiable in every patient from 18 weeks' gestation onwards with generally available real-time equipment.

Imaging the Great Arteries

The great arteries can be imaged in a variety of projections [1] only some of which will be described here. The horizontal views are usually the easiest views of the great arteries to find although they can be difficult to understand at first. The longitudinal views are more familiar to those used to postnatal cardiac imaging. Moving cranially from the four-chamber view, maintaining a horizontal projection, the aorta can be seen arising in the centre of the chest from the left ventricle (Fig. 8.4). This artery sweeps out to the right of the thorax at its origin. The horizontal section cranial to this plane will visualise the pulmonary artery. This artery arises anteriorly, close to the chest wall and is directed straight back towards the spine. The right ventricular outflow tract, the pulmonary valve, the main and right pulmonary artery and the ductal junction with the descending aorta can be seen in the longitudinal cut illustrated in Fig. 8.4. The ascending aorta is seen in cross-section as a circle in the centre. The aortic arch can be imaged in a longitudinal or horizontal section of the fetus. The important features to note in these views are:

1. Two arterial valves can always be seen.
2. The aorta arises wholly from the left ventricle.
3. The pulmonary artery at the valve ring is slightly bigger than the aorta.
4. The pulmonary valve is anterior and cranial to the aortic valve.
5. At their origins the great arteries lie at right angles to, and cross over, each other.
6. The arch of the aorta is of similar size to the pulmonary artery and duct and is complete.

If all these normal features are seen, major anomalies of the great arteries can be excluded.

Cardiac Malformations

Abnormalities of Connection

These can occur at the venous-atrial, the atrioventricular or the ventricular-arterial connection. Any of the connections may be absent (or atretic) or displaced (or inappropriate).

At the Venous-atrial Junction

The venous connection on the right side of the heart, the inferior vena cava, is rarely abnormal but when it is absent, this is usually associated with complex heart disease, which would be recognised from other features. Thus, it is not essential to identify this structure in a normal study.

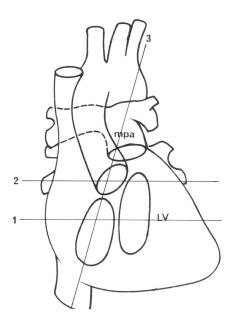

Fig. 8.4. Diagrammatic illustration of the cardiac skeleton. Section 1 images the four chamber view, section 2 the origin of the aorta can be seen in a horizontal section above the four-chamber view. Section 3 demonstrates the right heart connections: the inferior vena cava entering the right atrium, the tricuspid valve between the right atrium and ventricle, the origin of the pulmonary artery from the right ventricle and its connection to the duct.

Total Anomalous Pulmonary Venous Drainage. However, it is important to recognise the absence of the pulmonary venous connection. In the isolated form, it is an uncommon defect constituting about 2% of postnatal series [2]. The prenatal features include right ventricular dilatation and failure to identify the normal insertion of the pulmonary veins. There are three possible anatomical sites for anomalous venous drainage, supracardiac, infracardiac or the coronary sinus. The echocardiographic appearance will depend on the site of drainage. Where total anomalous pulmonary venous drainage occurs as an isolated anomaly, there is a good prognosis for corrective surgery if the infant is in good condition at operation. Thus, it would be a lesion which would particularly benefit from prenatal diagnosis and immediate postnatal therapy. However, the condition often occurs as part of the asplenia syndrome in combination with other cardiac anomalies where the outlook is poor.

Tricuspid atresia

In tricuspid atresia, there is no patent valve seen in the normal position between the right atrium and ventricle. The right ventricular chamber is small or indiscernible as a consequence of the lack of flow into the ventricle (Fig. 8.5). There is a ventricular septal defect of varying size. The great arteries can be

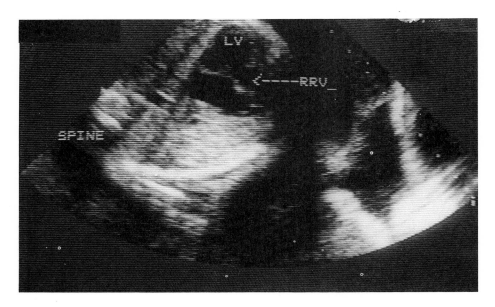

Fig. 8.5. The right ventricle (RRV) is much smaller than the left (LV) and there is no opening valve in the position of the tricuspid valve. This is tricuspid atresia.

normally connected or transposed. The arterial connection is important to identify as this will influence the prognosis. This form of heart disease constitutes about 3%–4% of both the infant and prenatal series, and is rarely associated with extracardiac anomalies.

Mitral Atresia

This is present when the mitral valve is not patent. It occurs in combination with a normally connected but atretic aorta in the hypoplastic left heart syndrome, or with double outlet right ventricle. The hypoplastic left heart will be described under aortic atresia. In mitral atresia with double outlet, there is no direct communication between the left atrium and ventricle, the floor of the left atrium being formed by muscular tissue. There is usually a ventricular septal defect and the great arteries arise from the right ventricle in abnormal parallel orientation. It is seen more frequently prenatally than in postnatal series and forms 5% of our series.

Atrioventricular Septal Defect

A common atrioventricular valve is found when there is a defect in the atrial and ventricular septa at the crux of the heart at the normal point of insertion of the two atrioventricular valves. The echocardiographic appearance is of the single valve opening astride the crest of the ventricular septum (Fig. 8.6). It is one of the commonest forms of heart disease seen in prenatal life [3] representing 17% of

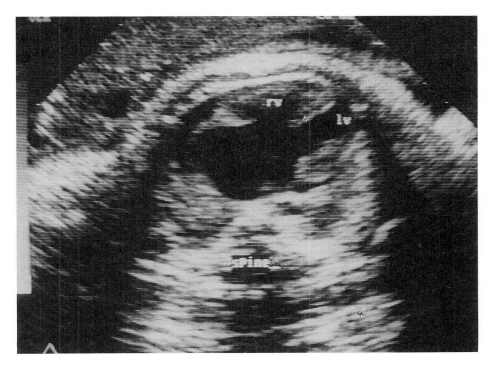

Fig. 8.6. There is a large defect in the atrial and ventricular septum with a common atrioventricular valve seen open in diastole in this frame.

our series compared with the expected rate of around 5% found in postnatal life. This type of defect is found in two situations: associated with complex other cardiac anomalies or associated with trisomy 21. An isolated defect in a normal child can occur but it is uncommon. Thus, when the diagnosis of an atrioventricular septal defect is made, the fetal chromosomes should be analysed. The prognosis will depend on the presence of associated cardiac or extracardiac lesions.

Double Inlet Connection

Double inlet connection is an unusual defect where both atrioventricular valves drain to one ventricle. The appearance is of absence of the interventricular septum. The origin of the great arteries is variable.

The Ventriculo-arterial Junction

Pulmonary atresia. In pulmonary atresia, the pulmonary root will not be found in its usual position or it will be very tiny in relation to the aorta. It occurs in three settings – with intact ventricular septum, with a ventricular septal defect, or as part of a complex of congenital heart disease. Pulmonary atresia with intact ventricular septum of the form seen most commonly postnatally is characterised by a small and hypertrophied right ventricle. This form represents about 2% of CHD in both the prenatal and postnatal series. However, pulmonary atresia is commonly found prenatally with a dilated right ventricle and severe tricuspid incompetence. This form is described later, under tricuspid dysplasia.

Aortic atresia. In this fatal condition, the aorta is tiny and the left ventricle small. Prenatally the left ventricle can be impossible to find or the left ventricular cavity can be small, thickwalled and echogenic as illustrated in Fig. 8.7. The aorta is hypoplastic from its origin to the site of entry of the ductus, where it becomes larger. Flow can be documented as reversed on pulsed or colour Doppler in the ascending aorta. It is a common form of heart disease, representing about 10% of postnatal and up to 20% of the prenatal series. It is rarely associated with extracardiac anomalies.

Transposition of the Great Arteries. In transposed great arteries, the normal positional arrangement of the arteries is lost. The aorta will arise anterior to the pulmonary artery and in parallel orientation to it, instead of the normal appearance of being at right angles to the pulmonary artery at its origin. It constitutes around 10% of infant series although it is infrequent in the prenatal series. This is because this lesion will not be detected by the four-chamber view alone and it is rarely associated with extracardiac anomalies. It can be a difficult diagnosis to make unless the operator is experienced in the interpretation of both the horizontal and the longitudinal views of the fetal heart. However, it is

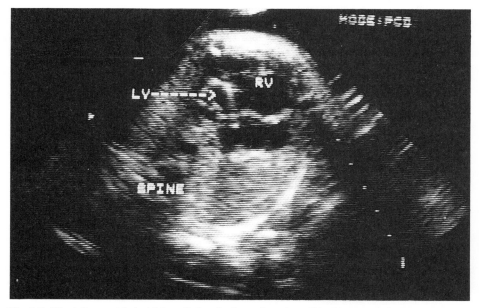

Fig. 8.7. The left ventricle is thick-walled and echogenic and small in this case of the hypoplastic left heart syndrome.

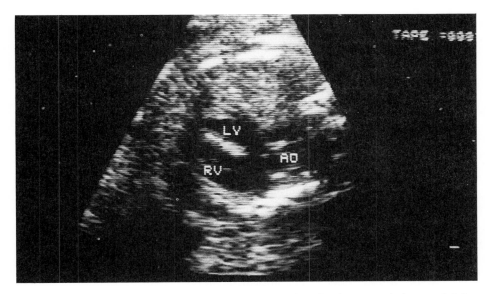

Fig. 8.8. The aorta arises astride the ventricular septum. There is a large ventricular septal defect positioned below the aorta. Examination of the pulmonary outflow tract will differentiate the three possible underlying diagnoses in this condition.

important to recognise prenatally as it has a relatively good prognosis if early corrective surgery takes place.

Aortic Override. In some forms of heart disease, the aorta is displaced anteriorly and arises astride the ventricular septum with a subaortic ventricular septal defect (Fig. 8.8). This finding is seen in Fallot's tetralogy, pulmonary atresia with a ventricular septal defect and with a common arterial trunk. The three conditions are differentiated by examination of the pulmonary outflow tract. The distinction is important as each carries a very different prognosis for corrective surgery. In the tetralogy, the pulmonary outflow is obstructed but patent; it is completely obstructed in pulmonary atresia and arises from the aorta in common arterial trunk. The tetralogy of Fallot is much the most common of the three conditions. All are seen prenatally with around the same incidence as is found postnatally, but in prenatal life, tetralogy of Fallot is associated in up to 50% of cases with extracardiac, particularly choromosomal, anomalies.

Double Outlet Right Ventricle. In double outlet right ventricle, both great arteries arise from the right ventricle anterior to the ventricular septum. Lesions which can occur in association with double outlet right ventricle include a ventricular septal defect, an atrioventricular septal defect, mitral atresia and pulmonary stenosis. The position of the great arteries relative to each other can vary but the normal "crossing over" of the two is usually lost.

Additional Abnormalities

Many other cardiac malformations can occur which do not involve abnormality of the connections of the heart. These tend to be less severe defects and more amenable to correction. Some can, however, be recognised prenatally. They include:

1. Valve stenosis
2. Ventricular septal defect
3. Valvular dysplasia or displacement
4. Cardiac tumour
5. Aortic arch abnormalities

Valve Stenosis

Any intracardiac valve can be stenosed or partially obstructed but in practice this mainly occurs at the aortic or pulmonary valve. The respective ventricular chamber and the valve itself may appear thickened.

When there is valvular obstruction prenatally, the affected artery will often be disproportionately small in size relative to the other artery. The Doppler sample volume placed in the stenosed artery will sometimes show a velocity of blood flow above the normal range but this is not such a consistent feature as that produced by a valvular stenosis postnatally. The left ventricle characteristically is globular, stiff and echogenic in appearance. The high left ventricular diastolic pressure in this condition will often lead to secondary left atrial hypertension and left to right shunting across the foramen ovale. In critical aortic stenosis, the prognosis is very poor even if the obstruction is relieved immediately after delivery, as the long-term damage to the left ventricle in the form of endocardial fibroelastosis is usually irreversible. In contrast, in critical pulmonary stenosis, early balloon valvotomy should be associated with a good result.

Ventricular Septal Defect

This is the most frequent form of heart disease seen in childhood, constituting nearly 20% of cases. However, it is relatively uncommon in the prenatal series and when it is found in the fetus, it is frequently associated with chromosomal anomalies. Ventricular septal defects (VSDs) can be present in various sizes and positions or be part of more complex anomalies. These factors will influence the prognosis. The majority of isolated VSDs close spontaneously after birth and therefore will not influence the child's future. Less than 5% of VSDs will require surgical correction.

Valvular Dysplasia

Any cardiac valve can be dysplastic but in practice in prenatal life this particularly affects the tricuspid valve. This valve may also be displaced from its normal position into the body of the right ventricle in Ebstein's anomaly. Both abnormalities will produce incompetence of the tricuspid valve and right atrial dilatation. This may be very gross and lead to secondary lung compression and hypoplasia. These conditions are both rare in paediatric practice but frequent in the prenatal series. This is probably because secondary lung hypoplasia results in early postnatal death. We have only two survivors in our series of 38 such cases where the right atrium was significantly dilated during gestation.

Aortic Arch Anomalies

The aortic arch can be interrupted completely, or partially obstructed as in coarctation. Coarctation can be a simple shelf region at the distal end of the arch or be associated with severe arch hypoplasia. Coarctation is a common form of heart disease, comprising about 10% of the total, but those cases recognised prenatally constitute the more severe end of the spectrum of this disease. The association of cystic hygroma, fetal hydrops, coarctation and Turner's syndrome is commonly recognised in early fetal life. The clues to the diagnosis of arch anomalies include the recognition of enlargement of the right ventricle and pulmonary artery relative to the left ventricle and aorta, in early pregnancy. A horizontal section of the arch of the aorta will show varying degrees of narrowing in coarctation and be incomplete in interruption. Where the coarctation is discrete and there are no other anomalies the prognosis for early corrective surgery is good. However, in complete interruption or severe arch hypoplasia it can prove difficult or impossible to reconstruct the arch adequately, and the mortality in those cases will be high.

Cardiac Scanning at Guy's Hospital

Electively, high-risk patients are booked for cardiac ultrasound scanning at 18 and 24–28 weeks' gestation. At 18 weeks all the connections are seen, and later, more minor defects are sought. Over 7000 such patients have been examined in our unit since early 1980. Over 500 anomalies have been accurately detected. In no patient has it proved impossible to visualise the atrioventricular connections and the two great arteries. Some minor defects, such as small ventricular septal defects, secundum atrial septal defects, and valve stenosis have been overlooked. No major false positive predictions have been made.

It is apparent from consideration of the forms of CHD seen prenatally that a different spectrum of disease is seen in prenatal life from those who survive to infancy [4]. This is reflected in the outcome of the pregnancies in which fetal heart disease has been detected (Fig. 8.9). Malformations detected tend to be the most severe forms of heart disease, and defects which are not commonly seen postnatally are frequently recognised, such as tricuspid dysplasia and cardiac tumours. Many of these fetuses do not survive intrauterine life, and this accounts partly for the discrepancy in disease patterns between prenatal and postnatal life. Up to 25% of continuing pregnancies with CHD resulted in spontaneous fetal

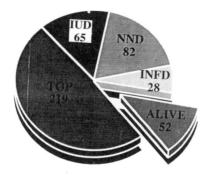

Fig. 8.9. The outcome of 446 cases of fetal cardiac anomaly detected to the end of 1989. Over half the total number of patients chose termination of pregnancy. Less than 12% of the group are alive. TOP, termination of pregnancy; IUD, intrauterine death; NND, neonatal death; INFD, infant death.

loss. A high proportion of parents, up to 75% when the diagnosis is made in time, will elect termination of pregnancy. Parents make their own decision concerning termination or continuation of the pregnancy, based on the information given by the paediatric cardiologist. A significant proportion (20%) of our detected cardiac abnormalities had associated chromosome defects. This possibility must be included in the counselling and the karyotype performed where this is appropriate. The presence of multiple congenital anomalies contributes to the high mortality found in our series of prenatally detected heart disease.

Prenatal Diagnosis of Fetal Abnormalities

Prenatal diagnosis of fetal abnormalities, like most new topics, has grown in a sporadic and haphazard way. It is now time to rationalise this specialty in order to improve the service provided and offer the optimum care for patients. The intrauterine diagnosis of any form of defect is a skill which should be concentrated in specialised centres. These centres should be organised to provide a comprehensive service which should include:

1. Experience in the diagnosis of all forms of fetal defect with access to shared care with the related paediatric specialist
2. Experience in fetal invasive techniques, such a drainage, shunt and needling procedures
3. Expert fetal pathology service for vital audit of diagnosis
4. Counselling for mothers with continuing pregnancies and for parents after termination of pregnancy
5. Delivery facilities for affected pregnancies under the management of a collaborating team.

Routine ultrasound departments should have easy access to expert advice and should be able to refer patients on an urgent basis. A videotape reporting service can work well for hospitals far from the specialist centre. Feed-back information is essential for these units, not only from the referral centre but from their own paediatricians if abnormalities which are detectable are being missed. A recent survey of approximately 80 infants presenting consecutively over a six-month period to Great Ormond Street Hospital with major heart disease, indicated that 50 should have been detectable on a four-chamber view scan. All but one mother had had a fetal scan. Thus, the technology and practice of fetal scanning is in place. It needs to be more organised and standardised if it is going to be maximally effective. However, effectiveness should be measured not only in disease prevention but in its ability to provide informed choice to all pregnant women.

References

1. Allan LD. Manual of fetal echocardiography. Lancaster, UK: MTP, 1986.
2. Anderson RH, Macartney FJ, Shinebourne EA, Tynan M, eds. Paediatric cardiology. London: Churchill Livingstone, 1987.
3. Machado MVL, Crawford DC, Anderson RH, Allan LD. Atrioventricular septal defect in prenatal life. Br Heart J 1988; 59:352–5.
4. Allan LD, Crawford DC, Anderson RH, Tynan MJ. Spectrum of congenital heart disease detected echocardiographically in prenatal life. Br Heart J 1984; 54:523–6.

Chapter 9

Doppler Ultrasound Studies and Fetal Abnormality

B. J. Trudinger

Introduction

The ability to study the fetal circulation with Doppler ultrasound has provided new information about fetal physiology and the pathophysiology of diseases of pregnancy. The fetal umbilical circulation has been most studied [1]. Such studies are simply performed. Studies have also been carried out on blood flow within the fetus, in particular in the aorta and the cerebral circulation.

In the circumstance of fetal abnormality such studies have been carried out to a limited extent. Most information is available about the relationship of the fetus and its placenta through the study of the umbilical artery flow velocity waveform in the presence of a spectrum of fetal anomalies. In the presence of congenital structural heart disease, Doppler assessment of blood flow in the cardiac chambers and across the valves is an integral part of a cardiac scan. Information about this is the subject of the preceding chapter and is excluded from the present chapter. Specific studies of blood flow to organs which are the site of malformation have been carried out to a very limited extent and little information is available. The recent availability of colour-flow Doppler has introduced the possibility of "tissue characterisation" of organs by flow patterns in vessels much smaller than those hitherto studied. Lack of information from clinical studies means that it is not yet possible to assess the value of this development, which involves far more costly equipment.

This chapter will:
1. Review the meaning of an abnormal Doppler flow velocity waveform in terms of the nature of the vascular bed under study, and the implications of disturbances in the waveform in terms of possible pathophysiology.
2. Examine the available information about Doppler studies in the presence of a chromosomal abnormality.
3. Examine the information about Doppler studies in the presence of structural fetal anomalies.
4. Analyse the possible use of Doppler screening programmes as an adjunct to the detection of fetal abnormality.

The Abnormal Doppler Study

Doppler studies of the umbilical artery have been analysed using indices of resistance (the systolic/diastolic ratio, the pulsatility index, and the resistance index) [2,3,4]. These reflect the downstream vascular bed – the umbilical placental circulation. Adverse fetal outcome has been associated with an abnormal umbilical artery flow velocity waveform in which the index of resistance is increased. Correlation studies have been performed in which the Doppler flow velocity waveform was matched to the placental vasculature studied after delivery. The "resistance vessels" in the placenta, the site of the arterial pressure drop, are the small arteries of the tertiary villi. The population of these vessels was reduced in placentas from pregnancies in which the Doppler umbilical flow velocity waveform was abnormal. In some placentas obliterative changes could be seen in these vessels [5]. Further evidence relating the size of the umbilical placental vascular bed to the Doppler flow velocity waveforms, is provided by animal experimental studies. Microsphere embolisation of the umbilical circulation caused an increase in the umbilical artery flow velocity waveform index of resistance [6,7].

Even more complete understanding of the nature of the placental vascular change existing when the flow velocity waveform index of resistance is increased, is provided from results of physical models of the placental vasculature using lumped electrical circuit equivalents [8]. In this approach detailed attention was paid to the branching of the umbilical arteries in the villus tree. Each vessel was assumed to have a resistance and a capacitance and values within the physiological range were given. The validity of this approach was confirmed by demonstrating that the model predictions of volume flow were realistic for input parameters in the physiological range and using typical values for pressure and placental size. The great strength of this approach is that the effects of various parameters on the flow velocity waveforms can be examined individually. Using this model, it can be shown that the pulsatility index (PI) of the flow velocity waveform for a given placenta size (number of primary and secondary branches) will increase as the fraction of terminal arterial vessels obliterated is increased, and this increase is not linear [9] (Fig. 9.1).

It is not until 50% or more of the terminal vessels are obliterated that the PI increases significantly. This is a fundamental property of vascular obliteration in the placental vascular tree. The Doppler umbilical flow waveforms are regarded

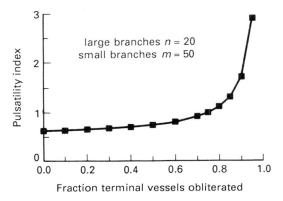

Fig. 9.1. Using a mathematical model of the umbilical placental circulation, the change in the umbilical artery pulsatility index was calculated in the presence of obliteration of an increasing fraction of the umbilical vascular bed.

as early predictors of vascular disease (relative to other clinical tests) in the placenta. Using our model it is demonstrated that extensive disease is present before waveform changes occur and that the placenta has a large reserve capacity.

Conceptually, the model is very valuable in emphasising the need to consider the various indices of resistance not as a measure of resistance, but rather as a measure of the number of resistance vessels in the vascular tree. The number of resistance vessels may be reduced by obliterative vascular disease and in such circumstances a progressive process of continuing obliteration may be expected to be associated with an increase in the indices of resistance (i.e. systolic/diastolic ratio, pulsatility index and resistance index). If the placenta (and so its vascular tree) are small then again the index of resistance will be high. This may occur in a "low growth potential" situation. In this situation there may be no process of vessel obliteration occurring in the placenta. Serial studies in these cases would reveal a decrease in the indices of resistance (which may, however, remain high or above the normal range) as the small placenta continues to grow.

Support for the existence of these two patterns among pregnancies with a high umbilical Doppler resistance index has been provided in a large clinical study [10]. A total of 794 high-risk pregnancies in which three or more umbilical Doppler studies had been performed were analysed. The trend in Doppler systolic/diastolic ratio was divided into three groups. In the normal group, results were within the normal range. The trend was designated abnormal if at least two of the last three values were above the 95th centile value and this group was further subdivided. An abnormal improving trend was present if the last two values, relative to earlier studies, were decreasing and moving down to the 95th centile value. An abnormal and worsening trend was present if these values were increasing relative to the 95th centile. It was demonstrated that both abnormal groups are strongly associated with small size at birth (corrected for gestation). Neonatal morbidity (admission to and duration of stay in neonatal intensive care) and perinatal mortality were, however, much greater in the abnormal and worsening group. It is in this group that the progressive obliterative vascular pathology is present in the placenta.

These considerations are important in analysing the significance of a high resistance pattern in the umbilical Doppler recordings of flow velocity waveforms from pregnancies with an abnormal fetus and in the analysis of these studies from other vessels.

Doppler Umbilical Studies in the Presence of Abnormal Fetal Karyotype

Review of the umbilical Doppler findings in the presence of major fetal abnormality from our centre (Table 9.1) demonstrates the association with the abnormal umbilical flow velocity pattern. These cases were referred for study because of a concern about the pregnancy and so do not necessarily equate to the results which might exist in a total population screen. Nonetheless, the abnormal fetal karyotype is the only group in which the association with an abnormal umbilical Doppler study is significant. Among our series of 22 cases an abnormal umbilical study was present in all five cases of trisomy 18, in eight of eleven cases of trisomy 21, in two of four cases of triploidy and in one case with a partial trisomy of chromosome 1. The Doppler study was normal in one case of D translocation. Some of these cases have been reported previously [11].

Table 9.1. Umbilical artery flow velocity waveform (FVW) and major fetal anomaly.
The distribution of umbilical artery flow velocity waveform systoic/ diastolic S/D ratio in a group of 88 fetuses in which major fetal abnormality was present. The 95th centile was used to classify the S/D ratio.

Anomaly group	Umbilical artery FVW		
	Normal	Abnormal	
All Cases	43	45	
Chromosomal	6	16	$P\ 0.02$
CNS	4	4	
Renal	8	2	$P\ 0.025$
GI Tract	7	7	
Heart	8	6	
Miscellaneous	10	10	

In a collected series of six reports totalling 95 cases [12] with absent end diastolic flow velocities, aneuploidy was reported in 12 cases (12.6%) and comprised five cases of trisomy 13, four cases of trisomy 18, two cases of trisomy 21, and one triploidy. The high frequency of the relatively uncommon trisomies 13 and 18 in this group was noted.

In the author's series, four of the five cases of trisomy 18 exhibited amniotic fluid volume which was normal or high (Fig. 9.2). The last two of these cases were identified before delivery because of a policy of karyotyping all cases of polyhydramnios associated with a high resistance pattern in the umbilical Doppler study. One of our cases presented with oligohydramnios, a small fetus

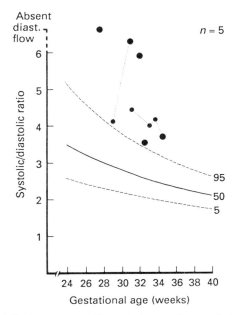

Fig. 9.2. The result of umbilical artery artery flow velocity waveform study in five fetuses with trisomy 18.

and abnormal fetal heart rate monitoring. A picture of "profound fetal compromise", usually with oligohydramnios, has been described by others [13]. It is suggested that a phase of "fetal demise", which is associated with absent growth and decreasing amniotic volume, may precede fetal death. Serial umbilical studies have demonstrated a pattern in which the index of resistance increases.

The trisomy 21 cases are divided between normal and abnormal Doppler results. The three perinatal losses in this group were all cases in which the umbilical Doppler study was abnormal. None of the eight cases in which the umbilical Doppler study was abnormal were associated with oligohydramnios. Serial studies from one patient are shown in Fig. 9.3. Serial studies in this group are consistent with the abnormal improving trend referred to above, and it is suggested that this group exhibits the true "low growth potential" picture. Data have been presented which demonstrated that the maternal angiotensin infusion sensitivity test (AIST) is negative in association with a high umbilical artery flow velocity waveform ratio which declines, in contrast to a positive result in those cases in which the ratio increases [14]. Fetal outcome was good in the negative AIST group. We have suggested that the positive AIST is associated with placental vascular obliteration.

Although two of the present four triploidy cases had an abnormal Doppler study these were close to the normal range. The placenta is large in these cases and indeed the large vascular spaces in the placenta strongly suggest the diagnosis on ultrasound scanning. Severe maternal hypertension is a frequent clinical feature combined with the above ultrasound findings.

Placental pathology in association with chromosomal abnormality has been little studied. Among 18 fetuses with autosomal trisomy (13 with either trisomy 18

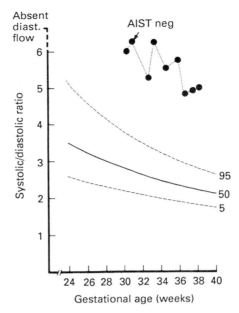

Fig. 9.3. Serial umbilical arter flow velociy waveform studies in one fetus with trisomy 21. An angiotensin infusion sensitivity test (AIST) was negative.

or 13, and five with trisomy 21) a reduction in small arteries in the tertiary villi of the placental vascular tree was noted [15]. This finding in a group in which trisomy 13 and 18 predominate agrees with the clinical findings presented above.

The possibility of fetal aneuploidy has been frequently used to justify percutaneous umbilical vein puncture and blood sampling in pregnancy associated with small fetal size for dates [16,17]. The value of blood gas analysis has been disputed in such cases [18]. The data presented here do not refute the need for such studies, but do introduce an element of discrimination in the need for such a study. The presence of a small fetus on ultrasound measurement with normal umbilical Doppler findings would not appear sufficient reason to justify fetal karyotyping.

Doppler Umbilical Studies in the Presence of Structural Fetal Abnormalities

The presence of abnormality in the umbilical Doppler flow velocity waveform in association with major fetal abnormality was first reported by Trudinger and Cook [11] and has been commented upon by others. The cases in the authors' laboratory are summarised in Table 9.1. The major correlate of the abnormal Doppler findings is small fetal size and this information is presented in Table 9.2. In all abnormality groups the association with birthweight less than the tenth centile is significant. In the 88 cases presented here, there were 50 perinatal

Table 9.2. Umbilical artery FVW and major fetal anomaly: birthweight <10th centile

Anomaly group	Umbilical artery FVW				
	Normal		Abnormal		
All cases	12/43	28%	28/45	62%	P 0.001
Chromosomal	1/6	17%	12/16	75%	
CNS	2/4	50%	3/4	75%	
Renal	3/8	38%	2/2	100%	
GI tract	3/7	43%	4/7	57%	
Heart	1/8	13%	3/6	50%	
Miscellaneous	2/10	20%	4/10	40%	

deaths. These were for the most part a consequence of the abnormality and the death rate did not differ between those with a normal (22/43) and abnormal (28/45) umbilical Doppler study. This has not, however, been the universal experience. In a study of 34 fetuses in whom congenital heart disease was detected between 18 and 32 weeks' gestation [19], ten revealed a pattern of absent diastolic flow velocities. Of these ten cases, eight died and the other two were aborted. This was a poorer outcome than those with a normal study. There did not appear to be any specific cardiac lesion associated with the abnormal Doppler study.

The examination of a group of small fetuses with major congenital malformation provides the opportunity to study growth in the presence of innate fetal disturbance. Fetal growth retardation has been broadly divided into two groups: those fetuses that are small (for fetal reasons) because of low growth potential, e.g. genetic abnormality or early infection; and those that are small because of deprivation of oxygen or nutrient supply (loss of growth support). Originally it was postulated that in those patients with a high systolic/diastolic ratio there was a process of vessel obliteration in the placenta triggered by the abnormal fetus [11]. We would now suggest that such a process is present only in those patients in whom the umbilical systolic/diastolic ratio or other index of resistance is high and increases. In the other cases the small fetus may be in balance with its small placenta (and vasculature) and such fetuses may exhibit a normal growth velocity. It remains to be determined whether the mechanisms triggering the placenta/vascular obliteration are set at a lower threshold in the presence of an abnormal fetus.

Doppler Umbilical Screening of a Total Obstetric Population

There has been much debate about the value of umbilical Doppler studies in screening a total obstetric population to detect cases of potential fetal growth retardation and compromise. The largest study, of 2097 patients from Belfast [20], concluded that such a programme was of "no value" at 28 weeks. An Australian study at 20 and 24 weeks reached a similar conclusion [21]. However, both of these studies and others [22] have noted that at varying gestation periods from 32 to 38 weeks there was a significant association between an abnormal umbilical Doppler result and small fetal size at birth. The author has been involved in a similar investigation of a total population of 500 screened with umbilical Doppler at 28 weeks (range 27–30 weeks). The systolic/diastolic ratio was used as the index of resistance and this was normal (<95th centile) in 465, elevated (95th–99th centile) in 22, and high (>99th centile) in 13. The results of this study are presented in Table 9.3. Previous reports have not used fetal abnormality and especially aneuploidy as an end point in the evaluation of umbilical Doppler screening. In our study the two fetuses with trisomy 21 were in the group with a high umbilical artery systolic/diastolic ratio. The importance of the diagnosis of fetal karyotype abnormality, and its strong association with an abnormal umbilical Doppler finding, makes this an important issue. This screening study was performed at 28 weeks, but it would seem relevant to evaluate this aspect at earlier gestational ages.

Table 9.3. Clinical features and pregnancy outcome of 500 studied patients: the results of screening a non-selected group of pregnancies between 27 and 30 weeks with umbilical Doppler examination

	Umbilical artery systolic/diastolic grouping		
	Normal (<95th)	Elevated (95–99th)	High (>99th)
Total cases	465	22	13
Maternal age (years) (mean ± SD)	26.3 ± 5.2	26.0 ± 4.3	27.1 ± 7.3
Parity (no. primip (% cases))	184 (40%)	14 (64%)*	3 (23.1%)
Gestation at delivery (weeks) (mean ± SD)	39.2 ± 1.6	39.5 ± 1.7	35.6 ± 3.5***
Premature delivery (no. <37 wk (% cases))	22 (2.4%)	1 (4.5%)	6 (46.2%)
Spontaneous	20	1	1
Elective	2	0	5
Birthweight (g) (mean ± SD]	3347 ± 482.8	3478 ± 548.4	2322 ± 807***
SGA fetuses (no. <10th centile (% cases))	27 (5.8%)	1 (4.5%)	3 (23.1%)**
Fetuses – Trisomy 21	0	0	2

SD, standard difference; SGA, small for gestational age.
*, $P<0.05$; **, $P<0.01$; ***, $P<0.001$.

Conclusions

Umbilical artery flow velocity waveforms recorded with Doppler ultrasound are often abnormal in the presence of major fetal abnormality. Such abnormal studies are strongly associated with small fetal size at birth.

Abnormal umbilical Doppler studies may result from a small placenta (vascular tree) in association with a small fetus (low growth potential) or a progressive obliterative vascular pathology in the placenta.

Fetal karyotype abnormality is very frequently associated with umbilical Doppler abnormality. An isolated finding of an abnormal umbilical Doppler study should prompt the assessment of fetal karyotype. Umbilical Doppler screening late in the second trimester may provide another means of assessing for cases of aneuploidy.

References

1. Trudinger BJ. The umbilical circulation. Semin Perinatol 1987; 11:311–21.
2. Giles WB, Trudinger BJ, Baird P. Fetal umbilical artery flow velocity waveforms and placental resistance: pathological correlation. Br J Obstet Gynaecol 1985; 92:31–8
3. McCowan LM, Mullen BM, Ritchie K. Umbilical artery flow velocity waveforms and the placental vascular bed. Am J Obstet Gynecol 1987; 157:900–2.
4. Bracero LA, Beneck D, Kirshenbaum N, Pieffer M, Stalter P, Schulman H. Doppler velocimetry and placental disease. Am J Obstet Gynecol 1989; 161:388–393.
5. Fok R, Pavlova Z, Benrischke K, Paul R, Platt LD. The correlation of arterial lesions with umbilical artery Doppler velocimetry in the placentas of small-for-dates pregnancies. Obstet Gynecol 1990; 75:578–583.
6. Trudinger BJ, Stevens D, Connelly A et al. Umbilical artery flow velocity waveforms and placental resistance: the effects of embolization on the umbilical circulation. Am J Obstet Gynecol 1987; 157:1443–9.
7. Morrow RJ, Adamson SL, Bull SB, Ritchie JWK. Effect of placenta embolization on the umbilical arterial velocity waveform in fetal sheep. Am J Obstet Gynecol 1989; 161:1056–60.
8. Thompson RS, Stevens RJ. A mathematical model for interpretation of Doppler velocity waveform indices. Med Biol Eng Comput 1989; 2:269–76.
9. Thompson RS, Trudinger BJ. Doppler waveform pulsatility index and resistance, pressure and flow in the umbilical placental circulation: an investigation using a mathematical model. Ultrasound Med Biol 1990; 16:449–458.
10. Trudinger BJ, Cook CM, Giles WB et al. Fetal umbilical artery velocity waveforms and subsequent neonatal outcome. Br J Obstet Gynaecol 1991 (in press).
11. Trudinger BJ, Cook CM. Umbilical and uterine artery flow velocity waveforms in pregnancy associated with major fetal abnormality. Br J Obstet Gynaecol 1985; 92:666–70.
12. Rochelson B. The clinical significance of absent and diastolic velocity in the umbilical artery waveforms. Clin Obstet Gynecol 1989; 32:692–702.
13. Schneider AS, Mennuti MT, Zackai EH. High Caesarean section rate in trisomy 18 births: a potential indication for late prenatal diagnosis. Am J Obstet Gynecol 1981; 140:367–70.
14. Cook CM, Trudinger BJ. Mothers with abnormal Doppler umbilical artery studies are angiotensin sensitive. Abstract No. 432, SGI Proceedings, San Diego. Br J Obstet Gynaecol 1991 (in press).
15. Rochelson B, Kaplan C, Guzman E, Arato M, Hansen K, Trunca C. A quantitative analysis of placental vasculature in the third-trimester fetus with autosomal trisomy. Obstet Gynecol 1990; 75:59–63.
16. Cox WL, Daffos F. Forestier F et al. Physiology and management of intrauterine growth retardation: a biologic approach with fetal blood sampling. Am J Obstet Gynecol 1988; 159:36–41.
17. Nicolaides KH, Economides DL, Soothill PW. Blood gases, pH, and lactate in appropriate and small for gestational age fetuses. Am J Obstet Gynecol 1989; 161:996–1001.
18. Nicolini U, Nicolaidis P, Fisk NM, Vaughan JI, Fusi L, Gleeson R, Rodeck CH. Limited role of fetal blood sampling in prediction of outcome in intrauterine growth retardation. Lancet 1990; 336:768–72.
19. Al-Gazali W, Chapman MG, Chita SK, Crawford DC, Allan LD. Doppler assessment of umbilical artery blood flow for the prediction of outcome in fetal cardiac abnormality. Br J Obstet Gynaecol 1987; 94:742–5.
20. Beattie RB, Dornan JC. Antenatal screening for intrauterine growth retardation with umbilical artery Doppler ultrasonography. Br Med J 1989; 298:631–5.

21. Newnham JP, Paterson LL, James IR, Diepereen DA, Reid SE. An evaluation of the efficacy of Doppler flow velocity waveform analysis as a screening test in pregnancy. Am J Obstet Gynecol 1990; 162:403–10.
22. Hanretty KP, Primrose MH, Neilson JP, Whittle MJ. Pregnancy screening by Doppler uteroplacental and umbilical artery waveforms. Br J Obstet Gynaecol 1989; 96:1163–7.

Discussion

Campbell: Dr Allan presented an important paper showing that we could detect the majority of serious cardiac malformations if we had a nationwide programme in existence.

Atkins: May I answer her very important challenge, which I think the Northern Region Study tries to answer, of how to translate something like the four-chamber view from a centre of expertise to the entire region or district.

I did not mention cardiacs in detail in chapter 2. The numbers of the last five years have only been coming off the computer over the last four weeks, but I have a fairly good feel for them. My feeling is that we are not doing as well as we hoped we were doing, or should be doing, but we can look at each case and see why. We can see in each case who the operator was, what the equipment was, what the gestation, what the build of the patient was, and so on.

As Dr Allan knows, in the Northern Region somebody from her department did visit, and we try and teach all the units to do this.

If we are to do audit, this surely is audit: this is not only looking at what happened, but is trying to see what can be done about it and how it can be changed. I would, therefore, make the plea that this sort of survey is spread.

I should like also to coin a phrase, almost. We keep talking about cost effectiveness: I should like to call it care effectiveness. Cardiac anomalies are one of the many where it really is care effective to make a prenatal diagnosis. It is not just the cases who are told the baby cannot survive, it is not just the cases who have to go somewhere else for delivery so that the baby can have immediate help, or the cases where the paediatrician has to be informed. It is the fact that these parents are prepared that makes such an enormous difference.

Allan: It took us years to begin to see a yield from the screening programme. That has been five years and we can demonstrate it is getting better every year. But it has been hard work, and repeated going out and teaching, and repeated reinforcement of the standards that we thought they would be able to apply immediately. Of course we have also looked at the factors that have had an affect and the equipment, timing, and so on are all very important. It will take a long time to change timing, to change equipment, to change staff training, but it does work in the end.

Campbell: In terms of NTDs, a multicentre study was carried out in my department in which training was supervised, and in a series of 10 000 reports there was 100% detection of spina bifida and anencephaly, which means that if one goes into the individual departments, shows them how to do it, and shows that we are really observing what they are doing, the results can be extremely good. But it is a lot of hard work.

Ward: There are two sides to the problem: the problem of detection and the problem of the advice that the couple are given. I was interested to hear that paediatric cardiologists are requesting more cardiologists to take care of these cases. I am sure that many of us who are involved in prenatal diagnosis of cardiac defects would usually involve our local paediatric cardiologist in advising the couple. The defects are complex: some can be repaired and some cannot. A lot of the advice that the parents are given by paediatric cardiologists is not all doom and gloom, and there may well be encouragement to continue with pregnancies.

Allan: The training of paediatric cardiologists is as important as the training of ultrasonographers. One of the things we have learned because we have such a big database, is that in general what is seen early in prenatal life does not bear much relation to what is seen in paediatric life, and although the paediatric cardiologists must be involved in order to get diagnostic accuracy, they must also be educated in prenatal cardiology. It is a different science from postnatal cardiology. There is a whole different spectrum of disease, and a developing spectrum, that they must know about if they are to be involved in this field. A wrong impression might be given by the paediatric cardiologist who is talking from an experience that is all postnatal, i.e. the survivors. That is as important as educating the ultrasonographer who makes the initial screen.

Donnai: How many centres like Dr Allan's will be necessary if 50% of the four-chamber view anomalies suspected at district general hospital level turn out not to have an anomaly screening in the expert centre?

A case has been made out for every baby having an examination in a specialist centre before a decision is made on the fate of that pregnancy by the family. If every pregnancy were screened this might involve huge numbers. It is obviously not convenient for huge numbers of people to get to her unit at Guy's Hospital, which is not at present funded by the Health Service, nor is it convenient for families from centres distant to that centre.

How would she suggest the organisation of specialist cardiac scanning?

Allan: I suggest that it is confirmed by a paediatric cardiologist who knows about the fetal spectrum of disease. The data are all there, available for them. There are perhaps ten paediatric cardiology centres in England and Wales that have experience in paediatric echocardiography, and one member of each of these groups should have expertise in fetal cardiology. That is how I think it should be organised in the long term. These are all supraregional centres, and that is how it should be organised in the future. Every diagnosis should be confirmed both by a paediatric cardiologist and, where this kind of audit is being performed, by the pathologist.

Lilford: That is exactly what happens in Yorkshire. There is one hospital which has two cardiologists who specialise in fetal cardiac scanning.

I am almost certain that nothing like half of all the anomalies they confirm are terminated. Dr Allan said they have had no totally false positive report, but there must be a number of pathology reports where although the presence of a lesion was confirmed, the extent and nature of the lesion was not what was expected.

Allan: There have been minor differences in the minutiae of a diagnosis, but in general the implications have not been any different at pathology from what we were expecting. At the beginning we were very cautious in our interpretation, but we have looked at 500 cases now. In the early days I did not know how some of the lesions would progress, I did not know the associations, and I did not know what the outcome would be, but on the database I now have I can be absolutely positive that X will not survive postnatal life. Maybe that makes a difference to how the patients choose because I am much more sure than I used to be. Whereas some of the paediatric cardiologists do not know all this, do not have that background and have to go through the same learning experience that I have been through of encouraging patients through the pregnancy with what looked like a correctable lesion but turned out in the end not to be.

Campbell: Could I follow up the statistics? Dr Allan gives a very low false negative rate, 1%, and yet she said that she would only expect to pick out 4–5 per 1000.

Allan: The 1% is at six-week follow-up. We sometimes hear about later atrial septal defects (ASDs) or pulmonary stenosis that turns up in the child, but in general we do not hear.

Six-week neonatal follow up is what we are talking about when we have a 1% miss rate. But there are other things like a mild aortic stenosis or pulmonary stenosis, which are congenital heart disease that will not be detected in prenatal life, and perhaps ASDs that do not present until later.

Drife: The four-chamber view is expected to pick up about half the potential abnormalities that one would expect?

Allan: Two out of three major defects that one would expect in infancy.

Drife: If this is introduced on a general basis, what sort of backlash will there be when people go through the scanning process and quite a substantial minority fail to have their abnormality diagnosed?

Allan: It should be clearly defined that we were only looking at one section and that there are major anomalies that will be missed. When we do what we call a complete fetal echocardiogram there are lesions that are missed. For example, ASD cannot be diagnosed prenatally unless it is huge, but that will need open heart surgery. It is important that the patient understands that there are minor defects – which may not seem too minor to them if open heart surgery is required, but in our terms they are minor defects that will be missed. The confidence limits can be clearly set out both at the screening stage and at the complete fetal echocardiogram stage. We have a written handout that we give to patients detailing what we cannot see and what we cannot do, and that is important for patients to understand. When it is clearly defined to them, I do not think they have any trouble.

In the long term, when we can have the time, the resources and the equipment to do a complete fetal echocardiogram, less will be missed. But at the present time we have a screening programme which we accept will miss some major abnormalities. That is all that is practical at the moment.

Campbell: Dr Whittle was wondering whether an early scan was as good as the 18–20 weeks scan. What percentage would be picked up if the scan was done at 12 or 13 weeks, as he does in his department?

Allan: We have been looking at vaginal scanning at 14 weeks and in the majority of cases – about 80% or 90% – we are able to define the connections. Unfortunately there are a few things that might not be fully developed at 14 weeks, so we are still using that as a research tool and looking at 18 weeks as well. That is perhaps not practical on a screening basis, but for the high-risk pregnancies a 14-week scan in the same way as other early diagnoses is very important in terms of patient management. If they can be offered early diagnosis, then so much the better.

Campbell: Two out of 500 Down's syndrome cases is above the normal prevalence but I suspect Professor Trudinger picked up the two Down's syndrome cases in his screening programme. Might this have a role in the future for screening for Down's syndrome?

Trudinger: It was an interesting association. Two is not a big number and neither is 500. We need much bigger numbers than that.

The chromosomal abnormalities do have abnormal umbilical Doppler studies and that they fell into that group is understandable, in terms of the fact that they do have small placentas.

I would now like to see data, and that is what we are attempting to see whether the patients can be recognised earlier in pregnancy. The problem is that there is a broad crossover zone between normal and abnormal which gets less as time goes on in pregnancy.

Nevin: Many chromosomal abnormalities have abnormal umbilical arteries. Is that why Professor Trudinger's group is picking up the abnormality of their results?

Trudinger: No, I do not think so. They may have abnormal umbilical arteries, they may have single umbilical arteries, but in cases with single umbilical arteries and no other fetal abnormality we have certainly seen normal wave forms.

Section III
DNA Analysis

Chapter 10

Overview of Linkage and Probes

M. E. Pembrey

Direct DNA analysis has revolutionised prenatal diagnosis for monogenic disorders in the last decade, principally because it can immediately exploit the rapid advances in gene mapping and also removes the need to examine tissue in which the gene in question is expressed. The latter means there is no theoretical limit to how early in pregnancy the diagnosis can be made. In practice of course reliable access to fetal material without endangering the viability of the fetus or embryo and the robustness of DNA analysis from as little as a single cell are the limiting factors.

With very rare exceptions (namely, carrier detection for X-linked mutations that distort random X-inactivation in heterozygous females), a chromosomal localisation of the disease specific locus is a prerequisite for any genetic prediction by DNA analysis. The disease locus has to be mapped to within a region no further in genetic distance than 5 cM from a polymorphic DNA marker, if prenatal diagnosis is to be offered as a clinical service. There is nothing magical about the recombination fraction of 0.05. It is just that once a disease gene is localised to a chromosome, a DNA marker at least this close is usually easy to discover by traditional linkage studies, and also errors in prenatal diagnosis above 5%–10% are unacceptable to most couples.

The International Human Genome Project and related research in molecular medicine are committed to mapping and eventually sequencing all the expressed human genes in the next decade or so. The challenge for those providing clinical genetic services is to develop generally applicable methods that can quickly translate these basic discoveries into robust and reliable tests for prenatal diagnosis and carrier detection. There are two fundamentally different approaches to DNA analysis for genetic prediction – gene tracking and mutation detection. Each has its advantages and disadvantages, but what is quite clear is that both will contribute to prenatal diagnosis for many years to come. It is with gene tracking that this chapter is concerned, particularly the problems posed by locus heterogeneity.

Gene Tracking

The great majority of DNA-based prenatal tests performed since the first examples about 10 years ago [1,2] have employed a genetic linkage approach rather than mutation detection. This is not surprising, since this strategy can be introduced as soon as close linkage of a DNA marker to a disease specific locus has been demonstrated and significant locus heterogeneity excluded (as discussed below). Furthermore, this approach is often the preferred strategy, even when the gene in question has been cloned and a gene-specific probe employed in the test. It was because of this latter point, and a need to draw the crucial distinction from mutation detection, that the term gene tracking was coined in 1983 to describe genetic prediction using nearby DNA sequence polymorphisms rather than detection of the disease-causing mutation itself.

In essence gene tracking is a family linkage analysis turned around to use a known linkage to make a genotype prediction. Traditionally linkage terminology concerns the genetic distance (the amount of meiotic recombination) between two gene loci. When thinking of gene loci in physical terms as any length of DNA sequence that is transcribed into RNA, it becomes somewhat confusing to talk of linkage analysis within a single gene locus and therefore a term that took account of these points seemed to me to be helpful. Gene tracking literally tracks the path of transmission of a mutation through the pedigree and addresses the question – has the fetus inherited the same chromosomal region as a previously affected or unaffected family member? It is independent of what particular mutation within the gene is causing the disease and this is the greatest advantage of this term.

Mutation detection is a search for any pathologically significant change in DNA sequence within the gene in question or detection of a particular DNA sequence change. As far as prenatal diagnosis is concerned it will usually be addressing the question – does gene X in the fetus have this particular DNA sequence change?

In reviewing the important issues that face both those offering a clinical service and working in human gene mapping research, I have chosen to follow the usual sequence of events from initial attempts to map a disease gene to description of the mutations found once the gene is cloned.

It should be emphasised that there may be many years between the mapping of a disease locus and actually identifying the gene in question so mutations can be characterised. For example, at the time of writing, the gene involved in Huntington's chorea has not been cloned, although it was mapped to the short arm of chromosome 4 several years ago [3]. Nevertheless, it is important to note that the gene tracking approach has allowed prenatal diagnosis in the meanwhile, as indeed it has with many other mutant genes that are mapped but not yet cloned. The rate of progress in mapping can be judged from Table 10.1 which lists the estimated frequency of the commoner serious Mendelian disorders in the general European population compiled by Carter in 1977 [4]. At that time none of the disease-specific loci listed were mapped for certain, although alpha globin (alpha thalassaemia) had been tentatively mapped to chromosome 16. Now

Table 10.1. Common monogenic disorders

Disorder (chromosomal location of gene)	Approximate birth incidence per 1000 in Europe
Autosomal dominant	
Familial hypercholesterolaemia (19p)	2.0
Huntington's chorea (4p)	0.5
Neurofibromatosis (17q)	0.3
Myotonic dystrophy (19q)	0.2
Hereditary motor and sensory neuropathy type I (17p)	0.2
Adult polycystic kidney disease (16p)	0.9
Familial adenomatous polyposis (5q)	0.1
Tuberose sclerosis (9q or 11q)	0.03
Retinoblastoma (13q)	0.05
Osteogenesis imperfecta (7q or 17q)	0.04
Marfan's syndrome (15q)	0.04
Dominant blindness	0.1
Dominant childhood deafness	0.1
Other dominants	2.0
Autosomal recessive	
Cystic fibrosis (7p)	0.5
Sickle cell anaemia (11p)	depends on ethnic origin
B thalassaemia (11p)	
Phenylketonuria (12q)	0.1
Neurogenic muscle atrophies (most 5q)	0.1
Congenital adrenal hyperplasia (21 hydroxylase deficiency) (6p)	0.1
Recessive severe congenital deafness	0.2
Recessive blindness	0.1
Recessive non-specific severe mental retardation	0.5
Other recessives	2.0
X-linked	
Duchenne muscular dystrophy (Xp21)	0.25 (males)
Haemophilia A (Xq28)	0.1 (males)
Haemophilia B (Xq27)	0.03 (males)
Ichthyosis (Xp22)	0.1 (males)
Fragile X mental retardation (Xq27)	0.75 (males)
Other X-linked	0.6 (males) All serious Mendelian disorders 10/1000 total livebirths

virtually all the common disorders have been mapped except for genes involved in autosomal recessive profound deafness, in which genetic (locus) heterogeneity causes great problems. Locus heterogeneity can bedevil gene mapping research, and in turn prenatal diagnosis. It becomes critically important to combine gene mapping and linkage expertise with clinical skills in delineation of the clinical subtypes of the disorder. The gene tracking approach calls for the closest coordination between the clinicians diagnosing and caring for the family and the clinical molecular geneticist performing the DNA analysis, a coordinating role well suited to the clinical geneticist.

Locus Heterogeneity

Genetic heterogeneity within a collection of families with what appears to be the same clinical disease, can be either due to different mutations at the same gene locus (allelic heterogeneity) or can involve different gene loci, usually on different chromosomes (locus heterogeneity). Locus heterogeneity is potentially one of the biggest problems in translating gene mapping discoveries into clinical service. With autosomal dominant and X-linked disorders, selection of very large families with many affected individuals for linkage analysis can relatively easily allow locus heterogeneity involving two (or three) well-separated loci to be resolved. Locus heterogeneity in autosomal recessive conditions tend to be more difficult to sort out but even here exploitation of large inbred families can lead to the loci being mapped. However, it is one thing to say that two mapped loci are involved in a disease, but quite a different matter to say which of the two genes is mutant in any particular small family seeking prenatal diagnosis.

Obviously when a disorder is first mapped it is most important to assess how many independent families have shown the same linkage results before proceeding to offer prenatal diagnosis to all families that present. It is easier to make this general assertion than decide at what point locus heterogeneity is excluded. Until a large body of data are brought together ruling out significant locus heterogeneity as has been done for cystic fibrosis for example, the clinical molecular genetics team has to incorporate this uncertainty into the prediction error (recombination rate) attached to the prenatal prediction. The international consortia that have often been set up following the localisation of a disease-specific locus have been very useful in quickly excluding (or establishing as the case may be) the existence of significant locus heterogeneity and in providing reliable recombination rates with specified DNA markers based on large numbers. These consortia in which the appropriate lay organisations often play a major role, have done much to enable the move from basic linkage research to prenatal diagnostic services. They come into their own when the gene is cloned and the range of mutations has to be defined.

When locus heterogeneity does exist, genetic prediction for all but the largest of families is hazardous, until one or both of the genes involved are identified. Tuberose sclerosis (TS) a serious disorder for which there would be a considerable demand for prenatal diagnosis, is one disease where locus heterogeneity has delayed the introduction of such a service. Studies in England mapped a TS locus to 9q34 [5] but research in America has implicated a locus on 11q [6].

It looks increasingly likely that at least two loci are involved. This poses no theoretical difficulties with respect to possible genetic mechanisms (there are several examples of mutations at different loci causing indistinguishable effects), but does mean little progress will be made until at least one of the genes is cloned. Osteogenesis imperfecta (OI) provides an example of how one can tackle the problem of locus heterogeneity, but also illustrates the difficulties. The autosomal dominant forms of OI (Type I and IV) are always due to a mutation in one of the two genes (COL1A1 on chromosome 17 or COL1A2 on chromosome 7) that encode the two collagen chains of type I collagen. This conclusion comes from linkage studies on a large number of multiplex families, in which it was clear that the disease was always coinherited with either the COL1A1 or COL1A2 gene [7]. To date type IV families have segregated with COL1A2. A small linkage study in any family requesting prenatal diagnosis may be able to indicate at which locus the mutation has occurred in that particular family. However, not all families have sufficient affected individuals to allow such a study and this is even more likely to be the case in the more severe forms of OI (Types II and III) where the demand for prenatal diagnosis is greater. Furthermore, it is not yet certain that all mutations underlying types II and III OI are confined to COL1A1 and COL1A2. The only way forward is mutation detection. Numerous mutations, including 53 different mutations involving glycine codons have been discovered in both Type I collagen genes, in keeping with what would be expected for a group of disorders that are predominantly autosomal dominant in inheritance. Sporadic cases usually represent dominant new mutations and some of the sibling recurrences are due to gonadal mosaicism rather than autosomal recessive inheritance. The clinical phenotype has not proved a reliable predictor of which gene is mutant or where the mutation is located within the gene [8].

It looks as if each child born with severe OI will need a combination of collagen biochemical analysis and a search of both genes for the causative mutation as part of their clinical work-up, if early DNA-based prenatal diagnosis is to be offered in future families. Type II OI is so severe that ultrasound scanning can offer second trimester diagnosis, but Type II OI is also nearly always lethal in the perinatal period. Type III is somewhat milder but can lead to severe crippling and here mutation detection is the only hope of providing useful prenatal diagnosis to parents in the face of a 5%–7% sibling recurrence risk.

Severe OI not only illustrates the problem of locus heterogeneity but the difficulties so many different mutations pose for mutation detection as an alternative to gene tracking. It serves to remind us that the haemoglobinopathies, the disorders that ushered-in DNA diagnosis a decade ago, are not typical when it comes to the ease of mutation detection. The beta and alpha globin genes are very small, less than 2 kb, with only three exons, so with the new methods of mutation detection it would be relatively easy to scan all the exons for any mutation. As it happens, although there are more than 40 different ways in which the beta globin gene can mutate to produce the disease β-thalassaemia, within a given population, one or a few mutations often predominate, as clearly demonstrated for subpopulations within the Mediterranean region [9]. α-Thalassaemia in the Far East is virtually all due to gene deletions that are very easily detected. Sickle cell disease is due to the same mutation throughout the world.

The common autosomal recessive disorder that seemed most likely to follow the pattern of the haemoglobinopathies was cystic fibrosis. However, by contrast the cystic fibrosis transmembrane conductance regulator (CFTR) gene is a fairly large gene with 27 exons and one year after cloning the gene, some 60 mutations have already been described [10]. As indicated in Chapter 5 it will be difficult to detect any more than 90% of mutations in the British families as part of a clinical service.

Limitations of Gene Tracking

Gene tracking requires a small family study and this can be a serious limitation. The appropriate family members, especially the affected individual(s), may be dead or unavailable or unwilling to provide a blood sample.

Involving other family members directly, particularly in autosomal dominant and X-linked disorders requires considerable organisation and coordination with general practitioners and other professionals. However, this extra activity although time-consuming and costly, does contribute to the counselling and education of the whole family with consequently more effective use of genetic services by them in the future.

One of the most disappointing aspects of the gene tracking approach is the discovery that the family, or certain branches of it, are "uninformative" for prenatal diagnosis or carrier detection, because the key individual is not heterozygous for any of the DNA sequence polymorphisms being used as a linked marker. Fortunately, there is the prospect of greatly reducing this problem in the future with the characterisation of a new set of highly variable sequences distributed throughout the genome, known as CA repeats [11]. Some 70%–90% of people are heterozygous enabling the two chromosomes of the pair to be distinguished.

Recombination between the mutation and the polymorphic marker used, leading to a wrong prediction, is a further limitation of gene tracking. This can be particularly troublesome when the gene is very large and/or is a region of increased recombination frequency as is the case with Duchenne muscular dystrophy (DMD) and Abbs et al. [12] estimated a recombination rate of 0.12 across the dystrophin gene which is 2 Mb long. This means that in the absence of a detectable deletion (which occurs in about 60% of DMD families) reliance on a single polymorphic marker within the dystrophin gene carries a prediction error rate of about 5%. The use of two DNA markers that flank the dystrophin gene can reduce this figure but increases the work involved.

At present fragile X mental retardation which affects up to 1 in 1000 males is waiting in the wings as it were. The fragile X locus has not yet been identified nor any mutations underlying the disorder elucidated. It is possible that this region shows a higher than normal rate of recombination. Recent DNA markers show only 2% recombination and can be employed for carrier assessment in females, in conjunction with cytogenetic studies. Most groups, however, still have reservations about using DNA analysis for first trimester prenatal diagnosis, given that cytogenetic detection of the fragile site is possible using chorionic villus material or fetal blood as discussed by Webb (Chapter 13). The genetics of fragile X are not well understood and it is likely that a significant proportion, perhaps up to 20% of males inheriting the same chromosome as a previously affected male, will not show any phenotypic abnormality. There is a faint hope that elucidation of the mutations involved might also go some way towards distinguishing so-called normal transmitting males from those who will be affected.

This last point on "penetrance" or the degree of expression is a general problem in many monogenic disorders, particularly autosomal dominant disorders where there can be considerable variation in severity of the disease within the family. DNA analysis, whether it be mutation detection or gene tracking can only predict whether the mutation has been transmitted. In years to come it may be possible to define some mutations where the phenotype is more variable than others but such genotype/phenotype correlation studies are in their infancy and, of course, limited to those conditions where the gene in question is cloned.

The Future

The ideal to which we are all working is, of course, mutation detection. Perhaps the one technical advance more than any other that has made this a more realistic aim than hitherto imagined has been the polymerase chain reaction (PCR), which allows a selective multimillionfold amplification of a target DNA sequence [13]. Once the sequence and exon/intron structure of a gene is known, direct sequencing after PCR and the use of techniques such as those based on single-strand conformation polymorphism [14] will permit mutation detection on a much wider scale. However, technical advances have to be matched with a much broader input of molecular genetics into all types of clinical practice, from the simplest practical procedure of banking a DNA sample from an affected individual or fetus, to the recognition that the "work-up" of a patient with a monogenic disease includes defining the mutation at the DNA sequence level. This will be a slow process, and in the meanwhile gene tracking will continue to play its role in helping couples restore family life in the face of known genetic risks by letting them try for healthy children, while avoiding the tragedies that genetic disorders can bring.

References

1. Kan YW, Dozy AM. Antenatal diagnosis of sickle cell anaemia by DNA analysis of amniotic-fluid cells. Lancet 1978; ii:910.
2. Old JM, Ward RHT, Petrou M et al. First trimester fetal diagnosis of haemoglobinopathies: three cases. Lancet 1982; ii:1413–16.
3. Gusella JF, Wexler NS, Conneally PM et al. A polymorphic DNA marker genetically-linked to Huntington's chorea. Nature 1983; 306:234–8.
4. Carter CO. Monogenic disorders. J Med Genet 1977; 14:316–20.
5. Fryer AE, Chambers A, Connors JM et al. Evidence that the gene for tuberous sclerosis is on chromosome 1. Lancet 1987; i:659–61.
6. Smith M, Smalley S, Cantor C et al. Mapping of a gene determining tuberous sclerosis to human chromosome 11q14–11q23. Genomics 1990; 6:185–94.
7. Sykes B, Ogilvie D, Wordsworth P et al. Consistent linkage of dominantly inherited osteogenesis imperfecta on the type I collagen loci: COL1A1 and COL1A2. Am J Hum Genet 1990; 46:293–307.
8. Sykes B. Bone disease cracks genetics. Nature 1990; 348:18–20.
9. Cao A, Pirastu M, Rosatelli C. Prenatal diagnosis of thalassaemia. Br J Haematol 1986; 63:215–20.
10. Davies K. Cystic fibrosis: complementary endeavours. Nature 1990; 348:110–11.
11. Weber JL, May PE. Abundant class of human DNA polymorphisms which can be typed using the polymerase chain reaction. Am J Hum Genet 1989; 44:388–96.
12. Abbs S, Roberts RG, Matthew CG, Bentley DR, Bobrow M. Accurate assessment of intragenic recombination frequency within the Duchenne muscular dystrophy gene. Genomics 1990; 7:602–6.
13. Erlich HA, Gelfand DH, Saiki RK. Specific DNA amplification. Nature 1988; 331:461–2.
14. Orita M, Suzuki Y, Sekiya T, Hayashi K. Rapid and sensitive detection of point mutations and DNA polymorphisms using the polymerase chain reaction. Genomics 1989; 5:874–9.

Discussion

Brock: A general point is that with the increasing awareness of heterogeneity in many of these diseases we may often go back from direct mutation detection into linkage analysis to get around this problem.

Pembrey: I agree.

Brock: So in CF this looks as though it may be the way forward for families which have an affected child?

Pembrey: Yes.

Williamson: In practice we do it case by case; if the mutation in that case is known then we actually do a mutation analysis.

Read: A lot also depends on the mode of inheritance of the disease. In autosomal recessive conditions, there is not a lot to choose between mutation analysis and linkage because with luck one can do it all in a very small number of tests. As soon as one considers a dominant condition like Huntington's or a serious X-linked condition like Duchenne, there is a very big difference between how easily it can be done by mutation analysis and the effort that is needed to do it by linkage. In those conditions it may be better to look very hard for mutations.

Chapter 11

Diagnosis of Genetic Defects in Eggs and Embryos

M. Monk

Introduction

Genetic risk for offspring of some couples is high. One in four of their children will be affected if both parents carry the same recessive gene, one in two if one parent carries a dominant disease mutation. For such couples, the genetic status of the fetus may be determined during pregnancy by diagnostic tests on samples of chorionic villus or on fetal cells obtained by amniocentesis. If the fetus proves to be affected, the couple has the option to terminate the pregnancy by abortion. Some couples may suffer the considerable trauma of repeated pregnancy terminations. Hence, attention has been focused on diagnosis in embryos before implantation.

Preimplantation diagnosis requires easy and safe access to embryos, efficient biopsy of one or a few cells from the embryo and highly sensitive diagnostic techniques for analysis at the single cell level. The diagnostic techniques include chromosome analysis [1,2], biochemical assays [3,4], in situ hybridisation [5] and DNA analysis [6-9]. Following diagnosis of the biopsied samples, only embryos free of the defect under test would be replaced in the uterus to initiate a pregnancy free from the anxiety of a possible termination at a later stage.

Access to Human Preimplantation Embryos

The access to human preimplantation embryos has come about due to the development of procedures for the treatment of infertility. For in vitro fertilisation (IVF) procedures, eggs are removed from the woman's ovaries, fertilised in vitro by the partner's sperm and between one and four embryos replaced in the uterus. An alternative approach to treatment of infertility is uterine lavage, a non-surgical process of flushing the woman's uterine cavity via the cervix (Fig. 11.1). In this case, a donor woman is artificially inseminated, the embryo recovered 5 days later by uterine lavage and transferred to the uterus via the cervix of the infertile woman.

The procedures for uterine lavage are described by Buster et al. [10], Sauer et al. [11,12] and Formigili et al. [13,14]. The efficiency of embryo recovery by uterine lavage may be as high as 40%–50%. For example, Sauer et al. [12,15] recovered 40 embryos in 90 insemination cycles from 27 women donors. However, the procedure was not without risk to the donor women two of whom showed retained pregnancies and three of whom suffered pelvic infection. Although embryo recovery is good, few embryos obtained by uterine lavage have reached the morula or blastocyst stage due to normal embryo wastage (see below). On the other hand, those that have developed will give rise to a pregnancy, after transfer of a single embryo, with high efficiency (about 50%) [12–14].

The development of uterine lavage procedures in the human is in its early stages. In contrast, procedures for IVF and embryo transfer [16,17] are widespread throughout the world and many thousands of babies have been born to infertile couples using this approach. In cases of infertility, the pregnancy rate is 15%–25% per replacement of up to four embryos and each embryo has a transfer efficiency of

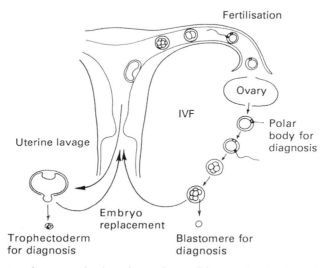

Fig. 11.1. Access to human preimplantation embryos. Diagram showing the normal course of fertilisation and development, IVF and uterine lavage. The stages for biopsy are the unfertilised egg, the 8-cell embryo and the blastocyst.

about 10%–15% [18]. The efficiency of IVF and embryo transfer may well be higher with couples seeking preimplantation diagnosis who are fertile (although the effects of biopsy on survival of the embryo must be considered).

Generally, embryos produced by IVF are replaced in the uterus when they are at the 4- to 8-cell stage. Many of these will be inherently defective. This embryo wastage is evident from the fact that only 30%–50% of IVF embryos reach the morula and blastocyst stages after 5 days of culture [18] and that many embryos obtained by uterine lavage show arrested development.

Karyotypic studies on human oocytes and embryos obtained by IVF indicate a high level of 20%–30% detectable chromosomal abnormalities [1,2,19,20]. The high rate of abnormality is mainly due to chromosome aberrations already present in the unfertilised eggs. There is no evidence that there is an increase in abnormality due to the IVF procedures [1].

Both the uterine lavage and IVF procedures are improving and it is not clear which will be the preferred route for access to embryos for preimplantation diagnosis in the future. Uterine lavage shows certain advantages over IVF in that the procedure is non-surgical and only a single embryo is replaced. Also the diagnostic procedures are on late morulae and blastocysts which have survived and which therefore have a much higher rate of implantation than do 4- to 8-cell embryos. In addition, more cells may be sampled from the blastocyst and the biopsied sample is extra-embryonic trophectoderm, thus avoiding the possible ethical objection to the use of cells from 6- to 8-cell stage embryos when all cells have both embryonic and extra-embryonic potential. However, the risks of uterine lavage have not as yet been properly evaluated; they may be unacceptably high.

Gametes

In addition to tests on the embryos themselves, diagnostic tests may also be possible on sperm and eggs. In the case of sperm, many attempts have been made to identify and separate X- and Y-bearing spermatozoa but to date these methods still appear to be unreliable. The value of sperm separation would be the avoidance of pregnancy with the male embryos where the mother is a carrier of an X-linked disease gene. X- and Y-bearing spermatozoa have been separated using discontinuous Percoll density gradients [21] but the reliability of the procedure needs to be further evaluated.

Tests on unfertilised eggs are possible through genetic analysis of the first polar body [9,22]. The polar body is the tiny cell lying next to the egg and one of the products of the first asymmetric meiotic division.

Biopsy

Cleavage Stage Embryos

Mouse embryos at the 8-cell stage are briefly incubated in acidic medium to remove the zona pellucida and transferred to medium without calcium and magnesium to loosen cell contacts. One blastomere is dislodged from each embryo by gentle pipetting. Following biopsy, about half the embryos transferred to foster mothers give rise to a fetus [3,23]. Krzyminska et al. [24] compared fetal development after biopsy of mouse embryos at the 4-cell, 8-cell and morula stages and concluded that biopsy of 8-cell embryos is the least detrimental to survival (52% giving rise to a fetus compared to 71% of controls).

In the human, the zona pellucida is left intact and the single blastomere is removed through a hole made in the zona. Handyside et al. [6] biopsied 25 human embryos at the 6- and 8-cell stage 3 days after IVF and 37% gave rise to the blastocyst by day 6 of culture. This frequency of ongoing development is similar to that of unoperated human cleavage stage embryos.

Blastocysts

Biopsies of trophoblast cells from mouse blastocysts are obtained by micromanipulation [25,26]. Clumps of 5–10 trophoblast cells are isolated from blastocysts in the process of hatching, or from blastocysts induced to herniate through a cut in the zona pellucida. The viability and implantation rate of the blastocysts following biopsy is only slightly less than that of unmanipulated embryos [25,26]. In the marmoset monkey, biopsies of up to 40% of trophectoderm cells have been taken from blastocysts which still produce normal young after replacement [27]. Trophectoderm biopsies of 30–50 cells can be grown to over 1000 cells in vitro.

The First Polar Body

Recently, the first polar body of the unfertilised egg has been removed for diagnosis. Monk and Holding [9] removed the zona from unfertilised human eggs and dislodged the polar body by gentle pipetting. For a clinical diagnosis, the polar body must be removed through a hole made in the zona by micromanipulation [22].

Non-Invasive Procedures on Whole Embryos

The ability to assess an embryo without the interference of removal of cells for diagnosis would be a considerable advantage. The approach is to incubate the preimplantation embryo briefly in serum-free medium and analyse metabolic processes such as release of synthesised factors or uptake of specific compounds into the embryonic cells. There are a few studies on the release of factors such as chorionic gonadtropin [28] and histamine-releasing factors [29] but such studies may be limited as they depend on early onset of embryonic gene activity.

The uptake of pyruvate and glucose has been measured in single mouse embryos at intervals throughout the preimplantation period followed by transfer of blastocysts to foster mothers to correlate uptake and viability [30,31]. Similar studies have been performed on human oocytes and preimplantation embryos

[32,33]. Degenerating embryos show decreased levels of uptake and such studies might be useful in evaluating the viability of preimplantation embryos and selection of the best embryos for transfer. However, it is doubtful at this stage whether the non-invasive approach will be applicable to diagnosis of disease.

Diagnostic Cytology

Human reproduction is very inefficient. It is estimated that the probability of a live term birth is only about 25% per ovulation cycle where intercourse occurred. Fertilisation failure occurs in 60% of cases and 15% of clinical pregnancies are lost by spontaneous abortion. The main reason for abortion is recognised to be due to lethal chromosome abnormalities in around 50% of cases [34] with autosomal trisomy as the most frequent abnormality. Cytogenetic studies on first trimester spontaneous abortions show that in the majority of cases of trisomy the abnormality arises due to non-disjunction at the first meiotic division of the oocyte and this is more frequent with increased maternal age.

Although most cases of autosomal trisomy result in abortion, some do survive, the most common being trisomy 21 which results in Down's syndrome. Cytogenetic analysis for chromosomal abnormalities is a well-established diagnostic procedure performed on cells obtained by chorionic villus sampling and amniocentesis, particularly when there is a risk of Down's syndrome. It requires good preparations of metaphase cells. Karyotyping procedures to detect numerical and structural chromosome abnormalities in preimplantation embryos might be practicable in cases of high risk of transmission of such defects. The procedures would be reliable at the level of 2–3 cells. This sensitivity is approached in human preimplantation embryos by a number of workers [1,2,19,20]. Successful karyotyping of cells biopsied from mouse preimplantation embryos has been reported by Severova and Dyban [35].

In Situ Hybridisation

In situ hybridisation allows detection of specific chromosomes in metaphase preparations or in interphase nuclei. The chromosome-specific probes are designed to detect a target sequence that is sufficiently repeated to provide a clear result. For diagnostic purposes, such probes allow detection of the number of copies of specific chromosomes in interphase nuclei and hence numerical chromosome aberrations (aneuploidy). Detection of the X and Y chromosomes may be used to sex embryos so as to avoid pregnancy with males where the mother risks passing on an X-linked genetic disease which is not amenable to specific diagnosis.

Sexing by in situ hybridisation has been applied successfully to whole human preimplantation embryos using a tritiated or biotinylated Y-specific probe which hybridises to sequences present in multiple copies on the Y chromosome [36–38].

The tritiated Y probe involves autoradiographic detection which takes 6–8 days [36] before the male embryos are clearly distinguished. Results may be obtained more rapidly (1–2 days) using a biotinylated probe and horseradish peroxidase [39], or alkaline phosphatase [38], immunocytochemistry. However, West et al. [39] found the biotinylated probe to be less reliable on the embryos. Using the tritiated probe it is estimated that reliable sexing could be achieved with 3–6 interphase nuclei.

Biochemical Microassay

Of those inherited diseases where the biochemical defect is known at the protein or enzyme level (around 200 cases) about 60 can be reliably diagnosed prenatally by chorionic villus sampling or amniocentesis [40]. Preimplantation diagnosis by biochemical microassay is only possible if the gene in question is normally expressed (i.e. transcribed and translated into active protein) by the embryo's own genome in the preimplantation stages of development. In addition, the level of the gene product inherited in the maternal cytoplasm of the egg must be sufficiently low so that the contribution of the embryo's own gene may be detected.

Sensitive biochemical microassays may be developed which detect the presence or absence of the gene product in a single cell, or in very few cells, taken from the preimplantation embryo for testing. In order that the level of the gene product in question may be evaluated, an internal control function is also screened in the same sample.

The procedures for preimplantation diagnosis were developed in a mouse model for Lesch-Nyhan syndrome (HPRT-deficiency) [3,25]. A double biochemical microassay for X-linked HPRT (hypoxanthine phosphoribosyltransferase, EC 2.4.2.8) and autosome-linked APRT (adenine phosphoribosyltransferase, EC 2.4.2.7) was developed to sufficient sensitivity to detect the presence or absence of HPRT activity in a single cell [41].

For diagnosis of the HPRT-negative male embryos of a mother heterozygous for X-linked HPRT-deficiency, single blastomeres biopsied from 8-cell embryos, or 5–10 trophoblast cells taken from blastocysts were assayed. Samples from the affected male embryos, with very low HPRT activity (and low HPRT:APRT ratios), were clearly distinguished. The diagnoses of HPRT-deficiency according to these single cell analyses were confirmed by transfer of the embryos and analysis of the fetuses.

Another microassay was developed to detect deficiency for adenosine deaminase (ADA, EC 3.5.4.4), the enzyme lacking in 30% of cases of severe combined immunodeficiency disease (SCID) [42]. This microassay is a quadruple microassay of very high sensitivity which is capable of detecting a deficiency of any one of four enzymes, ADA, HPRT, APRT and PNP (purine nucleotide phophorylase) in a single blastomere of a mouse 8-cell embryo.

The method of preimplantation diagnosis by biochemical microassay is clearly very successful in the mouse model systems. However, it appears that gene activation occurs later in human embryos than in mouse embryos. In an extensive study on human eggs and embryos, it was shown that the HPRT activity measured was inherited in the egg cytoplasm and that the embryonic gene was not detectably activated, at least not before the blastocyst stage [43]. This work emphasises the danger of extrapolating directly from animal models to the human and the importance of direct final analysis on human material.

Another two enzymes, glucose phosphate isomerase (GPI) [44] and hexosaminidase (I. Liebaers, personal communication), have been measured in preimplantation human embryos and, although the numbers and stages of embryos looked at were limited, there was no evidence of an increase in activity that might be indicative of onset of expression of the embryonic genes for these enzymes.

At this stage, direct biochemical microassay could not be applied to preimplantation diagnosis of Lesch-Nyhan disease (HPRT-deficiency), severe haemolytic anaemia (GPI-deficiency) or Tay–Sachs disease (hexosaminidase-deficiency). It is possible that expression of other enzymes (e.g. ADA, which is very low in the human egg) (Monk and Braude, unpublished) may be detectable earlier in the human. Also, further studies need to be done on blastocyst stage embryos.

DNA Analysis

The polymerase chain reaction (PCR) is a powerful technique capable of amplification of a specific DNA sequence from a single cell to a level where the DNA fragment can be readily analysed [45–47].

Preimplantation Diagnosis by PCR Amplification of Specific DNA Sequences in Mouse Model Systems

Preimplantation diagnosis by PCR amplification of a single copy gene sequence in a single blastomere isolated from an 8-cell embryo was first reported by Holding & Monk [8]. Mouse 8-cell embryos homozygous for a deletion of the entire beta-globin gene could be distinguished from normal embryos by PCR analysis of a single blastomere. Sensitivity and specificity of amplification of the beta-haemoglobin sequence was achieved with the use of nested primers and two sequential reactions. Over 90% of blastomeres tested (56 in all) give the correct result. Stringent precautions must be taken to ensure absence of contamination, the most important precautions being geographical isolation of the areas and equipment used for setting up the reactions and analysing the products.

Similarly, Gomez et al. [26] used the mouse mutant "shiverer" as a model. The gene encoding the myelin basic protein is deleted in shiverer mice. Trophectoderm biopsies taken from blastocysts were analysed for normal or mutant alleles of the gene by PCR using two sets of primers – one set for wild type mouse sequence not present in shiverer mice, and the other amplifying a fragment unique to shiverer DNA spanning the break points of the deletion. Gomez et al. [26] found that 96% of the blastocysts survived the biopsy procedure and 59% produced pregnancy (compared with 88% of unmanipulated controls). The diagnosis was correct in 25 out of 39 cases and the procedure distinguishes heterozygotes from homozygotes.

Preimplantation Diagnosis of Sex in the Human

PCR has been applied to the diagnosis of sex in single blastomeres of human 6-10-cell embryos [6]. The single blastomere was removed through a hole made in the zona and the sequence amplified was a repeat sequence on the Y chromosome present in about 800–1500 copies per cell. The first successful preimplantation diagnosis of sex followed by embryo transfer and pregnancy in the human has recently been reported [7]. The procedure was made available to five couples at risk of having a male child afflicted with severe X-linked genetic disease. Only diagnosed female embryos were replaced in each of the women resulting in two pregnancies and two sets of female twin babies. The results so far indicate a pregnancy rate of 35% per embryo which is higher than that reported previously in IVF programmes for treatment for infertility. This high pregnancy rate is achieved despite the removal of a cell for preimplantation diagnosis. This first report indicates that preimplantation diagnosis will be very successful.

Preimplantation Diagnosis of Single Copy Gene Defects in the Human

The determination of sex by PCR is facilitated by the presence on the Y chromosome of highly repeated DNA segments. However, PCR can be sufficiently reliable to detect single copy gene sequences in the mouse [8,26] and in the human [9,48,49].

In the human, amplification of the single copy sequences associated with cystic fibrosis and Duchenne muscular dystrophy in individual unfertilised human oocytes has been reported [48]. These preliminary experiments indicated that preimplantation diagnosis of single gene defects might be feasible with a single cell biopsied from the cleavage stage embryo.

PCR diagnostic procedures have been applied to the first polar body, the tiny cell lying next to the egg under the zona pellucida. The first polar body is a product of the first asymmetric meiotic division. Hence, in a heterozygous female, if the first polar body is mutant the egg will be normal and vice versa. However, if cross-over has occurred during meiotic prophase, both the egg and the polar body will be heterozygous for the condition.

Strom et al. [49] tested the first polar body of a single unfertilised human egg of a woman carrying a deletion, which, in the homozygous condition, causes cystic fibrosis (delta F508). Both she and her husband are heterozygous for this deletion and risk producing a child with the disease. The polar body was normal and hence it was deduced that the egg was mutant. This result was confirmed by fertilisation of the egg in vitro and biopsy followed by PCR analysis of the resulting preimplantation embryo which was not transferred.

Monk and Holding [9] analysed a series of unfertilised human eggs and the polar bodies separated from them for a beta-haemoglobin sequence which spans the site of the mutation which, in the homozygous condition, causes sickle cell anaemia. The specificity and sensitivity of amplification at the level of a single cell were again achieved by the two sequential reactions with two sets of primers. Control samples (from the final droplets of medium used to wash the cells) were negative, confirming absence of contamination. Overall, 6 of 11 unfertilised eggs following biopsy of the polar body gave a signal for both the egg and its polar body and all 11 controls were negative. Single blastomeres isolated from 8-cell human embryos also gave a strong positive signal.

To diagnose sickle cell anaemia, it is necessary to detect the presence or absence of the mutation in the amplified fragment. The mutation destroys a *Dde1* restriction enzyme site in exon 1 [50]; hence, a restriction digest of the amplified product positively identifies cells that are homozygous for the sickle cell mutation, those that are homozygous normal, and those that are heterozygous [9].

The reproducible amplification of a single copy gene sequence in a single first polar body demonstrates the feasibility of diagnosis of eggs before fertilisation. The first polar body of an unfertilised egg allows us to infer the genetic status of the oocyte. Only the oocytes without the defect would be fertilised and replaced in the mother. This approach avoids possible ethical difficulties of dealing with human embryos.

Currently, we are achieving close to 100% efficiency of amplification of single cells and a very low contamination rate of less than 3% (Mohadjerani and Monk, unpublished). The efficiency with which both beta-globin alleles are amplified in single buccal cells from an individual heterozygous for sickle cell trait is being determined (Mohadjerani and Monk unpublished). The ability to detect the heterozygous condition of a single cell is essential for diagnosis of unfertilised eggs using the first polar body. However, the reliability of detection of the mutant allele is already sufficient so as to allow preimplantation diagnosis of sickle cell anaemia using a single cell biopsied from the embryo.

Summary and Conclusions

Within the last three to four years the techniques of preimplantation embryo biopsy and diagnosis of genetic defects by non-invasive procedures, cytology, in situ hybridisation, biochemical and DNA analysis have been tried and evaluated. In the mouse it would appear that the removal of a few cells for testing will not jeopardise the future development of the embryo to term.

The preferred approach to preimplantation diagnosis will depend on the genetic defect under test. It may be possible to test for the absent or altered gene product directly, provided it is known that the product is synthesised under the direction of the embryonic gene. So far, the enzymes looked at in preimplantation embryos appear to be maternally inherited in the egg cytoplasm up to the morula stage. Diagnosis by DNA analysis using polymerase chain reaction amplification of the relevant gene sequence may be the more generally applicable approach provided the mutant DNA base sequence is known and contamination problems can be surmounted.

The first preimplantation diagnoses for sex in the human and transfer of female embryos are very encouraging. Preventive control of inherited X-linked disease not amenable to specific diagnosis is already in practice by identification and replacement of only female embryos.

In vitro fertilisation and subsequent development of human oocytes after removal of the first polar body has been reported and procedures for diagnosis for sickle cell anaemia in unfertilised eggs by PCR amplification of beta-globin sequences in the first polar body, and in single blastomeres of 8-cell embryos have been established.

The relatively easy access to gametes and preimplantation embryos makes preimplantation diagnosis a feasible proposition. Nevertheless, couples must be well counselled and they need to consider carefully whether preimplantation diagnosis, which incorporates the complex IVF procedures or the risks (at the present time) of uterine lavage, represents an advantage over a very early postimplantation diagnosis by chorionic villus biopsy.

References

1. Angell RR, Hillier SG, West JD, Glasier AF, Rodger MW, Baird DT. Chromosome anomalies in early human embryos. J Reprod Fertil 1988 (Suppl) 36:73–81.
2. Plachot M, Veiga A, Montagut J et al. Are clinical and biological IVF parameters correlated with chromosomal disorders in early lfe: a multicentric study. Hum Reprod 1988; 3:627–35.
3. Monk M, Handyside A, Hardy K, Whittingham D. Preimplantation diagnosis of deficiency of hypoxanthine phosphoribosyl transferase in a mouse model for Lesch-Nyhan syndrome. Lancet 1987; ii:423–6.
4. Monk M. Pre-implantation diagnosis. Bioessays 1988; 8:184–9.
5. West J. The use of DNA probes in preimplantation and prenatal diagnosis. Mol Reprod Dev 1989; 1:138–45.
6. Handyside AH, Pattinson JK, Penketh RJA, Delhanty JDA, Winston RML, Tiddenham EGD. Biopsy of human preimplantation embryos and sexing by DNA amplification. Lancet 1989; i:347–9.
7. Handyside AH, Kontogianni EH, Hardy K, Winston RML. Pregnancies from biopsied human preimplantation embryos sexed by Y-specific DNA amplification. Nature 1990; 344:768–70.
8. Holding C, Monk M. Diagnosis of beta-thalassaemia by DNA amplification in single blastomeres from mouse preimplantation embryos. Lancet 1989; ii:532–5.
9. Monk M, Holding C. Amplification of a B-haemoglobin sequence in individual human oocytes and polar bodies. Lancet 1990; 335:985–8.
10. Buster JE, Bustillo M, Rodi IA et al. Biologic and morphologic development of human ova recovered by non-surgical uterine lavage. Am J Obstet Gynecol 1985; 153; 211–17.
11. Sauer MV, Bustillo M, Rodi IA, Gorrill MJ, Buster JE. In vivo blastocyst production and ovum yield among fertile women. Hum Reprod 1987; 2:701.
12. Sauer MV, Bustillo M, Gorrill MJ, Louw JA, Marshall JR, Buster JE. An instrument for the recovery of preimplantation uterine ova. Obstet Gynecol 1988; 71:804–6.
13. Formigli L, Formigli G, Roccio C. Donation of fertilized uterine ova to infertile women. Fertil Steril 1987; 47:162–5.
14. Formigli L, Roccio C, Belotti G, Stangalini A, Coglitore MT, Formigli G. Non-surgical flushing of the uterus for pre-embryo recovery: possible clinical applications. Hum Reprod 1990; 5:329–35.
15. Sauer MV, Anderson RE, Paulson RJ. A trial of superovulation in ovum donors undergoing uterine lavage. Fertil Steril 1989; 51:131–4.
16. Edwards RG. Clinical aspects of in vitro vertilisation. In: Embryo research – yes or no. The Ciba Foundation Symposium, 1986; 39–42.
17. Braude PR. Fertilisation of human oocytes and culture of human preimplantation embryos. In: Monk M, ed. Mammalian development – a practical approach. Oxford: IRL Press, 1987; 281–306.
18. Edwards RG, Holland P. New advances in human embryology: implications of the preimplantation diagnosis of genetic disease. Hum Reprod 1988; 3:549–56.

19. Angell RR, Aitken RJ, Van Look PFA, Lumsden MA, Templeton AA. Chromosome abnormalities in human embryos after in vitro fertilisation. Nature 1983; 303:336–8.
20. Plachot M, Junca AM, Mandelbaum J, DeGrouchy J, Salat-Baroux J, Cohen J. Chromosome investigations in early life. 2. Human preimplantation embryos. Hum Reprod 1987; 2:29–35.
21. Iizuka R, Kaneko S, Aoki R, Kobayashi T. Sexing of human sperm by discontinuous Percoll density gradient and its clinical application. Hum Reprod 1987; 2:573–5.
22. Verlinsky Y, Ginsberg N, Lifchez A, Valle J, Moise J, Strom, C. Analysis of the first polar body: preconceptual genetic diagnosis. Hum Reprod 1990; in press.
23. Monk M, Handyside A. Sexing of preimplantation mouse embryos by measurement of X-linked gene dosage in a single blastomere. J Reprod Fertil 1988; 82:365–8.
24. Krzyminska UB, Lutjen J, O'Neill CO. Assessment of the viability and pregnancy potential of mouse embryos biopsied at different preimplantation stages of development. Hum Reprod 1990; 5:203–8.
25. Monk M, Muggleton-Harris A, Rawlings E, Whittingham D. Preimplantation diagnosis of HPRT-deficient male and carrier female mouse embryos by trophectoderm biopsy. Hum Reprod 1988; 3:377–81.
26. Gomez CM, Muggleton-Harris AL, Whittingham DG, Hood LE, Readhead C. Rapid preimplantation detection of mutant (shiverer) and normal alleles of the mouse myelin basic protein gene allowing selective implantation and birth of live young. Proc Nat Acad Sci USA 1990; 87:4481–4.
27. Summers PM, Campbell JM, Miller MW. Normal in vivo development of marmoset monkeys after trophectoderm biopsy. Hum Reprod 1988; 3:389–93.
28. Fishel SB, Edwards RG, Evans CJ. Human chorionic gonadotropin secreted in preimplantation embryos cultures in vitro. Science 1984; 223:816–18.
29. Cocchiara R, Di Trapani G, Azzolina A et al. Isolation of a histamine releasing factor from human embryo culture medium after in vitro fertilisation. Hum Reprod 1987; 2:341–4.
30. Gardner DK, Leese HJ. Non-invasive measurement of nutrient uptake by single cultured pre-implantation mouse embryos. Hum Reprod 1986; 1:25–7.
31. Gardner DK, Leese HJ. Assessment of embryo viability prior to transfer by the noninvasive measurement of glucose uptake. J Exp Zool 1987; 242:103–5.
32. Leese HJ, Hooper MAK, Edwards RG, Ashwood-Smith MJ. Pyruvate uptake by human oocytes and embryos. Hum Reprod 1986; 1:181–2.
33. Hardy K, Hooper MAK, Handyside AH, Rutherford AJ, Winston RML, Leese HJ. Non-invasive measurement of glucose and pyruvate uptake by individual human oocytes and preimplantation embryos. Hum Reprod 1989; 4:188–91.
34. Boue A, Boue J, Gropp A. Cytogenetics of pregnancy wastage. Adv Hum Genet 1985; 14:1–57.
35. Severova EL, Dyban AP. The live selection of mouse early embryos by sex and karyotype. Ontogenez 1984; 15:585–92.
36. West JD, Gosden JR, Angell RR et al. Sexing the human pre-embryo by DNA–DNA in situ hybridisation. Lancet 1987; i:1345–7.
37. Jones KW, Singh L, Edwards RG. The use of probes for the Y chromosome in preimplantation embryo cells. Hum Reprod 1987; 2:439–45.
38. Penketh RJA, Delhanty JDA, Vanden Barghe JA et al. Rapid sexing of human embryos by non-radioactive in situ hybridisation: potential for preimplantation diagnosis of X-linked disorders. Prenat Diagn 1989; 9:489–500.
39. West JD, Gosden JR, Angell RR et al. Sexing whole human pre-embryos by in-situ hybridisation with a Y-chromosome specific DNA probe. Hum Reprod 1988; 3:1010–19.
40. Patrick AD. Prenatal diagnosis of inherited diseases. In: Rodeck CH, Nicolaides KH, eds. Prenatal diagnosis. Proceedings of the XIth Study Group of the Royal College of Obstetricians and Gynaecologists. Chichester: Wiley, 1984; 121–32.
41. Monk M. Biochemical microassays for X-chromosome-linked enzymes HPRT and PGK. In: Monk M, ed. Mammalian development – a practical approach. Oxford: IRL Press, 1987; 139–61.
42. Benson C, Monk M. Microassay for adenosine deaminase, the enzyme lacking in some forms of immunodeficiency, in mouse preimplantation embryos. Hum Reprod 1988; 3:1004–9.
43. Braude PR, Monk M, Pickering SJ, Cant A, Johnson MH. Measurement of HPRT activity in the human unfertilised oocyte and pre-embryo. Prenat Diagn 1989; 9:839–50.
44. West JD, Flockhart JH, Angell RR et al. Glucose phosphate isomerase activity in mouse and human eggs and pre-embryos. Hum Reprod 1989; 4:82–5.
45. Saiki RK, Scharf S, Faloona F et al. Enzymatic amplification of B-globin genomic sequences and restriction site analysis for diagnosis of sickle cell anaemia. Science 1985; 230:1350–4.

46. Saiki RK, Gelfand DH, Stoffel S et al. Primer-directed enzymatic amplification of DNA with a thermostable DNA polymerase. Science 1988; 239:487–91.
47. Li H, Gyllensten UB, Cui X, Saiki RK, Erlich HA, Arnheim N. Amplification and analysis of DNA sequences in single human sperm and diploid cells. Nature 1988; 335:414–17.
48. Coutelle C, Williams C, Handyside A, Hardy K, Winston R, Williamson R. Genetic analysis of DNA from single human oocytes: a model for preimplantation diagnosis of cystic fibrosis. Br Med J 1989; 299:22–4.
49. Strom CM, Verlinsky Y, Milayeva S et al. Preconception genetic diagnosis of cystic fibrosis. Lancet 1990; 336:306–7.
50. Embury S, Scharf S, Saiki R et al. Rapid prenatal diagnosis of sickle cell anaemia by a new method of DNA analysis. N Engl J Med 1987; 316:656–61.

Discussion

Rodeck: The success rate with preimplantation diagnosis as far as fertilisation is concerned seems to be higher than is accepted in infertility circles. That is probably because for the first time people with normal fertility are having IVF.

Monk: Seven embryos out of 21 produced pregnancies.

Campbell: I agree that preimplantation in prenatal diagnosis looks very promising but it is an enormously expensive way of carrying out prenatal diagnosis. I doubt if the Health Service could ever bear the cost. Even in the most cost-effective IVF systems it will cost about £400 per cycle of treatment.

Williamson: One says that, but if we look at the CF cost, the overall cost of identifying that couple in the community who are at risk is of the order of £20 000 to £30 000. If we think it is justified to identify that couple in order to give them reproductive options, including the option of prenatal diagnosis, an additional £1000 on top of that is a very small proportion of the total that we are spending.

Campbell: One has to justify the first exercise, but accepting that, we are not just talking about CF, we are talking about the whole range of genetic diseases, and that will be very expensive.

Monk: People come in and say they want the new techniques, because they think the latest is best. Preimplantation diagnosis has had so much press coverage, and women have not been given all the facts. They have to go through all the complex procedures of IVF and the inefficiency of that, for preimplantation diagnosis versus a 1 in 4 risk with CVS and diagnosis and very early termination at nine weeks before the pregnancy is noticeable.

Lilford: There are many patients who are desperate to get early diagnosis especially after they have been through a few terminations.
Why confine the polar body biopsy to the prefertilisation period? My biologist tells me that she can see quite accurately which is the first and second polar body. Is there some danger of confusing the two?

Monk: The first polar body disintegrates quite quickly. The results I showed were from fresh eggs that were surplus to GIFT, not failed fertilisation eggs. We got four out of four successful amplifications and no contamination of both polar body and egg. But with the failed fertilisation eggs it was more difficult.

Pembrey: Whilst we are on error rates, we quote a 1% procedural error rate for CVS in counselling. Certainly in our lab, based on Southern blotting methods, in the first 100 prenatal diagnoses we had one false negative, predicted as a CF carrier, born as a CF affected, it was not a recombination, it was a rather cruel plasmid contamination which gave a band in the original affected channel and of the intensity just where we had expected a band from polymorphism.

John Old who receives samples from all over the world, has a 1% procedural error rate.

Brock: We do not really know what the PCR error rate is likely to be. Potentially there is a greater error rate there, because of contamination.

Section IV
Cytogenetic and Biochemical Disorders

Chapter 12

Fetal Karyotyping Using Chorionic Villus Samples

C. M. Gosden

Introduction

Since the introduction of amniocentesis in the early 1970s for fetal karyotyping to detect chromosomal abnormalities in older mothers, there has been a dramatic expansion in the use of prenatal chromosome studies and risk of chromosome disorder is the major indication for prenatal diagnosis at present. Although older mothers were the major group known to be at high risk for fetal chromosomal disorder, they were having only 20%–30% of the chromosomally abnormal children. Unfortunately, primary aetiological factors involved in the conception of chromosomally abnormal fetuses (other than advanced maternal age, a previous child or a parental chromosomal anomaly) have not yet been identified. This has made the development of screening methods for the recognition of abnormal pregnancies, either by maternal serum biochemical parameters (Wald, Chapter 4) or by the identification of structural markers on ultrasound scan [1] very important. Although secondary prevention is all that is possible at present for mutant chromosomal disorders, attention must be focused on the need to understand those factors which cause constitutional chromosomal abnormalities to occur, so that primary prevention becomes a realistic goal. Fetal karyotyping has a crucial role to play in the prevention of childhood handicap due to chromosomal abnormalities, since these occur in 1 in every 150 newborns, but it is essential to determine which pregnant women are at risk for chromosomal disorders and examine the methods available for karyotyping together with their safety and accuracy in order that rational approaches to chromosomal prenatal diagnosis can be developed.

Use of Chorionic Villus Sampling for Fetal Karyotyping

Prenatal chromosome studies can be undertaken on amniotic fluid cells and fetal blood, but it was the introduction of chorionic villus sampling (CVS) which made early first trimester prenatal diagnosis available for the first time on a relatively plentiful and accessible tissue. The advantages of early prenatal diagnosis were seen in the vastly increased uptake rates by the patients, but some problems in the use of trophoblast for diagnosis began to emerge, particularly trophoblastic divergence from the fetus, maternal cell contamination and the need to undertake further sampling in a significant proportion of cases. These findings raised questions about which methods of fetal sampling and karyotyping are the safest and most accurate when balanced against the need for early prenatal diagnosis for patient acceptability. Until recently, it was thought that the small volumes of amniotic fluid present in early pregnancy and the very low cell numbers were contraindications to first trimester amniocentesis. However, the need for early prenatal diagnosis, improved first trimester ultrasound and improved methods of amniotic fluid cell culture, have all combined to make early amniocentesis possible. Not only are randomised controlled trials to study the safety and accuracy of first trimester CVS against the "standard" of second trimester amniocentesis essential, but it is now imperative to find out whether CVS is safer and more accurate than early amniocentesis.

It is the sum of all the different risks involved (procedure-associated risks of early and late pregnancy losses, risk of test failure and misdiagnosis) which will influence the use of CVS. The methods of sampling and risks have been covered in previous chapters. The purpose of this chapter is to examine who is at risk for chromosome disorders, the chromosomal abnormalities involved and their diagnosis using early and late CVS against the background of the potential pitfalls of diagnosis using CVS.

Prenatal Karyotyping: Who is at Risk and What Are the Risks of Clinically Significant Chromosomal Disorders?

The largest group at present having cytogenetic prenatal diagnosis is that of older mothers. Two large collaborative studies showed that for women > 35 years old, the frequencies of clinically significant chromosomal abnormalities at amniocentesis were 1.84% to 2.01% [2,3] indicating the low risks in this group. Prenatal karyotyping is often regarded as a screening method for trisomy 21, but many other clinically significant chromosomal abnormalities also occur which include conditions causing severe handicap such as trisomies, deletions, duplications, de novo rearrangements and supernumerary marker chromosomes. The total risk for chromosomal abnormalities from amniocentesis data [3] at the age of 35 is 12.9 per 1000, whereas for Down's syndrome (DS) alone the risk is 3.5 per 1000 (only 27% of the total risk); at the age of 40, the overall risk is 23.6 per 1000, with DS 12.3 per 1000 (52% of the total); and at the age of 45, the overall risk has risen to 72.8 per 1000 with the risk for DS 45.3 per 1000 (62% of the total). Couples with a previous child with a mutant chromosomal abnormality and those with a strong family history of chromosome disorder also have a relatively low risk of recurrence (1%–2%). Parents who are carriers of a chromosomal rearrangement, translocation, or a genetic chromosomal syndrome have much higher risks of having an affected child varying from about 5%–35% depending on the abnormality involved, but relatively few people are carriers. Cytogenetic diagnosis for X-linked disorders (either as an adjunct to DNA analysis or used alone) involves a 50% risk for males. Other conditions where there is a very high risk of cytogenetic abnormality include genetic chromosomal syndromes such as fragile X-linked mental retardation, and the chromosomal instability syndromes such as ataxia telangiectasia, Fanconi's anaemia and Bloom's syndrome.

There are a growing number of screening tests carried out in the second trimester of pregnancy on women with no recognised increased risk of cytogenetic disorder. These include maternal serum alphafetoprotein (AFP) screening or triple testing (Chapter 4), and abnormalities on ultrasound scan [1]. The risks for those with abnormal maternal biochemical markers, are in general, relatively low, as karyotyping is usually offered when the risk is equal to or exceeds that of a woman of 35 years (> 1 in 280) (Chapter 4). The risks of fetal chromosomal abnormality for anomalies on scan may be as high as 35%–50%, depending on the fetal abnormality.

Possible Sources of Diagnostic Error in Karyotyping Using CVS

Fetal karyotyping has at least two different modes of approach. The first is to perform a screen for major chromosomal abnormalities (predominantly trisomies) in groups with relatively low risks (1%–2%) such as advanced maternal age, previous child with a mutant chromosomal abnormality and low maternal serum AFP. The second is to try to detect clinically significant chromosomal disorders which involve a variety of subtle but specific chromosomal anomalies in women with much higher risks (10%–35%) of cytogenetic disorder, such as those with genetic chromosomal syndromes or abnormalities on ultrasound scan. The difficulties of rapid karyotyping in pregnancies where there is a major structural abnormality of the fetus are exemplified by the problems encountered if the fetus has an omphalocele. This occurs in about 1 in 4000 pregnancies and about 25% of fetuses have major chromosomal disorders including trisomies 13 and 18, triploidy and a number of chromosomal deletions and duplications [4]. The Beckwith–Weidemann syndrome (BWS) occurs at a frequency of about 1 in

13 000 pregnancies and exomphalos is one of the cardinal signs in the syndrome. Many of the cases of BWS involve a microduplication in the short arm of chromosome 11. Thus up to 1 in 3 of all cases of omphalocele may be involve BWS and accurate karyotyping in cases of exomphalos would involve not just a search for an autosomal trisomy but a detailed search for a possible minute duplication of chromosome 11p15 in order to avoid the risk of obtaining a false negative result.

There are a number of reasons why the karyotype obtained from the placental villi may differ from that of the fetus. Chorionic villi are mitotic derivatives of the zygote but the cells which give rise to the cytotrophoblastic and extraembryonic tissues have already diverged from the epiblast at the 64 cell stage, so that some anomalies (particularly autosomal trisomies) may be restricted to trophoblast alone; these are described as confined placental abnormalities [5]. Other sources of apparent discrepancy between the fetus and the trophoblast include in vitro change, maternal cell contamination and twin pregnancy with death of the abnormal twin. Since the peak maternal age for dizygotic twin pregnancy is 38 years and at this age at least 1 in 40 pregnancies involves twins, this is by no means a rare cause of problems.

There were delays in identifying the scale of the problem of karyotype divergence of the placenta from the fetus in CVS. One of the principal reasons was that since only 27% of the chromosomal abnormalities diagnosed at CVS in a woman of 35 were expected to involve Down's syndrome, 73% would involve non-Down's syndrome anomalies so that abnormalities such as trisomy 18, 45,X, de novo translocations and tetraploidy were not unexpected. Furthermore, as the frequency of and severity of chromosomal abnormalities in CVS were expected to be greater than those found at second trimester amniocentesis (since these would contribute to the early spontaneous abortions in this group), it was not until discrepancies were shown to occur between the villi and the fetus, that the dangers of misdiagnosis in CVS were recognised.

Profound biological differences exist between the trophoblast and the fetus; for example, X-inactivation differs in the fetus and placenta. In a female fetus, X-inactivation occurs at random, so that in some cells it is the maternal X which is inactivated whereas in others it is the paternal. This is not the same in trophoblast, where it is always the paternal X which is inactivated [6]. The placenta also differs from the fetus when the fetus has pure trisomy 13 or 18, because the placenta almost invariably contains normal cells in addition to trisomic cells. The highest proportion of normal cells in the placenta in these cases is correlated with survival of the fetus until later in gestation, so that placental selection for normal cells aids fetal survival [7]. Arguments that the mechanism by which this is achieved is due to prolonged cell cycle times of trisomic cells, so that normal cells have a proliferative advantage in the placentas of trisomic fetuses, are difficult to support when considering the converse situation, that is the presence of trisomic lines (especially those for trisomy 18) which occur as confined placental abnormalities when the fetus itself has a normal karyotype. In this latter case, trisomic cell lines should be at such a proliferative disadvantage that they would rarely be seen unless there is some biological reason for their occurrence. Both these phenomena (the occurrence of normal cells in the placenta when the fetus is abnormal and abnormal cells in the placenta when the fetus is normal) are responsible for affecting the diagnostic accuracy of prenatal diagnosis using CVS and illustrate that more research is needed to extend our knowledge about mechanisms which exert control over the karyotype in fetus and placenta.

False Positive Results in First Trimester CVS

What are the risks of discrepancy between the CVS karyotype and the fetus itself when the CVS result is abnormal but the fetus is actually normal? Confined placental abnormalities are found most frequently in the cytotrophoblastic cells studied in the direct preparations [8,9]. Direct methods have other disadvantages, such as producing metaphase spreads which are of poorer quality, contain a high proportion of hypomodal cells and are more difficult to band than cultured cells, but they give a very rapid chromosome result and avoid significant maternal cell contamination [8,9]. Long-term culture which utilises mesenchymal cells (which belong to a cell lineage derived from the epiblast which is much more closely related to the embryo and has diverged later than the cytotrophoblast), take longer, but produce better metaphase spreads. These allow the identification of small chromosomal deletions and duplications (thus avoiding some false negative results) and they have lower rates of confined placental abnormalities but have high rates of maternal cell contamination. When CVS was in its infancy, most laboratories used only direct preparations for the speed and ease with which a karyotype could be obtained. Once the dangers of false positive results were recognised, a combination of both direct and culture methods were used to obtain diagnostic results for CVS, allowing the direct preps to monitor possible maternal cell contamination and cultures to minimise the risk of confined placental abnormalities. However, in some cases a chromosomal abnormality may be seen in both direct and cultured cells and this may not be present in the fetus. The vast majority of false positive results involve mosaicism and so the types of abnormality and their frequencies will be discussed with mosaic abnormalities.

Mosaicism in CVS and Abnormalities with a High Risk of Giving False Positive Results

Mosaicism is the presence of two or more cell lines with different karyotypes in a prenatal diagnostic sample or an individual. Most cases of true mosaicism arise as a result of an error in an early division in the zygote rather than to a parental nondisjunctional error. Although mosaicism in CVS may be associated with true mosaicism in the fetus, it is more often simply the consequence of confined placental abnormalities, in vitro changes, maternal cell contamination or the presence of cells from a deceased twin, but it is difficult, without further sampling, to distinguish those cases due to confined placental abnormalities from those which are real. The rates for chromosomal mosaicism in diagnostic CVS vary between 0.4% and 4.0% [8–20]. The results reported by Hook et al. [10] showed that even within one published series, rates for mosaicism varied between 12/925 and 2/2132 and the rates for trisomy 18 from 7/925 to 8/2132.

Four different types of chromosomal abnormalities are involved in false positive and mosaic results in CVS; autosomal trisomies, sex chromosome abnormalities, chromosomal rearrangements and polyploidy. The autosomal trisomies can be divided into the clinically recognised trisomies such as trisomies 8,9,13,18,21,22 where some may be associated with true mosaicism in the fetus, and trisomies not associated with clinically recognised syndromes, such as trisomies 3,6,7,14,15,16, and 20, usually representing confined placental abnormalities but may signal risk of adverse pregnancy outcome. No case of false positive diagnosis for pure trisomy 21 has yet been reported, although there are false positive results for trisomy 21 mosaicism, but reports of false positive results for other trisomies are frequent especially those for trisomy 18. The sex chromosome abnormalities include two major problem groups, 45,X cell lines, often not real in the fetus, and 46,XX/46,XY mosaicism which usually involves maternal cell contamination but is often viewed with the fear it might involve true hermaphroditism. Other forms of mosaicism involve 47,XXX, 47,XXY and 47,XYY.

Chromosomal rearrangements and supernumerary markers are a very heterogeneous group of abnormalities including de novo translocations, inversions, deletions and duplications as well as supernumerary marker chromosomes. The risks depend on the type of rearrangement, whether balanced and the chromosome regions involved. Many of the cases are unique and if after further sampling they are shown to be real, there are major problems in predicting what the prognosis will be. Polyploidy includes triploidy and tetraploidy, which have entirely different origins, probabilities of being real and consequences. Tetraploidy occurs frequently in CVS and is one of the leading causes of false positive results in CVS. Triploidy occurs less frequently, and in some cases involves twin pregnancy with death of the triploid twin but survival of its trophoblast.

In order to try and give some indication of the dangers of false positive results, a figure has been drawn up to show cases from the literature. These data have been accumulated from a number of different sources including the large collaborative studies, published series, and some individual cases [5–36]. In the early stages of CVS, termination was often carried out immediately an abnormal result was seen on the direct preparations, so that risk figures derived from the earlier studies on CVS represent a disproportionate number of cases derived from direct preparations alone. Confirmatory karyotyping on fetal tissue obtained after termination, is essential in the assessment of false positive risks, but this information is lacking on over 60% of all cases reported in the literature. The reasons vary from failing to identify fetal tissue in the products, the sample being placed in formalin, failure to grow or bacterial contamination in culture. Cumulatively these effects complicate the assessment of false positive rates. The highest risks for false positive results occurs for trisomies 3, 7, 8, 9, 12, 13, 15, 16, 18, 20, 21, 22, 45,X, polyploidy especially tetraploidy, 47,XXY and supernumerary markers and these can be seen, with indications of whether they occur in pure or mosaic form, in Fig. 12.1.

The Canadian randomised trial [37] indicated that the risks of confined placental abnormalities are not restricted to CVS, but demonstrated that the scale of the problem in CVS is much greater than that for amniocentesis. In the two randomised groups (of equal size), there were 45 chromosomal abnormalities in the CVS group, compared with only 22 in the group randomised to amniocentesis. Of the 45 chromosome abnormalities 19 (42%) in the CVS group were shown to be confined placental abnormalities; these included two non-mosaic and 12 mosaic aneuploidies, three polyploid mosaics and two unbalanced translocations. Of the 22 chromosomal abnormalities found in the amniocentesis

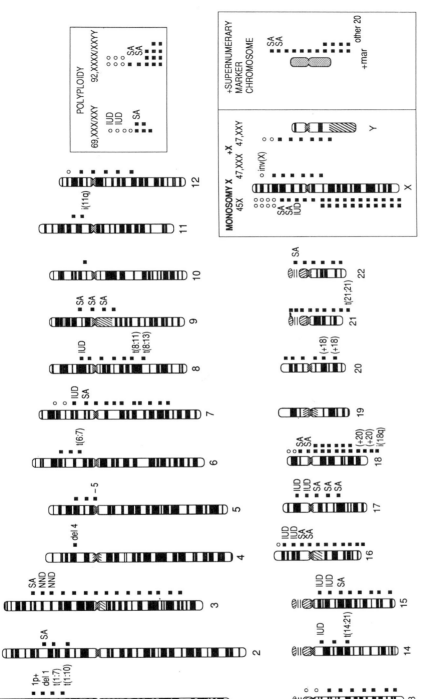

Fig. 12.1. Ideogram showing false positive results for chromosome abnormalities which have been reported in chorionic villus samples (CVS) in which the CVS sample showed the karyotypic abnormality indicated but the fetus was chromosomally normal. If the chromosome result has been associated with an adverse pregnancy outcome such as an intrauterine or neonatal death (IUD, NND) or spontaneous abortion (SA) this is indicated beside the specific abnormality, ○, pure (non-mosaic) trisomies; ■, mosaic abnormalities. Chromosome rearrangements are designated: translocations, e.g. t(1:7); deletions, e.g. del 4; isochromosomes, e.g. i(18q); inversions, e.g. inv(X).

Most cases involve trisomy or rearrangements, the exception is monosomy X where the 45X cases are to the left of the X chromosome and the 47,XXX and 47,XXY false positives are to the right of the X chromosome.

group, 3 of the 22 (14%), were abnormalities which were not present in the fetus, but were confined placental abnormalities all involving aneuploid mosaicism.

Adverse Pregnancy Outcome Associated with Confined Placental and Mosaic Karyotypes

Only recently has the association of certain adverse outcomes of pregnancy (such as spontaneous abortion, perinatal death or poor perinatal outcome) with certain types of mosaicism been recognised. In some reports, information on the outcome was limited and so adverse pregnancy outcome is almost certainly under-reported, but other cases of placental mosaicism have had detailed study and follow-up (5,20,23,29,31,32). Johnson et al [29] reported a series of 4319 pregnancies having first trimester CVS where chromosomal mosaicism was present in 55 cases (1.3%), but was confirmed in the fetus in only six cases (0.2%) and was confined to the placenta in the remainder (1.1%). In the group with confined placental mosaicism there were seven fetal losses (16.7%) compared with the loss rate of 2.7% overall.

There are specific chromosomal abnormalities which indicate a very high risk of adverse pregnancy outcome. These are shown in Fig. 12.1 and include trisomies 2,3,7,8,9,14,15,16,17,18,22, monosomy X, triploidy, tetraploidy and supernumerary markers. Of these abnormalities, trisomies 3,7,16,17,18 and polyploidy are associated with a higher risk of adverse outcome than the others.

False Negative Results in the First Trimester

False negative results, that is normal chromosome results in CVS but the birth of a chromosomally abnormal child, are relatively rare; the incidence is about 1 in 1000 cases of CVS, although they are rather more common in second trimester CVS than in the first trimester. False negative results may be grouped into three different subsets. The first occur because of biological selection in the placenta against abnormal cell lines so that the placenta is normal whereas the fetus is chromosomally abnormal. Thus if placental tissue alone is used for karyotyping abnormal cells will not be found. There is also a danger of a false negative result if the fetus is mosaic for the abnormality studied so that if the mosaic cell line is only found at a very low level in the placenta then it might not be found even if relatively large numbers of cells are studied. These are both forms of biological false negatives. There is also the problem of small chromosomal deletions such as those involved in the microdeletion syndromes which might be present in both fetus and chorion but might be overlooked in the CVS preparations (these could be described as technical false negatives).

False negative results have been reported for a number of different chromosomal abnormalities. Deletions are particularly common because they are difficult to detect in CVS, especially in direct preparations. Trisomies are also frequent because of placental selection against abnormal cell lines. The false negative cases reported in the literature include deletions including 4p-, 11q-, and deleted 18q, trisomies 18 and 21 and 47,XXY [24,26,38–44].

Maternal Cell Contamination

Cells of maternal origin are a possible source of diagnostic error in chromosome analysis from CVS. They are rarely a problem in direct CVS preparations and less than 1 division per 1000 cells is usually seen by the methods currently in use [9,45,46]. In cultured CVS samples they cause more problems and rates from 2%–14% have been reported [9,46]. Maternal cell contamination is a problem because of the risk of misdiagnosis (giving a maternal and not a fetal karyotype). If direct preparations with their low risk of maternal cells are used to assess the culture results, then the risk of misdiagnosis is minimised, but this requires extra time and resources. Testing for maternal cell markers is mandatory in all cases of 46,XX/46,XY mosaicism.

The Canadian MRC randomised trial [37] gave data for maternal cell contamination in the amniocentesis and CVS groups (which were of equal size) showed only four cases with problems in the amnio group compared with 26 in the CVS group. All four cases in the amnio group involved XX/XY mosaicism, but in none of these cases was any further testing done. In the 26 cases in the CVS group, 20 cases showed XX/XY mosaicism but there was no follow-up amniocentesis, four cases with a 46,XX karyotype and suspected maternal cell contamination had a follow-up amniocentesis and in two cases there was a discrepancy in sex diagnosis. All but one of these problem cases in CVS occurred in cultures.

How to Resolve Diagnostic Dilemmas: are Further Sampling or the Application of Different Techniques Necessary?

One of the principal advantages of first trimester CVS has always been thought to be that of allowing a first trimester termination if the fetus is abnormal. The dangers of false positive results from direct preparations alone now seem so great that the majority of centres wait for the culture result if the direct karyotype is abnormal. Unfortunately, even in these cases, false positive results still occur. There are thus three major questions which need careful consideration in cases of abnormal results in CVS. These are first, whether further sampling is justified and what proportion of patients will need further sampling. Second, what are the risks of two different sampling procedures and are there any remaining risks of false positive results if further testing is carried out. Third, it is important to question whether there are any other techniques which can be used in addition to cytogenetic analysis in these cases which would be of help.

At present there seems to be no way of distinguishing cases of confined placental mosaicism from those pregnancies in which the abnormality is real without further testing. Since a repeat of CVS would have the same risk of misdiagnosis and further studies on the original CVS sample, such as those analysing more cells or the use of DNA probes would be expected to give the same results as those seen initially and it is important to recognise that in the vast majority of cases the problem lies in the sample itself and not in the analysis. It is therefore essential that more detailed studies should be done on a different tissue offering some significant advantage in accuracy of diagnosis. In the American collaborative study [47], 3% of CVS patients required a subsequent amniocentesis and in the Canadian trial [37] 9.9% had further testing. If an amniocentesis is to be undertaken after CVS for a suspected false positive result, then the greatest chance of accuracy and the way of minimising risks of confined anomalies is to wait until the amniotic fluid contains sufficient cells of fetal origin to give the greatest chance of karyotypic accuracy. The cell types which predominate in an early amniocentesis specimen are those from trophoblastic and extraembryonic membranes and it is not until about 16–17 weeks gestation that with an increasing turnover of amniotic fluid from fetal swallowing and urination, the proportion of cells from the fetal surfaces starts to increase. Few centres have yet reviewed the risks for those patients who have had both CVS and amniocentesis in the same pregnancy. In a study of 82 patients who underwent both CVS and amniocentesis in the same pregnancy, a fetal loss rate of 2.5% was found by Brandenburg et al. [48], which is similar to the rates in this centre for each of the procedures separately. This centre has some of the lowest reported procedure related risk, but they suggest that the risks for a patient having two procedures in the same pregnancy are, at minimum, the sum of the risks for the two separate procedures.

Are There Other Methods, New Techniques or Other Studies in the Pregnancy Which Might be of Help When There are Problems?

The use of recombinant DNA methods as an adjunct to chromosome studies can be very important in defining the exact nature, at a molecular cytogenetic level of some chromosomal disorders. However, the use of DNA probes in cases of mosaicism must be undertaken with great care if major diagnostic errors are to be avoided. DNA probes can be used to assess dosage (as for example in autosomal trisomies) or to establish the identity of a particular chromosome or chromosomal segment, but there are problems with quantitation or other difficulties which must be recognised. We have used a variety of different DNA probes (either chromosome, gene or sequence specific to investigate metaphase and interphase cells in CVS, AFC and fetal blood cells, using radioactive and biotinylated probes of different types using conventional microscopy and the confocal (laser) microscope. We have been unable to achieve levels of hybridisation above 80% in interphase cells although every single metaphase examined shows clear hybridisation [49]. Thus although the majority of cells from a pure trisomy 21 individual will show metaphase cells with three labelled chromosome 21s, less than 80% of the interphase cells show three blocks of hybridisation. It may be possible to undertake analysis cell by cell using the polymerase chain reaction (PCR), and thus if it were simply a problem of technical artefact of inaccessibility to probes or other factors, this might help. If, however, it is due to biological properties of interphase cells, then new methods for resolving problems of false positive results and mosaicism need to be developed.

Detailed ultrasound scan in cases of mosaicism can be misleading. The presence of the normal cell lines may dilute the phenotypic effect and ameliorate the effects of abnormal development in structural anomalies such as heart defects, so that the absence of structural abnormalities must not be taken as a sign of reassurance. Many of the individuals with true mosaicism are severely mentally and developmentally handicapped even in the absence of abnormalities detectable on scan in utero. It is also difficult to predict the clinical effects of the abnormality from the proportion of abnormal cells in CVS because the cardinal tissues for predicting the severity of the phenotypic abnormalities are those such as brain in cases of trisomy 21 or disorders causing mental handicap, or the proportion of 45,X cells in the fetal gonads in 45,X/46,XY mosaicism where the phenotype may be either normal male, Turner-like female or intersex depending on the proportion of 45,X cells.

Advantages and Problems of Second and Third Trimester CVS (Placental Biopsy)

Risk of chromosomal abnormality is often only recognised relatively late in the second trimester for indications such as abnormalities on scan or oligohydramnios [50,51], when the need for urgent karyotyping is compounded by parental anxiety and the gestational age at which termination is still possible. Specific chromosomal abnormalities such as trisomies 13 and 18 are frequently found when the fetus has major organ system abnormalities such as renal or cardiac malformations. For more generalised disorders, such as severe intrauterine growth retardation, anomalies may be distributed throughout the karyotype, varying from fetal triploidy to almost all described deletions and duplications, since these all cause disruptions in the patterns of fetal growth. Many obstetricians have expertise in the sampling procedure (since CVS is a less specialised technique than fetal blood sampling) but its use must be evaluated by considering the benefits and problems of the procedure. The initial figures suggest that the accuracy is less than that for first trimester CVS, although indications, risks of abnormalities and misdiagnosis differ so that it is difficult to make valid comparisons with other procedures. The results from the collaborative registry for late CVS [51], show that the patients have a very high risk of chromosomal abnormality; 21% of cases showed a major karyotypic anomaly. However, fetal losses were 10% and perinatal deaths 6.8%; although these loss rates seem very high, this is a very unusual group with a high background rate of structural abnormalities in the fetus and a high termination rate for structural abnormalities in the fetus as well as fetal chromosome disorders. In 3.3% of cases karyotyping was unsuccessful.

The reason that false negative results occur more often in second trimester CVS than in the first trimester is probably twofold. The first is that one of the major reasons for undertaking CVS in the second trimester is to obtain an urgent result if the fetus has a major structural abnormality on scan. Many of these cases involve trisomies 13 and 18 where the chance of normal cells being present in the placenta is high. Two false negative diagnoses in late CVS have been reported, both for trisomy 18 [52]. The second is that the quality of the direct preparations may be poor and there is a substantial chance of failing to detect small chromosomal deletions.

Other problems also occur with late CVS. There may be problems of tissue specificity; for example if the fetus has diaphragmatic hernia, the major risks of chromosomal abnormality are those of trisomy 18 or isochromosome 12p (Pallister–Killian syndrome PKS). There is a very high risk of false negative results for trisomy 18 in late CVS. For PKS the abnormality is usually only seen in fibroblasts; it is not expressed in fetal blood cells and little is known about its expression in cytotrophoblast. Thus fetal karyotyping in cases of diaphragmatic hernia should try to avoid CVS because of the risk of false negative results for trisomy 18 but include a tissue culture providing fibroblastic lines (i.e. either amniotic fluid or CVS) for PKS. Chromosomal imprinting also affects the phenotype according to whether the abnormality is inherited from father or mother, may affect cardinal features and in fetal triploidy affects the risk of invasive trophoblastic disease [53].

Is CVS More Labour Intensive Than Other Methods of Fetal Karyotyping?

Fetal karyotyping is important as a public health resource for many hundreds of thousands of patients at risk. If the screening levels for maternal age and AFP levels depend on the number of samples which can be handled, then questions of whether CVS is more labour intensive than amniocentesis are important. Direct chromosomal preparations from chorionic villi have the advantages of rapid results and low risk maternal cell contamination, but high risk of false positive results. Cultures from CVS avoid many problems of confined placental abnormalities but have a high risk of maternal contamination. A combination of both methods achieves maximum accuracy with the least chance of misdiagnosis. A workload measurement system was devised by the Association of Clinical Cytogeneticists (ACC Working Party, 1989). Their studies showed that one trained cytogeneticist (TC) could karyotype 325 postnatal blood samples per year (unit value of 250 units). Amniotic fluid samples were more labour intensive, with a value of 520 units so that a TC could handle 158 amniotic fluid samples per year. The unit value for direct preparations from chorionic villi was 450 units, with an additional 550 units if culture techniques were used, giving a total of 1000 units for both methods for CVS. Since mistakes can occur if either the direct method or culture methods are used alone, then only 82 CVS samples can be fully karyotyped (using both direct and culture methods) per TC per year, compared with 158 amniocentesis specimens per TC per year.

Conclusions: assessment of risks, accuracy and problems

Over 60 000 diagnostic CVS samples have already had successful chromosome analysis. The use of CVS depends on the factors which include the sum of the procedure associated risks and the accuracy of diagnosis. In some pregnancies at low risk of fetal chromosomal abnormality, where the principal risk is that of autosomal trisomy the use of new methods for targeting trisomies and other major aneuploidies such as DNA analysis, in situ hybridisation using DNA probes on interphase and metaphase cells, fluorescence activated cell sorting or computerised chromosomal analysis might soon be possible. Detailed chromosome analysis requiring special cytogenetic skill and expertise to investigate pregnancies with a substantial risk of small chromosomal rearrangements such as small deletions and duplications or requiring specialised karyotypic techniques, would be reserved for those pregnancies at high risk of these anomalies. If CVS is more labour intensive than amniocentesis then this too must be considered in the context of the number of women who can be offered prenatal karyotyping and the effects of this on the proportion of children who will be born with chromosomal abnormalities. Those women at risk who do request fetal karyotyping at present are predominantly those who are well educated, caucasian, with higher than average family incomes who live in large towns and cities and attend younger obstetricians for prenatal care [54]. If prenatal karyotyping for those at risk is to have impact as a public health measure, then the methods of sampling must have low risks, and the chromosome results should have a high degree of accuracy with the lowest possible risks of misdiagnosis as these are all correlated with acceptability to the patients and uptake rates.

References

1. Nicolaides KH, Rodeck CH, Gosden CM. Rapid karyotyping in non-lethal fetal malformations. Lancet 1986; i:283–7.
2. Schreinemachers D, Cross P, Hook E. Rates of trisomies 21, 18, 13, and other chromosomal abnormalities in about 20,000 prenatal studies compared with estimated risks in live births. Hum Genet 1982; 61:318–24.
3. Ferguson-Smith MA, Yates JRW. Maternal age specific rates for chromosomal aberrations and factors influencing them; a report of a collaborative European study on 52,965 amniocenteses. Prenat Diagn 1984; 4:5–44.
4. Gosden CM. Prenatal diagnosis of chromosomal anomalies in prenatal diagnosis. In: Lilford RJ, ed. Prenatal diagnosis and prognosis. London, Boston, Tokyo, Wellington: Butterworths, 1990; 104–64.
5. Kalousek DK, Dill FJ, Pantzar T, McGillivray BC, Yong SL, Wilson RD. Confined chorionic mosaicism in prenatal diagnosis. Hum Genet 1987; 77:163–7.
6. Harrison KB. X-chromosome inactivation in the human cytotrophoblast. Cytogenet Cell Genet 1989; 52: 37–41.
7. Kalousek DK, McGillivray B. Confined placental mosaicism and intrauterine survival of trisomies 13 and 18. Am J Hum Genet 1988; 44:54–60.
8. Sachs ES, Jahoda MGJ, Los FJ, Pijpers L, Reuss A, Wladimiroff JW. Interpretation of chromosome mosaicism and discrepancies in chorionic villi studies. Am J Med Genet 1990; 37:268–71.
9. Ledbetter D. Cytogenetic results of chorionic villus sampling: high success rates and diagnostic accuracy in the USA Collaborative study. Am J Obstet Gynecol 1990; 162:495–501.
10. Hook EB, Cross PK, Jackson L, Pergament E, Brambati B. Maternal age-specific rates of 47, +21 and other cytogenetic abnormalities diagnosed in the first trimester of pregnancy in chorionic villus biopsy specimens: comparison with rates expected from observations at amniocentesis. Am J Hum Genet 1988; 42:797–807.
11. Czepulkowski BH, Heaton DE, Kearney LU, Rodeck CH, Coleman DV. Chorionic villus culture for first trimester diagnosis of chromosome defects. Prenat Diagn 1986; 6:271–82.
12. Hogge WA, Schonberg SA, Golbus MS. Prenatal diagnosis by chorionic villus sampling: lessons of the first 600 cases. Prenat Diagn 1985; 5:393–400.
13. Hogge WA, Schonberg SA, Golbus MS. Chorionic villus sampling: experience of the first 1,000 cases. Am J Obstet Gynecol 1986; 154:1249–50.
14. Martin AO, Simpson JL, Rosingky BJ, Elias S. CVS in continuing pregnancies II. Cytogenetic reliability. Am J Obstet Gynecol 1986; 154:1353–62.
15. Therkelson AJ, Jensen PKA, Hertz JM, Smidt-Jensen S, Hahnemann N. Prenatal cytogenetic diagnosis after transabdominal sampling in the first trimester. Prenat Diagn 1988; 8:19–31.

16. Jahoda MGJ, Pijpers L, Reuss A, Los FJ, Wladimiroff JW, Sachs ES. Evaluation of transcervical chorionic villus sampling with a completed follow-up in 1550 consecutive pregnancies. Prenat Diagn 1989; 9:621–8.
17. Leschot NJ, Wolf H, Van-Prooijen-Knegt AC et al. Cytogenetic findings in 1250 chorionic villus samples obtained in the first trimester with clinical follow-up of the first 1000 pregnancies. Br J Obstet Gynaecol 1989; 96:663–70.
18. Lilford RJ, Linton G, Irving HC, Mason MK. Transabdominal chorion villus biopsy: 100 consecutive cases. Lancet 1989; i:1415–16.
19. Sachs ES, Jahoda MGJ, Kleijer WJ, Pijpers L, Galjaard H. Impact of first trimester chromosome, DNA and metabolic studies on pregnancies at high genetic risk: experience with 1,000 cases. Am J Med Genet 1988; 29:292–303.
20. Callen DF, Korban G, Dawson G et al. Extra embryonic/fetal karyotypic discordance during diagnostic chorionic villus sampling. Prenat Diagn 1988; 8:453–60.
21. Mikkelsen M. Cytogenetic findings in first trimester chorionic villi biopsies; a collaborative study. In: Fraccarro M, ed. First trimester fetal diagnosis. Berlin: Springer–Verlag, 1985; 110–20.
22. Mikkelsen M, Ayme S. Chromosomal findings in chorionic villi; a collaborative study. In: Vogel F, Sperling A, eds. Human genetics. Berlin: Springer–Verlag, 1987; 597–606.
23. Simoni G, Gimelli G, Cuoco C et al. Discordance between prenatal cytogenetic diagnosis after chorionic villi sampling and chromosomal constitution of the fetus. In: Fraccarro M, Simoni G, Brambati B, eds. First trimester fetal diagnosis. Berlin, Heidelberg: Springer-Verlag, 1985; 137–43.
24. Crane JP, Cheung SW. An embryogenic model to explain cytogenetic inconsistencies observed in chorionic villus versus fetal tissue. Prenat Diagn 1988; 8:119–29.
25. Delozier-Blanchet CD, Engel E, Extermann P, Pastori B. Trisomy 7 in chorionic villi: follow-up studies of pregnancy, normal child, and placental clonal anomalies. Prenat Diagn 1988; 8:281–6.
26. Eiben B, Hansen S, Knipping J, Massenberg R, Goebel R, Hammans W. Translocation trisomy 21 in CVS not found in embryoblast: three different cell lines in CVS, amnion- and placental culture. Prenat Diagn 1989; 9:365–7.
27. Greene JE, Dorfman A, Jones SL, Bender S, Patton L, Schulman JD. Chorionic villus sampling; experience with an initial 940 cases. Obstet Gynecol 1988; 71:208–12.
28. Jackson LG. CVS Newsletter, 1987, 1988, 1989, 1990. Philadelphia, Thomas Jefferson University.
29. Johnson A, Wapner RJ, Davis GH, Jackson LG. Mosaicism in chorionic villus sampling: an association with poor perinatal outcome. Obstet Gynecol 1990; 75:573–7.
30. Reddy KS, Blakemore KJ, Stetton G, Corson V. The significance of trisomy 7 mosaicism in chorionic villus cultures. Prenat Diagn 1990; 10:417–23.
31. Schwartz S, Ashai S, Meijboom EJ, Schwartz MF, Sun CC, Cohen MM. Prenatal detection of trisomy 9 mosaicism. Prenat Diagn 1989; 9:549–54.
32. Schwinger E, Seidl E, Klink F, Rehder H. Chromosome mosaicism of the placenta – a cause of developmental failure of the fetus? Prenat Diagn 1989; 9:639–47.
33. Schulman LP, Tharapel AT, Simpson JL et al. Three different, non-mosaic sex chromosome abnormalities (direct cytotrophoblasts, mesenchymal core cultures, and abortus skin fibroblasts): implications for elucidating chorionic villi mosaicism. J Med Genet 1989; 26:791–2.
34. Vejerslev LO, Mikkelsen M. The European collaborative study on mosaicism in chorionic villus sampling: data from 1986 to 1987. Prenat Diagn 1989; 9:575–88.
35. Verjaal M, Leschot NJ, Wolf H, Trefers E. Karyotypic differences between cells from placenta and other fetal tissues. Prenat Diagn 1987; 7:343–8.
36. Wegner RD, Hohle R, Karkut G, Spurling K. Trisomy 14 mosaicism leading to cytogenetic discrepancies in chorionic villi sampled at different times. Prenat Diagn 1988; 8:239–43.
37. Canadian Collaborative CVS-Amniocentesis Clinical trial Group. Multicentre randomised clinical trial of chorion villus sampling and amniocentesis; first report. Lancet 1989; i:1–6.
38. Eichenbaum SZ, Krumins EJ, Fortune DW, Duke J. False negative finding on chorionic villus sampling. Lancet 1986; ii:391.
39. Miny P, Basaran S, Holzgreve W, Horst J, Pawlowitzki IH, Ngo TK. False negative cytogenetic result in direct preparations after CVS. Prenat Diagn 1988; 8:633.
40. Simoni G, Gimelli G, Cuoco C et al. First trimester fetal karyotyping: one thousand diagnoses. Hum Genet 1986; 72:203–9.
41. Leschot NJ, Wolf H, Weenink GH. False negative findings at third trimester chorionic villus sampling (C.V.S.). Clin Genet 1988; 34:204–5.
42. Eiben B, Leipolt M, Schibbe I, Ulbricht R, Hansmann I. Partial deletion of 4p in fetal cells not present in chorionic villi. Clin Genet 1988; 33:49–52.

43. Bartels I, Hansmann I, Holland U, Zoll B, Rauskolb R. Down syndrome at birth not detected by first-trimester chorionic villus sampling. Am J Obstet Gynecol 1989; 34:606–7.
44. Linton G, Lilford RJ. False negative finding on chorionic villus sampling. Lancet 1986; ii:630.
45. Blakemore KJ, Samuelson J, Breg WR, Mahoney MJ. Maternal metaphases on direct chromosome preparation of first trimester decidua. Hum Genet 1985; 69:380.
46. Roberts E, Duckett DP, Lang GD. Maternal cell contamination in chorionic villus samples assessed by direct preparations and three different culture methods. Prenat Diagn 1988; 8:635–40.
47. Rhoads GG, Jackson LG, Schlesselman SE et al. The safety and efficacy of chorionic villus sampling for early prenatal diagnosis of cytogenetic abnormalities. N Engl J Med 1989; 320:609–17.
48. Brandenburg H, Jahoda MGL, Pijpers L, Reuss A, Kleyer WJ, Wladimiroff JW. Fetal loss rate after chorionic villus sampling and subsequent amniocentesis. Am J Med Genet 1990; 35:178–80.
49. West JD, Gosden CM, Gosden JR et al. Sexing the human fetus and identification of polyploid nuclei by DNA–DNA in situ hybridisation in interphase nuclei. Mol Reprod Dev 1989; 1:129–36.
50. Nicolaides KH, Soothill PW, Rodeck CH, Warren RC, Gosden CM. Why confine chorionic villus (placental) biopsy to the first trimester? Lancet 1986; i:543–4.
51. Holzgreve W, Miny P, Schloo R, and Participants of the "Late CVS" registry. Late CVS International registry: compilation of data from 24 centres. Prenat Diagn 1990; 10:159–67.
52. Wirtz A, Seidel H, Bruisis F, Murken J. Another false-negative finding on placental sampling. Prenat Diagn 1988; 8:321.
53. Bagshawe KD, Lawler SD, Paradinas FJ, Dent J, Brown P, Boxer GM. Gestational trophoblastic tumours following initial diagnosis of partial hydatidiform mole. Lancet 1990; 335:1074–6.
54. Naber JM, Huether CA, Goodwin BA. Temporal changes in Ohio amniocentesis utilisation during the first twelve years (1972–1983), and frequency of chromosomal abnormalities observed. Prenat Diagn 1987; 7:51–65.

Chapter 13

Prenatal Diagnosis of the Fragile-X Syndrome

T. Webb

Background

The Martin-Bell syndrome is a common form of X-linked mental retardation, characterised by co-segregation with a fragile site at Xq27.3 (FRAXA). With a birth frequency of approximately 0.4–0.8/1000 males and 0.2–0.6/1000 females [1–4], this syndrome is second only to Down's as the most common form of heritable mental retardation. Penetrance in males in whom the fragile-X can be demonstrated (as opposed to the usually FRAXA negative yet gene-carrying transmitting males) is considered to be approximately 100%, while about one-third of carrier females are mentally impaired to some degree. The fragile site is nearly always found in affected female heterozygotes but is detected in only between a half and two-thirds of those of normal intelligence. In general those who are handicapped express the fragile Xq27.3 more easily than those who are not, but approximately 10% of normal female carriers do express FRAXA at a high level [5,6]

Due to the high prevalence of the Martin-Bell syndrome there is an urgent need for both accurate carrier detection and prenatal diagnosis. Theoretically this should be feasible both cytogenetically due to the co-segregation of the fraXq27.3 marker chromosome and, because the disease locus is accurately mapped, by employing linked probes. However, difficulties have been encountered using either method. Cytogenetically, demonstration of FRAXA requires special culture methods involving manipulation of the thymidine pool [7], but even then it is present in only a percentage of mitoses, a level of 4% being considered to be diagnostic in lymphocytes. Molecular biological methods have proved to be unsatisfactory as there is a high level of crossing over between linked probes. The fragile-X has been designated a hot spot for recombination.

Demonstration of fragile-X is achieved either by employing a medium with a very low concentration of folic acid or no folate at all, or by using folate antagonists such as methotrexate (MTX), which blocks the enzyme tetrahydrofolate reductase [8], or FUdR, which blocks thymidylate synthetase [9]. In each case the result is to reduce the amount of thymidine available to the cell. An excess of thymidine will also serve to induce FRAXA as shown by Sutherland et al. [10]. Fragile-X was first demonstrated in amniocytes retrospectively by Jenkins et al. [11] who found 12/60 (20%) of FRAXA after treatment with 0.4 µM-FUdR. A prospective prenatal diagnosis was first made in fetal lymphocytes by Webb et al. [12] and in amniocytes by Shapiro et al. [13], who supported the finding of 3/106 (2.8%) of FRAXA in these cells with a concurrent finding of 2/58 (3.4%) in fetal lymphocytes. The first diagnosis of fragile-X in a heterozygous female was in 1984 when Wilson and Marchese [14] detected 4/81 (5%) of FRAXA-positive amniocytes treated with MTX but none (0/243) in amniocytes cultured in TC199 without antagonists.

Difficulties were being experienced due to the poor chromosome morphology, low mitotic index and long culture times in cultures treated with MTX or FUdR, and low levels of fragile-X being induced in non-treated cultures. This problem was highlighted by Rocchi et al. [15] who found discordant expression of the fragile site in a pair of male monozygotic twins. Treatment of fetal lymphocytes with 0.1µM-FUdR revealed 32% and 34% of fragile-X respectively but treatment of amniocytes with 0.05µM-FUdR allowed expression in 2/243 cells from twin 1 and 11/41 cells from twin 2.

In 1986 three papers summarised experience in prenatal diagnosis so far. Jenkins et al. [16] reported 79 successful diagnoses of which 27 were FRAXA positive, Shapiro and Wilmot [17] found four fragile-X positive males from 74 amniotic fluid specimens and Tommerup et al. [18] reported five positive cases from a total of 23 pregnancies at risk.

Despite these successes all three groups advised caution when employing amniotic fluid cells. Jenkins et al. [16] found variation in fragile-X levels of between 0.5 and 15.5% in different cultures from a single case. Shapiro and Wilmot [17] found a similar variation, between 0 and 22%, and in their report a pair of twins had 1.4% and 12% of FRAXA-positive cells respectively. Very low or borderline positive levels of 1.4% and 1.9% were alo being found – levels comparable to those of Tommerup et al. [18], who found a fetus with 3/220 fragile-X-positive cells to be negative after termination whereas a level of 1/355 subsequently proved to be positive.

The possibility of an easier diagnostic method was offered by Sutherland et al. [19], who induced the fragile-X in amniocytes with an excess of thymidine, a method earlier proved to be successful for lymphocyte cultures [20]. When compared to the fragile-X levels demonstrated in parallel cultures treated with methotrexate, an excess of thymidine increased the demonstrable level of fragile-X severalfold.

A further improvement in technique came when the fragile site was shown to be inducible in chorionic villus sampling (CVS) cells. Using TC199 and 0.05 µg/ml of MTX as an inducing agent, Tommerup et al. [21] found 5% of fragile-X in long-term CVS cultures, a result confirmed after termination. The fragile site, however, was found neither in direct preparations nor in a 24 hour culture in Chang medium.

The application of molecular biological techniques offered one approach to the resolution of some of the problems. Tommerup et al. [21] used the flanking probes F1X and St14 but found a crossover between them in the family under investigation. Oberle et al. [22], using the same approach, suggested that the method was not yet accurate enough to use by itself although it could help to decrease the false negative rate. Shapiro et al. [23] used the proximal probe DXS10 in preference to F1X, but pointed out that for one case in particular detection was further complicated by the possibility of confusion between a double crossover and a transmitting male, in a fetus who was positive according to DNA typing but who was cytogenetically negative.

By 1988 comparisons were being drawn between the different available methods. Jenkins et al. [24] confirmed the prenatal diagnosis of fragile-X in 35 cases, 26 males and 9 females, but there were false negatives using amniocytes, and once again large between-culture discrepancies were encountered in a single case (0 and 19.2%). One fragile-X-positive male was also detected out of seven CVS samples. These authors did not find that an excess of thymidine was superior to FUdR as an inducer of the fragile site. Shapiro et al. [23] described several approaches to diagnosis including amniotic fluid, CVS, fetal blood sampling and molecular methods. Their total of 160 cases included 140 amniotic fluids, 90 of which yielded a result: 56 were males with seven fragile-X positives, but also included were two false negatives. Five out of 38 females were also fragile-X positive. In 13 cases studied after CVS sampling, of five fragile-X negative males, two were in fact false negatives, whereas three fetal blood samples yielded a positive male and a positive female. Molecular methods employing DXS10 and DXS52 (St14) as flanking probes suggested that four cases were RFLP positive but the chromosome results were negative in three of them. Purvis-Smith et al. [25] performed an inter-laboratory study on 50 potential cases of the fragile-X syndrome, finding ten to be FRAXA positive. There were, however, four cases of between-laboratory discrepancy and one false negative. There were some encouragingly high levels of fragile-X-positive CVS cells after culture in 600 mg/l of thymidine.

There appeared to be a general consensus [17,23–25] that reliability of diagnosis is maximised if at least three different tissue culture methods are employed, duplicate cultures are studied for each induction system, and at least 100 mitoses are scored for each case. Where amniocytes or CVS cells demonstrate low levels of the fragile site, molecular methods or fetal blood samples should be used.

Although the first reported prospective prenatal diagnosis was made in fetal lymphocytes [12], due to an increased risk of fetal loss fetal blood sampling has been recommended only for high risk pregnancies [25] or as an aid in the resolution of difficult cases [24]. Demonstration of the fragile-X is easier in fetal lymphocytes and higher percentages are generally obtained [26,27]. Despite this, there has been a false negative reported [28]. Orlandi et al. [29] summarised the risks of fetal blood sampling, finding 16 fetal losses (4.3%) out of 370 cordocenteses. The fetal loss rate correlated neither with operator experience nor with gestation over the gestation period 12–18 weeks, falling to half the original level between 19 and 21 weeks of gestation.

The Prenatal Diagnosis of Fragile Xq27.3

In this study 64 pregnancies considered to be at risk for the Martin–Bell syndrome underwent prenatal diagnosis for detection of fraXq27.3. Of these pregnancies 51 were subjected to fetal blood sampling via cordocentesis as the main diagnostic method, two had amniocentesis alone and 11 had only first-trimester CVS. Amniotic fluid cells were studied together with the fetal blood samples for the earlier subjects [27], but later subjects had fetal blood sampling either as the sole diagnostic method or subsequent to a negative CVS. The majority of cordocenteses (36/57) were carried out at King's College Hospital. All but one of the samples were received within 24 hours of sampling. In every case but three, the fetal sample was accompanied by a blood sample from the mother in order to determine whether or not she manifested the fragile-X chromosome. Her family history was also taken in order to provide an estimate of her carrier status. Cultures were set up in a variety of different media designed to demonstrate the presence of the fragile site fraXq27.3 (Table 13.1). Cultures were maintained at 37°C for 48 hours before the addition of either folate antagonists or an excess of thymidine for 24 hours. Harvesting, slide-making and Giesma banding were all by routine methods. CVS samples and amniocytes were cultured as described previously [30].

Care was taken to maintain the "balance" of the cytogenetic results. Were possible, approximately equal numbers of mitoses were recorded from each culture so that there was no undue bias towards any one method. This was often difficult as some samples responded poorly to the folate antagonists FUdR and MTX. As demonstration of FRAXA was often facilitated by these additives, the positive cells were not allowed to become "diluted out" by over-representation of a kinder culture with a higher mitotic index but a low demonstration of FRAXA.

Table 13.1. Media used for detection of fraXq27.3 (FRAXA) in fetal lymphocytes

1. TC199 + 2% fetal calf serum (FCS)
2. TC199 + 2% FCS + 10^{-7}M methotrexate (MTX)
3. TC199 + 2% FCS + 10^{-7}M FUdR
4. RPMI 1640 + 10% FCS + 300 mg/litre thymidine
5. RPMI 1640 + 10% FCS + 600 mg/litre thymidine
6. FXI

Results

Of the 64 pregnancies studied, 52 had male outcomes and 12 had female. Fifteen of the males (15/52) and five of the females (5/12) were FRAXA positive (Table 13.2).

Table 13.3 indicates the distribution of the pregnancies and their outcomes according to the type of diagnostic samples received. Out of 51 prenatal blood

Table 13.2. Prenatal diagnosis of the Martin-Bell syndrome

Total number of pregnancies	64
Normal males	37
Martin–Bell syndrome males	15
Normal females	7
Fragile-Xq27.3 positive females	5

Table 13.3. Cytogenetic outcomes of prenatal diagnosis for Martin–Bell syndrome

Fetal blood	Total 51		
Normal males	33		
FRAXA males	13	11 terminated	2 live born
Normal females	3		
FRAXA females	2	1 live born	1 terminated
CVS	Total 11		
Normal males	2		
FRAXA males	2	Both terminated	
Normal females	4		
FRAXA females	3	2 terminated 1 ongoing as yet	
Amniocentesis	Total 2		
Normal males	2		
FRAXA males	0		

samples, 14 were found to be fragile-X positive, 12/46 males and 2/5 females. All the results were confirmed either postnatally or after termination and one male was found to have been missed, giving 13/46. Two FRAXA-positive males came to term, one which was missed and one whose mother refused termination. Of the two FRAXA-positive females, one was terminated and one was liveborn. Both had very low levels of fraXq27.3 but the diagnosis was confirmed in both cases. Where the fragile site was detected in the first trimester CVS samples, both positive males were terminated as were two of the three females.

The distribution of mothers according to their carrier status and pregnancy outcome is shown in Table 13.4 which shows that 10/16 mothers who were FRAXA positive themselves and who had already had a FRAXA-positive child were found to be carrying a FRAXA-positive male, whereas only 1/12 of FRAXA-positive mothers who had not previously had a FRAXA-positive child was carrying a FRAXA-positive male. When all the carrier mothers are considered together, then out of 28 with male outcomes, 11 were FRAXA positive, showing an expected deficiency (11/28 = 39%) [5,6].

Those mothers who had not had a previous FRAXA-positive pregnancy and who were FRAXA negative themselves, i.e. those who were referred because of family history alone, had only a 1/18 chance of carrying a FRAXA-positive male (Table 13.4). Only those mothers known to be carriers of the Martin-Bell syndrome are considered in Table 13.5. This group includes all those who are FRAXA positive themselves and those who are FRAXA negative but who have had a previous affected child. In total, 14 of the 33 males (42%) born to carrier

Table 13.4. Outcome of pregnancy when mothers are divided into groups according to their status as carriers

Carrier status of mother	Number	FRAXA positive males	FRAXA positive females
Previous FRAXA +ve child and mother FRAXA +ve	21	10/16	2/5
FRAXA −ve mother of FRAX +ve child	6	3/5 (one missed)	1/1
FRAXA +ve mother, no previous FRAXA +ve child	14	1/12	1/2
FRAXA −ve mother, no previous FRAXA +ve child (FM only)	19	1/18	0/1
Mothers' chromosomes not done	3	0/1	1/2 (previous FRAXA +ve child)
Total FRAXA +ve mothers	35	11/28	3/7

Table 13.5. Outcome of pregnancy of known FRAXA carriers

Type of tissue	Number	FRAXA +ve males	FRAXA +ve females
Fetal blood only	32	10/28	2/4
CVS only	8	2/3	3/5
CVS + fetal blood	2	2/2 (1 missed)	0
Amniocentesis	0	0	0
Total	42	14/33	5/9

mothers were FRAXA positive but so were 5/9 females. Approximately equal numbers of males and females were also observed by Shapiro et al. [23].

The numbers of FRAXA cells scored for each of the 13 positive males is shown in Table 13.6 for the individual culture media, and the range for each type is shown in Table 13.7.

The media without additives yield many more mitoses (average = 137 per sample) than those with MTX (average = 32) or FUdR (average = 29) but an excess of thymidine is slightly less harsh (average = 56). All the media with additives are more efficient at inducing the fragile site than those without, but results from all three methods are very comparable (Table 13.7). Table 13.6 illustrates the danger of placing too much emphasis on large numbers of mitoses from low folate media alone, when the addition of MTX, FUdR or an excess of thymidine causes a marked reduction in mitotic index. This may reduce the observable level of FRAXA-positive cells disproportionately. The average percentage of FRAXA cells scored per positive male diagnosis in this study was 12% with a range of 0–32% (Table 13.7). The average number of cells scored per positive male diagnosis was 224 with a range of 105–390. Of the 33 fetal blood samples with normal 46XY male karyotypes, 7545 mitoses were examined for FRAXA, an average of 236 per subject. A total of 26 mitoses (0.3% average) were found to carry a fragile-X chromosome. This is probably attributable to the presence of the common fragile site FRAXD at Xq27.2 [31].

Table 13.6. Detection of fraXq27.3 in fetal blood samples from males

Subject	Number and percentage of FRAXA-positive cells in various culture media								
	Low or no folate media (%)		Media 10^{-7}M MTX (%)		Media with 10^{-7}M FUdR (%)		Media with excess thymidine (%)		Total (%)
1	6/88	(7)	12/28	(43)	15/28	(54)	ND		33/144 (23)
2	50/211	(24)	14/60	(23)	5/22	(23)	ND		69/293 (24)
3	21/227	(9)	10/24	(42)	1/2		ND		32/253 (13)
4	25/175	(14)	1/3		3/5		ND		29/183 (16)
5	3/130	(2.3)	4/13	(31)	–		ND		7/143 (5)
6	6/317	(2)	5/67	(7)	2/6		ND		13/390 (3)
7	0/123	(0)	6/22	(27)	27/148	(18)	ND		33/293 (11)
8	2/78	(3)	2/22	(9)	6/15	(40)	2/2		12/117 (10)
9	2/88	(2)	2/64	(3)	0/11		0/2		4/165 (4)
10	7/93	(8)	2/28	(7)	8/103	(8)	19/100	(19)	36/324 (11)
11	0/162	(0)	0/22	(0)	0/23	(0)	0/50	(0)	0/257 (0)
12	0/30	(0)	23/50	(46)	2/6		12/30	(40)	37/116 (32)
13	1/53	(2)	1/17	(6)	1/9		31/150	(21)	34/229 (15)

Table 13.7. Range of FRAXA-positive cells obtained from males when lymphocytes were cultured in different media

Culture medium	Range of mitoses obtained	Average	Range of FRAXA-positie mitoses	Average
Low or no folate medium	30–317	137	0/162–50/211 (0–24%)	123/1775 (7%)
Medium + MTX	3–67	32	0/22–23/50 (0–46%)	82/420 (19.5%)
Medium + FUdR	0–148	29	0/23–15/28 (0–54%)	70/378 (18.5%)
Medium with excess Thymidine	2–150	56	0/50–12/30 (0–40%)	64/334 (19%)
Total	105–390	224	0/257–37/116 (0–32%)	339/2907 (12%)

The two FRAXA males detected after first trimester CVS had a total of 13/146 and 5/115 fragile-X positive cells respectively, an average of 7% (M. McKinley, personal communication).

Of the five FRAXA-positive female heterozygotes detected, two were diagnosed after fetal blood sampling, one being terminated. The three others were detected after first trimester CVS and two were terminated. Levels of FRAXA in the three samples ranged from 4/72 to 14/101 (6%–14% positive) with an average of 10%. This is considerably higher than the levels induced in the two fetal blood samples (1.5% and 2.5% respectively).

Problems Encountered During the Study

1. Two of the pregnancies spontaneously aborted after fetal blood sampling, resulting in the loss of three male fetuses. Neither of these abortions was considered to be a consequence of the procedure [27] and they occurred in 1983 and 1985. There has been no spontaneous loss in the last 40 pregnancies sampled.
2. One false negative diagnosis was made after fetal blood sampling [28].
3. The blood sample from one patient came divided into two paediatric tubes which were not mixed, separate cultures being set up for each sample. The outcome from the first was a normal male karyotype 46XY, with 0/312 cells FRAXA positive, the second sample was a mixture of male and female cells, the male cells being 46XY and the female cells 46XX fraXq27.3, 11% positive. It was concluded that the fetus was a karyotypically normal male and that the second sample was contaminated with maternal blood.
4. One sample from a pregnancy previously found to be male after direct CVS examination had the karyotype 46XX. This was adjudged to be a maternal sample. The cordocentesis was repeated and a male with a 46XY karyotype was diagnosed.
5. One sample contained very few blood cells, appearing to be largely amniotic fluid. No mitoses were obtained and when the sampling was repeated a 46XY karyotype was diagnosed.

In order to determine whether it is possible to anticipate whether or not a potentially positive male fetus will be easily detectable in terms of proportions of positive cells, the percentages of FRAXA cells present in the blood samples from eleven positive males were correlated with the percentages present in their mothers. A correlation coefficient of $r = 0.14$ indicates that the level in the mothers was not indicative of the potential level in fetal cells.

Discussion

Prenatal diagnosis for the Martin–Bell syndrome by means of the cytogenetic demonstration of FRAXA has resulted in the detection of 15 positive male pregnancies and five positive female pregnancies out of a total of 52 males and 12 females. It is believed that there were no false positives (all except one were examined after birth or post-termination) but of 34 fetal blood samples from males found to be FRAXA negative one positive was subsequently found to have been missed.

The false-negative rate in CVS samples is still under investigation but both of the terminated FRAXA-positive female fetuses were confirmed as heterozygotes after detection of the fragile site in fibroblasts.

Despite the fact that only one-third of female carriers are mentally handicapped, one of the two detected by fetal blood sampling and two of the three detected by CVS were terminated. There is a need for a reliable method of distinguishing between the affected and the normal fragile-X-positive carrier. In general, handicapped carriers have higher proportions of FRAXA-positive cells but the relationship between the levels in fetal blood samples or CVS samples and degree of potential handicap is unknown. Once the gene for the Martin–Bell syndrome has been isolated it may be possible to distinguish between affected and normal heterozygotes.

The unexpected finding that carrier mothers with an affected or FRAXA-positive child are seven times more likely to have a FRAXA-positive male than are carrier mothers without such a previous child (10/16 or 62.5% as opposed to 1/12 or 8%) may be a consequence of small numbers, but if real, this implies clustering and that certain types of normal carrier are more likely to have affected children than are other types. Such differences could possibly be visualised as X-inactivation differences or imprinting differences, depending on the position of the mother in the pedigree.

An expected shortfall in FRAXA-positive males born to known carriers was found in this study as only 11 were detected in 28 FRAXA-positive mothers instead of 14. The "missing" three account for 21% of the expected number – the percentage of FRAXA-negative transmitting males believed to be present in Martin–Bell syndrome pedigrees [6].

The fact that there have been no pregnancy losses after cordocentesis over the past five years indicates it to be a relatively safe procedure. A positive result, however, means that any termination will be in the second trimester. Until first trimester CVS biopsies have been proved to yield an acceptably low level of false negative results, fetal blood sampling will be required for confirmation [24] even if it is not the diagnostic method of choice.

Despite variations in diagnostic techniques aimed at improving accuracy [32,33], amniocentesis has not generally been found to be very reliable for the prenatal diagnosis of the fragile-X syndrome, as even when a successful diagnosis is achieved, the percentage of FRAXA-positive cells can be low when compared to that in other tissues and termination is sometimes late [34]. When exposed to agents such as MTX or FUdR the cells grow slowly, reducing the mitotic index and lowering the number available for analysis. This extends the culture time even further. The use of low or no folate media may not be sufficient to induce a high enough expression of the fragile site to facilitate diagnosis. Not only do low frequencies of the fragile site imply the possibility of false negatives, but because of the large numbers of mitoses needed there is the chance of obtaining a false positive due to the detection of the common site FRAXD or to non-specific telomeric changes in the region Xq27–8. Occasionally there is a spontaneous low level of expression of fragile Xq27.3 in amniotic fluid samples which have been referred to the laboratory for other reasons. Shapiro et al. [23] have found that these are usually artifacts.

The use of flanking DNA markers provides an alternative test in cases where the cytogenetic results are equivocal [35]. Sutherland and Mulley [36] suggest a strategy for obtaining the most accurate information, but point out that molecular methods are still of limited use for prenatal diagnosis, although they are considered to be helpful for carrier detection, where between one-third and half of the normal heterozygotes never demonstrate the cytogenetic marker under any culture conditions currently employed. It is envisaged that eventually first trimester CVS will be found to yield accurate diagnoses both positive and negative, eliminating the need for fetal blood sampling at 18–19 weeks of pregnancy. The relative robustness of the cells coupled with easier induction of fraXq27.3 make this the method of choice for the prenatal diagnosis of the Martin-Bell syndrome, especially when supported by molecular methods using closely linked probes.

References

1. Gustavson KM, Blomquist HK, Holmgren G. Prevalence of the fragile-X syndrome in mentally retarded children in a Swedish county. J Med Genet 1986; 23:581–7.
2. Turner G, Robinson H, Laing S, Purvis-Smith S. Preventive screening for the fragile-X syndrome. N Engl J Med 1986; 315:607–9.
3. Webb T, Bundey S, Thake A, Todd J. The frequency of the fragile X chromosome among school children in Coventry. J Med Genet 1986; 23:396–9.
4. Kahkonen M, Alitalo T, Airaksinen E et al. Prevalence of the fragile X syndrome in four birth cohorts of children of school age. Hum Genet 1987; 77:85–7.
5. Sherman SL, Morton NE, Jacobs PA, Turner G. The marker (X) syndrome: a cytogenetic and genetic analysis. Ann Hum Genet 1984; 48:21–37.
6. Sherman SL, Jacobs PA, Morton NE et al. Further segregation analysis of the fragile X syndrome with special reference to transmitting males. Hum Genet 1985; 69:289–99.
7. Sutherland GR. Heritable fragile sites on human chromosomes 1. Factors affecting expression in lymphocyte culture. Am J Hum Genet 1979; 31:125–35.
8. Mattei MG, Mattei JF, Vidal I, Giraud F. Expression in lymphocyte and fibroblast culture of the fragile X chromosome: a new technical approach. Hum Genet 1981; 59:166–9.
9. Glover TW. FUdR induction of the X chromosome fragile site: evidence for the mechanism of folic acid and thymidine inhibition. Am J Hum Genet 1981; 33:234–42.
10. Sutherland GR, Baker E, Fratini A. Excess thymidine induces folate sensitive fragile sites. Am J Med Genet 1985; 22:433–43.
11. Jenkins EC, Brown WT, Duncan CJ et al. Feasibility of fragile X chromosome prenatal diagnosis demonstrated. Lancet 1981; ii:1292.
12. Webb T, Butler D, Insley J, Weaver JB, Green S, Rodeck C. Prenatal diagnosis of Martin–Bell syndrome associated with fragile site at Xq27–28. Lancet 1981; ii:1423.
13. Shapiro LR, Wilmot PL, Brenholz P et al. Prenatal diagnosis of Fragile X chromosome. Lancet 1982; i:99–100.
14. Wilson MG, Marchese CA. Prenatal diagnosis of fragile-X in a heterozygous female fetus and postnatal follow-up. Prenat Diagn 1984; 4:61–6.
15. Rocchi M, Pecile V, Archidiacono N, Monni G, Dumez Y, Filippi G. Prenatal diagnosis of the fragile X in male monozygotic twins: discordant expression of the fragile site in amniocytes. Prenat Diagn 1985; 5:229–31.
16. Jenkins EC, Brown WT, Wilson MG et al. The prenatal detection of the fragile X chromosome: review of recent experience. Am J Med Genet 1986; 23:297–313.
17. Shapiro LR, Wilmot PL. Prenatal diagnosis of the fra(X) syndrome. Am J Med Genet 1986; 23:325–40.
18. Tommerup N, Aula P, Gustavii B et al. Second trimester prenatal diagnosis of the fragile X. Am J Med Genet 1986; 23:313–24.
19. Sutherland GR, Baker E, Purvis-Smith S, Hockey A, Krumins E, Eichenbaum SZ. Prenatal diagnosis of the fragile X syndrome using thymidine induction. Prenat Diagn 1987; 7:197–202.
20. Sutherland GR, Baker E. Effects of nucleotides on expression of the folate sensitive fragile sites. Am J Med Genet 1986; 23:409–17.
21. Tommerup N, Sondergaard F, Tonnesen T, Kristensen M, Arveiler B, Schinzel A. First trimester prenatal diagnosis of a male fetus with fragile X. Lancet 1985; i:870.
22. Oberle I, Mandel JL, Boue J, Mattei MG, Mattei JF. Polymorphic DNA markers in prenatal diagnosis of fragile-X syndrome. Lancet 1985; i:871.
23. Shapiro LR, Wilmot PL, Murphy PD, Breg WR. Experience with multiple approaches to the prenatal diagnosis of the fragile X syndrome: amniotic fluid, chorionic villi, fetal blood and molecular methods. Am J Med Genet 1988; 30:347–54.
24. Jenkins EC, Brown WT, Krawczun MS et al. Recent experience in prenatal FRA(X) detection. Am J Med Genet 1988; 30:329–36.

25. Purvis-Smith S, Laing S, Baker E, Sutherland GR. Prenatal diagnosis of the fragile X – the Australasian experience. Am J Med Genet 1988; 30:337–45.
26. Webb T, Gosden CM, Rodeck CH, Hamill MA, Eason PE. Prenatal diagnosis of X-linked mental retardation with fragile (X) using fetoscopy and fetal blood sampling. Prenat Diagn 1983; 3:131–7.
27. Webb T, Rodeck CH, Nicolaides KH, Gosden CM. Prenatal diagnosis of the fragile X syndrome using fetal blood and amniotic fluid. Prenat Diagn 1987; 7:203–14.
28. Webb TP, Bundey S, McKinley M. Missed prenatal diagnosis of fragile-X syndrome. Prenat Diagn 1989; 9:777–81.
29. Orlandi F, Damiani G, Jakil C, Lauricella S, Bertolino O, Maggio A. The risks of early cordocentesis (12–21 weeks): Analysis of 500 procedures. Prenat Diagn 1990; 10:425–8.
30. McKinley MJ, Kearney LU, Nicolaides KH, Gosden CM, Webb T, Fryns J-P. Prenatal diagnosis of fragile X syndrome by placental (chorionic villi) biopsy culture. Am J Med Genet 1988; 30:355–68.
31. Sutherland GR, Baker E. The common fragile site in band q27 of the human X chromosome is not coincident with the fragile X. Clin Genet 1990; 37:167–72.
32. von Koskull H, Aula P, Ammala P, Nordstrom AM, Rapola J. Improved technique for the expression of fragile-X in cultured amniotic fluid cells. Hum Genet 1985; 69:218–23.
33. Tejada I, Boue J, Gilgenkrantz S. Diagnostic prenatal sur les cellules du liquid amniotique d'un fetus male, porteur du chromosome X fragile. Ann Genet 1983; 26:247–50.
34. Schmidt A, Passarge E, Seemanova E, Macek M. Prenatal detection of a fetus hemizygous for the fragile X-chromosome. Hum Genet 1982; 62:285–6.
35. Forster-Gibson CJ, Mulligan LM, Simpson NE, White BN, Holden JJA. An assessment of the use of flanking DNA markers for fra(x) syndrome carrier detection and prenatal diagnosis. Am J Med Genet 1986; 23:665–83.
36. Sutherland GR, Mulley JC. Diagnostic molecular genetics of the fragile X. Clin Genet 1990; 37:2–11.

Discussion

R. Harris: Could I ask about the correct protocol now for a woman at risk of producing a child with fragile X? What is the best sequence of events for prenatal diagnosis?

Webb: I think CVS should be the first test. If it is positive, fine. If it is negative and the woman is at high risk, one would refer her for fetal blood sampling. In CVS samples the cells seem to be much more robust than amniotic fluid cells and it is much easier to demonstrate the fragile X in them.

R. Harris: And having to proceed if the fetus is female? That was the real point.

Webb: What is happening is when the CVS is positive in females the mothers are aborting, and this poses a problem. We need to be able to tell which females are intellectually affected and which are not, and even DNA will not tell us.

Pembrey: It does hold up in general that if levels are high – 20% fragile-X in a female – she is more likely to be mentally retarded than if the level is 2% or 3%. But it is by no means cut and dried.

Donnai: If we are saying one third of fragile-X females have some intellectual handicap, that is a much higher risk of intellectual handicap than in males with Klinefelter's or females with triple X syndrome. However, in some instances, it is thought a reasonable decision to abort a child with a sex chromosome abnormality, whereas in a female with fragile-X a lot of reservations are being expressed.

Webb: Yes. That is right.

Lilford: One of the reasons is that there is a high chance of the same thing happening again with fragile-X, whereas with triple X it is unlikely to happen next time.

But I am distrustful of this one-third. Surely what is really happening is that the IQ scale has shifted over to the left and there is a general diminution of intelligence in girls with fragile X.

Webb: That is true. But one-third actually receive special education.

Pembrey: The decision to abort can be influenced by the family. We know of families where there is a very severely affected girl and they are more likely to want to abort an affected pregnancy whether it is male or female.

Donnai: I was somewhat disturbed that people suggested that if a fetus was female, one should not do the test for the carrier state of the disorder. Given the range of possible problems in carrier females, it is the families who should make the choice on the next course of action, rather than us.

J. Harris: There are two issues here. One is that of taking on oneself the task of judging what is an acceptable risk for a particular family. Surely that is their decision.

Second, there does seem to be a dangerous tendency here to judge the quality of parental decisionmaking about abortion. We, as a society, do not do that for abortions in general, when there is no question of fetal abnormality. It seems a very dangerous tendency to start picking and choosing among parents as to their quality of decisionmaking given our assessment of what is a reasonable risk for them to run. Either we provide them with the information and let them make choices, or we do not, but we should not be in the business of nudging them one way or another depending on what we feel about the desirability of aborting females at less risk, and so on.

Rodeck: My undersanding is that the reluctance to perform prenatal diagnosis in female fetuses was due to the quality of the information being inadequate.

Lilford: There must be quite a strong argument for going straight to fetal blood sampling if the parents are of a mind that they would wish to terminate a carrier fetus, a carrier female fetus.

Webb: Yes. Except that CVS is earlier, and a decision can be made on a positive result at that point.

Donnai: But if there was a negative result, rather than accepting that as the final answer should one then proceed to fetal blood sampling?

Webb: Yes.

Donnai: Professor Gosden highlighted some enormous problems with CVS. One can hope eventually that DNA might solve the problem of fragile-X, but DNA will not help us with our difficulties with CVS for cytogenetic reasons.

Whittle: Professor Gosden (Chapter 12) underlines several of the problems that laboratories are facing in trying to sort out CVS material.

When the Fetoscopy Society were abroad last year we heard from laboratories in America many of the things that Professor Gosden has commented on here. We went back to our own lab to see if they were finding the same sort of problems, and their comment was that they could usually sort them out. I wonder, therefore, whether some of the things she was discussing are problems for the laboratory in its learning curve. Certainly our experience would not be so much doom and gloom, although I appreciate what she is saying.

Gosden: I take the point exactly. I think there is a learning curve. The price that has to be paid is that CVS is labour intensive. To cut down on maternal contamination specimens have to be sorted, effectively. If one wishes to avoid confined placental abnormalities one must do direct and cultured preparations. Diagnostic error can be kept to an absolute minimum with a false-negative rate of about 1 in 1000. The labs will do their very best to give the most accurate results, but the cost is that they will say they can do only a limited number.

I am sure most people will be able to get their false-positive rate down to <1.5%, something like that, and with further sampling they could get it down to almost nothing. The patients want first trimester termination and they ask about the chances of a positive result being accurate, and they are told perhaps only 50%. If any of us was in that position as a parent trying to make the choice, would we go for the early first trimester termination or would we have the amniocentesis on the grounds there is a 50% risk it might not be real? It is one of the most agonising decisions.

Donnai: I think that is right. And because of that what I think we have moved towards – in Manchester anyway – is keeping CVS for the high-risk group, partly because of the resource implications and partly because of the problems. CVS is not turning out to be the simple solution that will completely and utterly replace amniocentesis for women at low, medium or high risk. It may be that CVS is an expert procedure for people at high risk and that we must look toward the triple test for general screening.

Lilford: I entirely agree with all that and that CVS really comes into its own at high risk.

There is one point I must make. The chances of false positive and false negative really become very small when we have both the direct and the culture preparation. I have looked at the literature very carefully and I can find very few cases of either false positive or false negative on both culture and direct.

Gosden: Some such cases were given in Chapter 12.

Chapter 14

Advances in Diagnosis of Biochemical Disorders

H. Galjaard

Introduction

At present some 4500 disorders in man are known or assumed to be of Mendelian inheritance. Most of the genetic disorders have initially been defined by clinicians on the basis of clinical and pathological manifestations [1–3]. The development of chromatographic methods in the 1950s stimulated the use of chemical analyses in the diagnosis of genetic disorders and led to an exponential discovery rate of hereditary amino acidopathies during the 1960s. The important work on normal and abnormal haemoglobins by Pauling, Ingram and others led the way to the search for the responsible (enzyme) protein defects in other genetic diseases and during the past three decades nearly 400 genetic protein defects have been identified. This has offered new perspectives in health care and in basic research.

The demonstration of a protein defect in cell material from a patient enables an early and accurate diagnosis of a genetic disease. In some instances understanding of the biochemical pathology has led to therapeutic approaches which may prevent severe handicap. Examples are the administration of factor VIII to haemophilia patients, of vitamins to patients with certain types of aminoacidopathies or organic acidaemias or of a phenylalanine-free diet to patients with phenylketonuria [2,4,5].

Unfortunately, most genetic disorders cannot be treated and the main emphasis has, therefore, been on strategies to prevent the birth of an affected child. In populations with a high prevalence of a specific gene mutation, large-scale screening for carriers, followed by genetic counselling of couples at risk before reproduction is the most efficient approach to prevention. The availability of a simple, accurate and cheap test is of course mandatory as well as optimal attention for the psychosocial and organisational aspects of the screening programme. The first example of carrier screening is the hexosaminidase A assay in leukocytes or serum of young adults of Ashkenazi Jewish ancestry because of the high incidence (1 in 30 carrier) of the severe neurological disorder G_{m2}-gangliosidosis type Tay–Sachs [6,7]. Up to 1989 nearly 800 000 individuals had been tested, nearly 30 000 carriers had been identified as were some 1000 couples at risk. By providing genetic counselling to all couples at risk and by performing prenatal monitoring of pregnancies at risk the incidence of Tay–Sachs disease among American Ashkenazi Jews has decreased by 95% since 1970 (Kaback, personal communication). The results of these carrier screening programmes have been a model for those on β-thalassaemia [8,9] and hopefully will be for cystic fibrosis in the future (see Chapter 5).

In most instances the birth of a patient with a genetic metabolic disease is unexpected and efforts should be focused on prevention of the birth of further affected people in the same family. This requires an early and accurate diagnosis of the index patient and genetic counselling of all relatives at risk. The options for couples at increased genetic risk are: accepting the risk; refraining from pregnancy; artificial insemination with donor germ cells; or prenatal monitoring with the possibility of interrupting the pregnancy if the fetus is found to be affected [3,10]. The expression of some 100 (enzyme) protein deficiencies in cultured skin fibroblasts formed the basis for the antenatal diagnosis of genetic metabolic diseases using cultured amniotic fluid cells [3]. The introduction of chorion villus sampling, during the early 1980s, meant a further step forward: most antenatal biochemical diagnosis could now be established within a few days during the first trimester of pregnancy [11–13]. The biochemical analysis of chorionic villi has gained wide application during the last few years and will largely replace amniocentesis in the prenatal diagnosis of genetic metabolic disease [13–15].

The use of human mutant cells with a defined (enzyme) protein defect has also contributed to various areas of research. The genetic background of clinically different variants of "the same" metabolic disease could be investigated by biochemical analysis after somatic cell hybridisation using cultured skin fibroblasts from different patients [16]. Electrophoretic studies of interspecies hybrids with defined human chromosome contents has enabled rapid progress in human gene mapping [17]. Also, human mutant cells with a specific protein defect have served as a model in studies on correction of metabolic disturbances by introducing a normal protein or gene [18].

During the past decade radioimmunolabelling of cultured skin fibroblasts and other cell types from normal individuals and patients with different variants of a disease has enabled a refined analysis of the molecular defect underlying an enzyme deficiency [19,20]. A large array of different post-translational defects has been found and more insight has been gained in to the molecular aetiology of clinical heterogeneity [2].

Purification of a protein involved in a genetic disease also enables its (partial) amino acid sequencing and thereby the preparation of oligonucleotides which can be used to identify the encoding gene. Once the normal gene sequence is known, the gene mutations involved in different patients can be identified and further research can be conducted, aimed at understanding the pathogenesis of different clinical variants of a genetic disease. Also experiments in cell culture and in (transgenic) animals may lead to therapeutic strategies.

Despite all these advances and the fact that new genetic metabolic diseases are being identified continuously, still 90% of the Mendelian disorders are of unknown molecular aetiology. As a consequence little can be done in terms of early diagnosis, carrier detection and prevention. The development of recombinant DNA technology and its application in the direct, and indirect demonstration of gene mutations meant an important breakthrough in this context. During the last decade several dozen genetic disorders with an unknown protein defect became eligible for (prenatal) molecular diagnosis [14,21,22].

Yet, there seems to be a long way from the finding of a disease-linked polymorphism to real understanding of the pathogenesis. Future insight into the exact way an abnormal protein deranges the molecular and cellular biology and ultimately the organ's functions [5] will require expertise in biochemistry, cell biology, physiology and pathology in addition to molecular genetics. In the following sections a brief overview will be given of the present scope of prenatal and postnatal biochemical diagnosis and of the advances in the analysis of various types of molecular defects.

Biochemical Diagnosis

Postnatal Diagnosis

At the level of the individual, early diagnosis of a genetic disease first of all depends on the alertness of parents in recognizing that something might be wrong with the development of their child. Unfortunately, too few parents have sufficient knowledge about the normal milestones in child development and about major signs of disturbances. Also, there may be emotional hindrances in accepting the possibility of a disease or handicap, irrespective of knowledge.

A second important step in early diagnosis is a thorough clinical investigation by the medical doctor who should also spend sufficient time to listen to the parent's or child's experiences and worries. Since nearly 4500 different Mendelian disorders are known today it is, of course, impossible to review the major signs and symptoms. There are, however, clinical features that should raise suspicion of the existence of a genetic metabolic disease (Table 14.1). Neither of the clinical features mentioned in Table 14.1 is pathognomonic for a given disease but a combination of several symptoms and signs should certainly be considered a reason to refer the patient to an expert paediatrician. Also the loss of acquired functions may point to a severe genetic disease.

Several academic centres have laboratory facilities for chemical analysis of abnormal metabolites in body fluids. The most common methodologies are quantitative and qualitative chromatographic analyses of amino acids [22], gas chromatography sometimes combined with mass spectrometry for the identification of (abnormal) organic acids [23] and electrophoresis and thin-layer chromatography for the detection of abnormalities in mucopolysaccharides or oligosaccharides. The development of high-performance liquid chromatography

Table 14.1. Clinical features possibly associated with genetic metabolic disease

Early symptoms and signs	Later during child development
Lethargy	Retarded growth
Feeding problems	Delayed psychomotor development
Failure to thrive	Loss of acquired functions
Vomiting	Skeletal abnormalities
Diarrhoea	Coarsening facial features
Dehydration	Hepatosplenomegaly
Acidosis	Muscular hypo- or hypertonia
Hypoglycaemia	Ophthalmological abnormalities
Persistant icterus	Speech- or hearing impairment
Convulsions without fever	Ataxia, seizures, myoclonus
Hypo- or hypertonia	Spastic quadriplegia
Unusual odour	Recurrent respiratory infections
Abnormal hair	

has meant an important step forward in the analysis of purine and pyrimidine metabolism.

Altogether the chemical analysis of blood and urine with these techniques allows the diagnosis of some 100 genetic metabolic diseases most of which are rare. In our experience 3%–5% of the suspected patients, who undergo extensive chemical testing, are found to suffer from one of these genetic diseases. Usually, a follow-up by enzyme assay(s) is necessary to establish the responsible protein defect.

In some 300 genetic diseases with a known protein defect no abnormalities can be detected by chemical analysis of metabolites in body fluids. In these instances the diagnosis requires direct (enzyme) protein analysis on cell material from patients with a suspected genetic disorder. Genetic defects of (enzyme) proteins that occur in all cell types can most easily be demonstrated by biochemical analysis of white blood cells. If a protein is expressed in a few cell types only, an organ biopsy, usually of liver or muscle, may be required. Quite a number of genetic disorders can be diagnosed by biochemical analysis of erythrocytes or of blood proteins [2].

In all instances where prenatal diagnosis might be considered in the future, a skin biopsy should be taken and cultured fibroblasts should be analysed and stored in liquid nitrogen. If the necessary facilities are not available, one of the national or international cell repositories may assist. (The Dept. Cell Biology and Genetics, Erasmus University Rotterdam, Netherlands, P.O. Box 1738 has a large human mutant cell line collection and the same is true for the Human Genetic Mutant Cell Repository, Institute for Medical Research, Copewood and David Streets, Campden NJ 018103 USA.) Cultured skin fibroblasts may sometimes also be needed in postnatal diagnoses where the assay is based on the incorporation and metabolism of radiolabelled compounds in living cells [3].

Sometimes it is useful to store cultured fibroblasts from a patient with an unexplained (genetic?) disorder. If a "new" (enzyme) protein defect is being described in a patient with similar clinical features or a new methodology becomes available, the stored cells can be tested and sometimes a biochemical diagnosis can be established even after a patient has died. A post-mortem diagnosis can still be valuable in genetic counselling of the parents or of other relatives.

Because of the large variety of biochemical disorders and the complexity of many of the laboratory assays the referring clinicians must be aware of the existing network of biochemical experts within their country and sometimes abroad [3]. Laboratories involved in (enzyme) protein assays on cellular material are often focused in their research on one area of cell metabolism. The biochemical diagnosis of particular categories of disorders can best be centralised in such laboratories because it usually guarantees reliable diagnoses. Centralisation has the additional advantage that cell lines from patients with different rare variants of a particular category of diseases are more easily accessible also for research purposes.

Since many (enzyme) protein assays, especially on cultured cells are rather complex, time consuming and costly, the clinician should be selective in his referral. Concentration of patients with rare metabolic diseases in a limited number of clinical centres and regular contacts between the referring clinicians and laboratory experts are prerequisites for a proper selection. If the necessary requirements are fulfilled a genetic protein defect is found in 15%–20% of the cell samples referred. Each year a total of about 500 "new" patients are diagnosed using biochemical methods in the Netherlands with a population of 14 million and 180 000 newborns (see also Chapter 22).

Prenatal Diagnosis

The indication for prenatal biochemical analysis is an increased risk of a fetal genetic disease which is known to be demonstrable in amniotic fluid or fetal cells. In some instances the presence of an abnormal level of specific metabolites in amniotic fluid is a reliable indication of a fetal abnormality (Table 14.2). The

Table 14.2. Chemical analysis of amniotic fluid supernatant in fetal diagnosis

Disorder	Biochemical parameter
Congenital adrenal hyperplasia	17-Hydroxyprogesterone
Congenital nephrosis	Alphafetoprotein
Meckel's syndrome	Alphafetoprotein β-glycoprotein SP1
Polycystic kidneys	Alphafetoprotein
Argininosuccinic aciduria	Argininosuccinic acid
Congenital erythropoieitic porphyria	Porphyrin
Citrullinaemia	Citrulline
Glutaric acidurias I and II	Dicarboxylic acids
Isovaleric acidaemia	Isovalerylglycine
Methylmalonic aciduria	Methylmalonic acid
Multiple acyl-CoA dehydrogenase deficiency	Isovalerylglycine
Multiple carboxylase deficiency	Methylcitric acid
Non-ketotic hyperglycinaemia	Glycine, serine
Pituitary dysgenesis	Prolactin
Tyrosinaemia	Succinylacetone

advantage of this approach is that the results are available within 1–2 days after amniocentesis in week 16 of pregnancy. The methodologies are similar to those used in postnatal biochemical diagnosis. A possible disadvantage is the fact that abnormal metabolites are a secondary effect which may not necessarily correctly reflect the supposed primary genetic (enzyme) protein defect. The choice between (enzyme) protein analysis in cultured amniotic fluid cells and metabolite studies in amniotic fluid supernatant has now become rather academic in cases where chorionic villus sampling in the first trimester of pregnancy is feasible.

The demonstration of the primary protein defect in cultured amniotic fluid cells has the advantage of great reliability, but the disadvantage of a rather long waiting period for the parents since the cultivation of sufficient fetal cells usually takes 2–3 weeks depending on the type of assay [3]. The use of microchemical techniques has made it possible to shorten this period to 7–10 days for a few dozen enzymes, mostly involved in lysosomal storage disorders [24]. Again, this problem hardly exists if chorionic villus samples are being used for biochemical analysis.

Direct biochemical assays on chorionic villus samples obtained during the first trimester of pregnancy are feasible for nearly all (enzyme) protein defects that are eligible for biochemical diagnosis in cultured amniotic fluid cells (for reviews see references 11–13). Most enzyme assays are performed on cell homogenates after careful inspection under the microscope that the samples to be analysed are indeed chorionic villi without maternal cell contamination. Also enzyme defects which have to be demonstrated by incorporation studies of radiolabelled substrates in living cells can be diagnosed directly (1–2 days) by using uncultured, intact chorionic villus cells [25,26]. Table 14.3 lists the genetic metabolic diseases which have been diagnosed prenatally by biochemical analysis of cultured amniotic fluid cells or by direct assay of uncultured chorionic villi.

In some instances the activity of a particular enzyme in uncultured chorionic villi is too low to allow for a reliable interpretation of a decreased activity in case of an affected fetus. Cultivation of chorionic villus cells prior to biochemical analysis or the use of cultured amniotic fluid cells will solve this problem. Table 14.4 lists a few diseases where cultivation of chorion cells is mandatory for a reliable first-trimester diagnosis.

Prenatal diagnosis of genetic metabolic disease comprises only a few per cent of all prenatal diagnoses. Usually it concerns rare diseases for the clinician whereas the laboratory has to adapt to a rather large variety of complex biochemical assays, often at unexpected times. Early communication between the referring obstetrician and the laboratory is necessary to enable adequate preparation. Cultured skin fibroblasts from an index patient and possibly from the heterozygous parent(s) have to be analysed in parallel with control cells and the fetal cells from the pregnancy at risk. Sufficient fetal cell material must be available (usually 10–20 mg wet weight of chorionic villi and in the case of enzyme assays with natural substrate 20–30 mg may be required) and transportation must be performed under sterile conditions, at room temperature and during a period not longer than 1–2 days. A clinical diagnosis of the index patient is not acceptable as the only basis for prenatal biochemical analysis, but data demonstrating a genetic (enzyme) protein defect in cultured skin fibroblasts should be available.

If these conditions are being fulfilled prenatal biochemical diagnosis using cultured amniotic fluid cells, uncultured or cultured chorionic villus cells is highly

Table 14.3. Genetic metabolic diseases diagnosed prenatally by biochemical analysis of fetal cells

Lysosomal Storage Diseases
Lipid metabolism

Fabry's disease	CV	Metachromatic leukodystrophy	CV, VC
Farber's disease	AC, CV?	Mucolipidosis II/III	VC, AF
Galactosialidosis	CV	Mucolipidosis IV	CV, AC
Gaucher's disease	CV	Multiple sulphatase deficiency	CV
G_{m1}-gangliosidosis	CV	Niemann–Pick disease A/B	CV, VC
G_{m2}-gangliosidosis		Niemann·Pick disease C	AC, VC?
Tay–Sachs disease	CV	Sialidosis (ML I)	VC, AC
Sandhoff's disease	CV	Wolman's disease	CV
Krabbe's disease	CV		

Carbohydrate/glycoprotein metabolism

Aspartyl-glucosaminuria	CV, AC	β-Mannosidosis	CV?
Fucosidosis	CV	α-NAGA-deficiency	CV?
Glycogenosis II (Pompe)	CV	Salla and sialic acid	CV, VC,
α-Mannosidosis	CV	Storage disease	AF, AC

Mucopolysaccharide metabolism
MPS type:

I (Hurler)	CV, VC, AF	IV (Morquoi A/B)	CV, VC, AF
II (Hunter)	CV, AF	VI (Maroteaux-Lamy)	CV, VC, AF
III (Sanfilippo A, B, C, D)	CV, VC, AF	VII (Sly)	CV?

Peroxisomal Disorders

Adrenoleukodystrophy		Hyperoxaluria I	FL
Neonatal ALD	CV, VC	Pseudoneonatal ALD	CV, VC
X-linked ALD	CV, VC	Pseudo-Zellweger syndrome	CV, VC
Chondrodysplasia punctata	CV, VC	Refsum's disease	CV, VC
		Zellweger syndrome	CV, VC

Other Disorders
Carbohydrate metabolism

Congenital haemolytic anaemia	AC	Glycogenosis type:	
		I (Von Gierke)	FL
Galactosaemia	CV	III (Cori)	CV
Galactokinase deficiency	AC, CV?	IV (Anderson)	CV
		Pyruvate carboxylase	CV
Glycerol kinase deficiency	CV, CVC	deficiency	

Amino acid metabolism

Argininosuccinic aciduria	CV, AF	Leucinosis	CV, VC
Carbamoyl-P synthetase	FL	3-Methylcrotonyl CoA	CV, VC
		carboxylase deficiency	
Citrullinaemia	CV, AF	Methylmalonic acidaemia	AF, VC, AC
Cystinosis	CV	MTHF-reductase deficiency	AFC, CV?
Glutaric aciduria I	CV, AF, AC	Multiple carboxylase	CV, AF
Glutaric aciduria II	AF, AC	deficiency	
Homocystinuria	AC, VC?	Propionic acidaemia	CV, AF
HMG-CoA-lyase deficiency	CV, AF, AC	Tyrosinaemia	CV, AF
Isovaleric acidaemia	CV, VC, AF		

Continued next page

Table 14.3 (*continued*)

Nucleic Acid Metabolism

Adenosine deaminase deficiency	CV	Purine nucleoside phosphorylase deficiency	CV
Ataxia telangiectasia	VC	Xanthine-sulphite oxidase	AC, CV?
Cockayne's syndrome	VC	Xeroderma pigmentosum	VC
Lesch–Nyhan syndrome	CV		

Miscellaneous

Acute intermittent porphyria	AC	Hypercholesterolaemia	AC, VC
		Hypophosphatasia	CV
Adrenal hyperplasia	AF, CV	Menkes' disease	CV, AC
Ceroid lipofuscinosis	AC, CV	X-linked ichthyosis	CV
Cystic fibrosis	AF, CV[a]		
Cytochrome b_5 reductase	AC, CV?		
Cytochrome c oxidase	CV?, VC?		

AC, amniocytes; AF, amniotic fluid; CV, uncultured chorionic villi; FL, fetal liver; VC, cultured chorionic villus cells
[a] Now by DNA analysis.

Table 14.4. Diseases where first-trimester fetal diagnosis requires chorion cell cultivation

Peroxisomal disorders	Xeroderma pigmentosum
"I-cell" disease	Cockayne's syndrome
Niemann–Pick disease type C	Ataxia telangiectasia
Sialidosis	In cases of very low activity in direct chorion villus assays

reliable. Misdiagnoses that have been reported were usually caused by too little cell material, contamination with maternal cells or by insufficient communication between the clinician taking care of the index patient, the obstetrician performing fetal cell sampling and the laboratory [14].

In various genetic diseases the responsible metabolic defect is not expressed in amniotic fluid cells or chorionic villi. The best alternative is, of course, (in)direct demonstration of the gene mutation by DNA analysis. If this is not possible biochemical analysis of fetal blood or of fetal tissue biopsies has been used. Examples are fetal liver biopsy for the diagnosis of ornithine transcarbamylase deficiency [27], glycogenosis I [28] and primary hyperoxaluria I [29] and fetal skin biopsies for (electron) microscopic diagnosis of genetic skin diseases [30,31]. A disadvantage of this approach is the late stage of pregnancy, the risk for fetus and mother and the limited number of obstetricians capable of performing such sampling. The introduction of ultrasound-guided fetal blood sampling from the umbilical cord has improved this situation [32,33] (Chapter 15). The largest experience with fetal blood analysis has been gained with pregnancies at risk for haemoglobinopathies or haemophilia [30,32].

So far, no prenatal diagnoses have been reported based on biochemical analysis of dissected cells from a preimplantation embryo. This approach would technically be feasible in the few cases where ultramicrochemical methods are available permitting reliable enzyme assays in one or few cells [34] (see also Chapter 11). In vitro fertilisation, followed by manipulation, analysis and reimplantation of the remaining pre-embryo seems to me no useful alternative for the present methodology of chorion villus sampling and analysis during the first trimester.

Analysis of Genetic Protein Defects

For diagnostic purposes the demonstration of a functional impairment of a protein is usually sufficient. In the case of enzyme deficiencies this is often expressed as a percentage of the normal conversion rate of artificial or natural substrate by measuring in the test tube the formation of specific chemical products [3]. In some instances differences in residual enzyme activity can best be assessed by studying specific metabolic pathways in (cultured) living cells after uptake of radiolabelled natural substrates [34,35]. Genetic abnormalities of non-enzymic proteins, such as haemoglobinopathies can be demonstrated by a different electrophoretic mobility or by chromatographic evaluation of globin chain synthesis [36]. Such procedures also form the basis of carrier screening for β-thalassaemia or sickle-cell anaemia.

During the past few decades it has become more and more evident that there is considerable clinical and biochemical heterogeneity within many of the genetic metabolic diseases. Clinically, severe, rapidly fatal early infantile forms of a disease exist as well as late onset, milder juvenile forms and still less severe adult variants. In some instances, like in glycogenosis II, a relationship between clinical severity and residual enzyme activity could be established [37], but in many other diseases such a relationship has not been observed. A well-known and still unexplained example is the complete α-iduronidase deficiency both in patients with the clinically very severe form of mucopolysaccharidosis I (MPS) (Hurler disease) and in patients with the very mild form of this disease (Scheie syndrome). Another example of clinical heterogeneity is asymptomatic adults in their 40s and 50s who were found to have very low hexosaminidase A activities as measured with artificial substrate during Tay–Sachs screening programs [38].

In some genetic metabolic diseases the biochemical heterogeneity has been intriguing. In patients with different types of lysosomal storage disorders a marked discrepancy was found between the results of in vitro enzyme assays and in vivo studies using natural substrate. Sometimes a severe enzyme deficiency has been observed in a clinically normal parent of a patient with a genetic metabolic disease; in other instances a nearly normal enzyme activity is found in a severely affected person [2,3,34,38]. Also, the discovery of multiple enzyme deficiencies as in certain lysosomal storage disorders [39,40] and organic acidurias [41] pointed to the need of more refined analyses of genetic protein defects.

An important step forward in this context was the use of radioimmunoprecipitation, first applied by Hasilik and Neufeld [42] in studies on the biosynthesis of lysosomal enzymes. Incorporation of radioactive amino acids into living cultured cells will label all newly synthesised proteins; precipitation with specific antibodies followed by electrophoretic separation of different molecular weight forms in a pulse–chase experiment will then enable a follow-up of the biosynthesis and post-translational modifications of normal and mutant proteins (Fig. 14.1). With similar techniques the phosphorylation and glycosylation of proteins can be

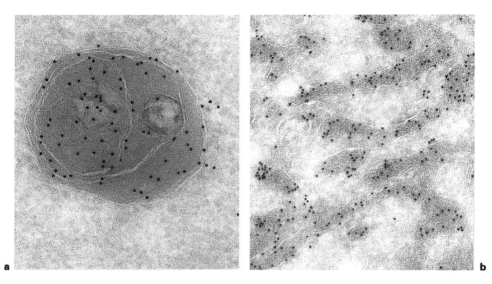

Fig. 14.1. Immunoelectronmicroscopy using gold labelling. **a**, normal intralysosomal localisation of the protective protein. **b**, site directed mutagenesis causes the precursor of the protective protein to accumulate in the endoplasmic reticulum. (By kind permission of Dr. A. d'Azzo and R. Willemsen.)

studied. Once specific antibodies are available immunoelectronmicroscopical studies also allow in situ studies of the intracellular compartmentation in normal and mutant cell material (Fig. 14.2).

During the last decade a variety of different types of protein defects also within one disease, have been delineated [2]. Some genetic disorders appear to be based on the complete absence of synthesis of a particular protein. In other instances there is a normal rate of synthesis but a genetic abnormality of the precursor protein results in a post-translational defect. Depending on the type of defect the result in molecular or cell biological terms will vary. Table 14.5 gives examples of different types of post-translational defects with different consequences for the ultimate (enzyme) protein function and hence for the pathological and clinical manifestations.

Detailed understanding of the nature of a genetic protein defect is mandatory to gain insight into the pathogenesis and to devise adequate strategies for replacement therapy.

A good example is the combined β-galactosidase–neuraminidase deficiencies observed in patients with the severe neurological disorder galactosialidosis [39]. Initially, a deficiency of either of these enzymes was thought to be the primary defect; subsequent radioimmunological studies revealed, however, that a defective third, not yet described, lysomal protein designated as "protective protein" is responsible for the deficiency of the two other lysosomal enzymes which lead to the pathological manifestations [45–47]. Therapy in such patients should be aimed at the replacement of the protective protein or of its encoding gene [48]. Replacement with normal β-galactosidase or sialidase would not have any clinical effect since these proteins would be rapidly degraded or remain inactive in the patient's cells as long as the protective protein is deficient.

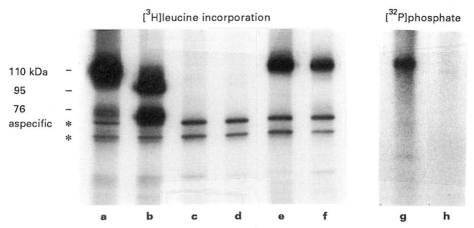

Fig. 14.2. Radioimmunoprecipitation studies of the biosynthesis of acid α-gluosidase in fibroblasts from patients with different variants of glycogenosis II. **a**, α-glucosidase in normal fibroblasts after pulse labelling (110 kDa precursor mainly). **b**, α-glucosidase in normal fibroblasts after 16 hours chase (95 kDa intermediate form and 76 kDa mature lysosomal enzyme). **c**, Patient with infantile form of glycogenosis II without any synthesis of precursor form (pulse). **d**, Same after 16 hours chase. **e, f**, Pulse and chase experiment show normal synthesis of precursor form but defective processing into mature lysosomal enzyme. **g**, [^{32}P]phosphate labelling in control fibroblasts shows phosphorylation of precursor form of α-glucosidase. **h**, Defective phosphorylation of α-glucosidase in cells from a patient with infantile glycogenosis II leading to extralysosomal compartmentalisation of the precursor. (With kind permission from the work by Dr A.J.J. Reuser, Department of Cell Biology and Genetics, Erasmus University, Rotterdam.)

Table 14.5. Genetic diseases associated with different types of post-translational modifications

Disease	Post-translational modification	Diagnostic (enzyme) protein defect
G$_{M2}$-gangliosidosis Tay–Sachs variant	Precursor protein (α-chain) insoluble, sticks to endoplasmic reticulum	Hexosaminidase A
Glycogenosis II variant Tay–Sachs variant	Precursor protein unstable and rapidly degraded	Acid α-glucosidase Hexosaminidase A
Methylmalonic acidaemia	N-terminal deletion of precursor with loss of target-signal to enter mitochondrion	Methylmalonyl-CoA mutase deficiency
"I-cell" disease	Defective N-acetylglucosamine phosphostransferase preventing transfer of mannose-6-phosphate marker for lysosomal compartmentalisation	Multiple lysosomal enzyme deficiency
Galactosialidosis	Defective protective protein resulting in rapid intralysosomal degradation of β-galactosidase and in defective activation of lysosomal neuraminidase	β-galactosidase and neuraminidase
Variants of metachromatic leukodystrophy and G$_{M2}$-gangliosidosis	Defective activator protein(s) preventing proper interaction between enzyme and substrate	Arylsulphatase A impaired G$_{M2}$-ganglioside degradation with normal hexosaminidase activity in vitro

Another recent advance in biochemical diagnosis is the elucidation of the genetic and molecular nature of the so-called activator proteins, involved in the interaction between various lysosomal hydrolases and their glycolipid substrates [49]. Protein studies and DNA analyses have recently shown that four different activator proteins are derived from a common precursor molecule [50,51]. Understanding of the molecular nature of those proteins and of their physiological role has led to a better insight in the pathogenesis of patients with hitherto unexplained discrepancies between clinical features and the activity of particular lysosomal enzymes. [2].

Nearly every year "new" (enzyme) protein defects responsible for hitherto unexplained Mendelian disorders are reported. Often this is the result of step by step pathological and biochemical studies. In other instances routine testing of patient material for an enzyme for which no deficiency in man is known, leads to the discovery of a "new" disease as in the lysosomal storage disorder due to N-acetyl-α-galactosaminidase deficiency [52]. Of course the delineation of new protein defects often provides new perspectives for (prenatal) diagnosis.

Sometimes new categories of genetic biochemical disorders are defined which at the same time lead to new insights into normal molecular and cell biological processes. An example is the delineation of membrane transport proteins that are instrumental in energy-dependent transport of low-molecular-weight compounds out of the lysosome. Diseases like cystinosis [53], Salla disease and other sialic acid storage disorders are due to a genetic defect of one of these proton-driven carrier proteins [54] and other yet unexplained disorders are likely to join this group in the future.

Once a responsible protein defect has been defined and the protein has been purified for normal cells, partial amino acid sequencing will enable the synthesis of specific oligonucleotides which can be used to screen gene libraries. In the last few years many genes have been identified using this approach [17]. Knowledge of the gene sequence provides information about the primary protein structure and DNA technology will enable the production of sufficient quantities of a single human protein to enable studies of its 3-dimensional structure. Site-directed mutagenesis, gene transfer in cultured cells and experiments with transgenic animals will provide more insight into the pathogenesis in different variants of a genetic metabolic disease. Also, new strategies for treatment can be tested once normal and mutant proteins and their encoding genes are available. Finally, the mutations involved in different clinical and biochemical variants of a disease can be defined at the level of DNA. Rapid progress is already made in this area and in the future most patients will be classified on the basis of the nature of their mutation.

Despite all advances, in some 90% of the Mendelian disorders the responsible protein defect is not known [1]. The approach of reverse genetics has been fruitful in delineating mutations involved in such diseases or of closely linked polymorphisms [14, 21, 55]. Up to now some three dozen Mendelian disorders of unknown molecular aetiology can be diagnosed prenatally by DNA technology. In terms of understanding the pathogenesis, however, the identification of a gene is only the beginning of what turns out to be a rather long path. The amino acid sequence of an unknown protein may give certain clues to its 3-dimensional structure but the ultimate isolation and characterisation of a protein and certainly the elucidation of its function will often be difficult. It requires expertise in different fields of biochemistry, cell biology, physiology and pathology. The same is true for studies on the pathogenesis of Mendelian disorders with a known protein defect and for devising successful strategies for treatment. If real advances are to be made in human genetics in the future it is important to maintain know-how in all the disciplines mentioned above, not just in the popular field of molecular biology.

References

1. McKusick VA. Mendelian inheritance in man, 8th edn. Baltimore: Johns Hopkins University Press, 1988.
2. Scriver CR, Beaudet AL, Sly WS, Valle D, eds. The metabolic basis of inherited disease, 6th edn. New York: McGraw Hill, 1989.
3. Galjaard H. Genetic metabolic disease: early diagnosis and prenatal analysis. Amsterdam: Elsevier, 1980.
4. Benson PF. Screening and management of potentially treatable genetic metabolic disorders. Lancaster: MTP Press, 1984.
5. Bickel H, Guthrie R, Hammersen, eds. Neonatal screening for inborn errors of metabolism. Berlin: Springer, 1980.
6. Kaback MM, ed. Tay–Sachs disease: screening and prevention. New York: Alan Liss, 1977.
7. Kaback MM. Heterozygote screening. In: Emery AEH, Rimoin D, eds. Principles and practice of medical genetics. Edinburgh: Churchill Livingstone, 1990.
8. Cao A. Results of programmes for antenatal detection of thalassemia in reducing the incidence of the disorder. Blood Rev 1987; 1:169–76.
9. WHO report. Community control of hereditary anemias. Bull WHO 1983; 61:63–80.
10. Harper PS. Practical genetic counselling, 2nd edn. Bristol: Wright, 1984.
11. Brambati BM, Simoni G, Fabro S, eds. Chorionic villus sampling. Clinical and biochemical analysis. Vol. 21. New York: Marcel Dekker, 1986.
12. Galjaard H. Fetal diagnosis of inborn errors of metabolism. In: Rodeck CH, ed. Fetal diagnosis of genetic disease. Baillières Clin Obstet Gynaecol 1987; 1:547–68.
13. Kleijer WJ. Prenatal diagnosis. In: Fernandez J, Saudubray JM, Tada K, eds. Inborn metabolic diseases. Berlin: Springer, 1990; 683–95.
14. Galjaard H. World-wide experience with first-trimester fetal diagnosis by molecular analysis. In: Vogel F, Sperling K, eds. Human genetics. Berlin: Springer, 1987; 611–21.
15. Poenaru L. First-trimester prenatal diagnosis of metabolic diseases: a survey in countries from the European Community. Prenat Diagn 1987; 7:331–41.
16. Galjaard H, Reuser AJJ. Genetic aspects of lysosomal storage disease. In: Dingle JT, Dean RT, Sly W, eds. Lysosomes in biology and pathology. Amsterdam: Elsevier, 1984; 315–46.
17. Human Gene Mapping 9.5. Cytogenet Cell Genet 1988; 49:1–258.
18. Desnick R. Treatment of inherited metabolic diseases. In: Emery AEH, Rimoin D, eds. Principles of medical genetics. Edinburgh: Churchill Livingstone, 1990.
19. Hasilik A, Neufeld EF. Biosynthesis of lysosomal enzymes in fibroblasts. J Biol Chem 1980; 255:4937–45.
20. Barranger J, Brady RO, eds. Molecular basis of lysosomal storage disorder. New York: Academic Press, 1984.
21. Cooper DN, Schmidtke J. Diagnosis of genetic disease using recombinant DNA, 2nd edn. Hum Genet 1989; 83:307–34.
22. Bremer HJ, Duran M, Kamerling JP, Przyrembel H, Wadman S. Disturbance of amino acid metabolism: clinical chemistry and diagnosis. Munich: Urban and Schwarzenberg, 1981.
23. Goodman SI, Markey SP. Diagnosis of organic acidemias by gas chromatography – mass spectrometry. Laboratory and research methods in biology and medicine, vol. 6. New York: Alan Liss, 1981.
24. Galjaard H. Miniaturisation of biochemical analysis of cultured (amniotic fluid) cells. In: Latt SA, Darlington GJ, eds. Methods in cell biology, vol. 26. New York: Academic Press, 1982; 241–68.
25. Kleijer WJ, Thoomes R, Galjaard H, Wendel U, Fowler B. First-trimester (chorion biopsy) diagnosis of citrullinaemia and methylmalonic aciduria. Lancet 1984; ii:1340.

26. Kleijer WJ, Horsman D, Mancini GMS, Fois A, Boué J. First-trimester diagnosis of maple syrup urine disease on intact chorionic villi. N Engl J Med 1985; 313:1608.
27. Rodeck CH, Nicolaides KH. Fetoscopy and fetal tissue sampling. Br Med Bull 1987; 39:332–7.
28. Golbus HS, Simpson TJ, Koresawa M, Appelman Z, Ropers C. The prenatal determination of glucose-6 phosphatase activity by fetal liver biopsy. Prenat Diagn 1988; 8:101–4.
29. Danpure CJ, Jennings PR, Penketh RJ, Wise PJ, Cooper PJ, Rodeck CH. Fetal liver alanine: glyoxylate amino transferase and the prenatal diagnosis of primary hyperoxaluria type 1. Prenat Diagn 1989; 9:271–81.
30. Status report on fetoscopy and fetal tissue sampling. Prenat Diagn 1984; 4:79–81.
31. Epstein CJ, Cox DR, Schonberg SA, Hogge WA. Recent developments in the prenatal diagnosis of genetic diseases and birth defects. Ann Rev Genet 1983; 17:49–83.
32. Weiner CP. The role of cordocentesis in fetal diagnosis. In: Williamson RA, ed. Fetal diagnosis. Clin Obstet Gynaecol 1988; 31:285–92.
33. Nicolaides KH, Rodeck CH. Fetal blood sampling. In: Rodeck CH, ed. Fetal diagnosis of genetic defects. Baillières Clin Obstet Gynaecol 1987; 623–48.
34. Shapiro LJ, Aleck KA, Kaback MM et al. Metachromatic leucodystrophy without arylsulfatase A deficiency. Pediat Res 1979; 13:1179–81.
35. Mancini GMS, Hoogeveen AT, Galjaard H, Mansson JE, Svennerholm S. Ganglioside G_{M1} metabolism in living human fibroblasts with β-galactosidase deficiency. Hum Genet 1986; 73:35–8.
36. Weatherall DJ, Clegg JB. The thalassemia syndromes, 2nd edn. Oxford: Blackwell, 1981.
37. Reuser AJJ, Kroos M, Willemsen R, Swallow D, Tager JM, Galjaard H. Clinical diversity in glycogenosis II: biosynthesis and in situ localization of acid α-glucosidase in mutant fibroblasts. J Clin Invest 1987; 79:1689–99.
38. Sandhoff K, Conzelmann E, Neufeld EF, Kaback MM, Suzuki K. The G_{M2}-gangliosidosis. In: Scriver CR, Beaudete AL, Sly WS, Valle D, eds. The metabolic basis of inherited disease, 6th edn. New York: McGraw Hill, 1989; 1807–39.
39. Wenger DA, Tarby TJ, Wharton C. Macular cherry-red spots and myoclonus with dementia: coexistent neuraminidase and β-glactosidase deficiences. Biochem Biophys Res Commun 1978; 82:589–95.
40. Leroy JG, Ho MW, McBrinn MC, Zeelke K, Jacob J, O'Brien JS. I-cell disease: biochemical studies. Pediat Res 1972; 6:752–7.
41. Robinson BH. Lactic acidemia. In: Scriver CR, Beaudet AL, Sly WS, Valle D, eds. The metabolic basis of inherited disease, 6th edn. New York: McGraw Hill, 1989; 869–88.
42. Hasilik A, Neufeld EF. Biosynthesis of lysosomal enzymes in fibroblasts: synthesis as precursors of higher molecular weight. J Biol Chem 1980; 255:4937–45.
43. Dingle JT, Dean RT, Sly W, eds. Lysosomes in biology and pathology. Amsterdam: Elsevier, 1984.
44. Barranger J, Brady RO, eds. Molecular basis of lysosomal storage disorders. New York: Academic Press, 1984.
45. d'Azzo A, Hoogeveen AT, Reuser AJJ, Robinson D, Galjaard H. Molecular defect in combined β-galactosidase and neuraminidase deficiency. Proc Natl Acad Sci 1982; 79:4535–9.
46. Hoogeveen AT, Verheijen FW, Galjaard H. The relation between human lysosomal β-galactosidase and its protective protein. J Biol Chem 1983; 258:12143–6.
47. Verheijen FW, Palmeri S, Hoogeveen AT, Galjaard H. Human placental neuraminidase. Am J. Biochem 1985; 149:315–21.
48. Galjart NJ, Gillemans CHAM, Harris A et al. Expression of cDNA encoding the human protective protein associated wth lysomal β-galactosidase and neuraminidase: homology to yeast proteases. Cell 1988; 54:755–64.
49. Li YT, Li SC. Activator proteins related to the hydrolysis of glycosphingolipids catalyzed by lysosomal glycosidases. In: Dingle JT, Dean RT, Sly W, eds. Lysosomes in biology and pathology. Amsterdam: Elsevier, 1984; 99–118.
50. O'Brien JS, Kretz KA, Dewjyu N, Wenger DA, Esch F, Fluharty AL. Coding of two sphingolipid activator proteins by the same genetic locus. Science 1988; 241:1091–101.
51. Nakano T, Sandhoff K, Stümper J, Christomanou H, Suzuki K. Structure of full-length cDNA encoding for sulfatide activator. J Biochem 1989; 105:152–4.
52. Diggelen OP van, Schindler, D, Willemsen R et al. α-N-acetylgalactosaminidase deficiency: a new lysosomal storage disorder. J Inherited Metab Dis 1988; 11:349–57.
53. Gahl W, Baskan N, Tietze F, Bernardini I, Schulman JD. Cystine transport is defective in isolated leukocyte lysosomes in patients with cystinosis. Science 1982; 217:1263–5.

54. Mancini GMS, Jong HR de, Galjaard H, Verheijen FW. Characterization of a proton-driven carrier for sialic acid in the lysosomal membrane: evidence for a group-specific transport system for acidic monosaccharides. J Biol Chem 1989; 264: 15247–54.
55. Weatherall DJ. The new genetics and clinical practice, 3rd edn. Oxford: Oxford University Press, 1990.

Discussion

Donnai: Professor Galjaard has made a persuasive plea to keep the protein biochemists employed as well as the DNA scientists.

R. Harris: What he stresses so very correctly is that all the specialties are needed, the obstetricians, the geneticists, the biochemists, the molecular biologists, the haematologists.

Read: One interesting question is now many balls can anyone keep in the air at once in a laboratory. In a way it is more of a problem for biochemical laboratories because techniques are more specific to each particular problem. How many diseases could they accept referrals for in Professor Galjaard's laboratory?

Galjaard: At this moment two-thirds of our prenatal and postnatal diagnoses are from foreign countries and one-third comes from Holland.

Read: I was not thinking so much in terms of the number of specimens as the number of different enzyme tests that a laboratory would do.

Galjaard: There is certainly a limit to that. Probably the diagnostic activities should be divided between laboratories that have a specific interest in a particular pathway because of their research. The disadvantage of that is decentralisation.

Brock: This may come under Professor Galjaard's second title (Chapter 22) and it may not be a problem in Holland, but in Britain there has been a distinct physical separation of biochemical geneticists from molecular geneticists. In a lot of places it is very difficult to make any communication at all; they have more or less stopped talking to each other and each have pushed their own claims.
  I quite agree that they must get back together, but it is difficult to see how it can happen on an organisational basis.

Galjaard: But they will because the research molecular biologists are not far enough yet. Once they are far enough, once they have the gene and they have the sequence, then they will turn to the protein people. They will go for the proteins. Diagnostics are different. In my view (and this will be discussed further in Chapter 22), diagnostics should come under the umbrella of clinical genetics with its counselling, chromosomes, biochemical diagnosis, DNA diagnosis.

Pembrey: With regard to the point that Professor Brock mentioned and also bringing other things together, it is the clinical geneticists who can play a very useful role in acting as a conduit for communication. We should, where possible, have the DNA services embedded in research centres, have the two feed off each other, but they have to be organised separately, have their own organisation. But the clinical geneticists can act to bring together these disparate groups.

Section V
Special Techniques: 2

Chapter 15

Cordocentesis

K. H. Nicolaides

Access to the fetal circulation was originally achieved by exposing the fetus at the time of hysterotomy [1]. Subsequently with the development of fibreoptics, fetoscopy was used to visualise and sample vessels on the chorionic plate [2] and the umbilical cord [3]. More recently, improvements in imaging by ultrasonography have made fetoscopic guidance unnecessary and fetal blood can be obtained by ultrasound guided puncture of the fetal heart (cardiocentesis; [4]), intrahepatic umbilical vein (hepatocentesis; [4]) or an umbilical cord vessel (cordocentesis; [5]). Currently the most widely used method both for fetal blood sampling and transfusion is cordocentesis.

Technique

Cordocentesis can be performed by a single operator in an outpatient setting in the ultrasound department without need for maternal fasting, sedation, antibiotics, tocolytics or fetal paralysis [6]. However, various centres use different techniques, which may include: maternal hospitalisation, fasting and sedation; pre- and/or postoperative administration of antibiotics and tocolytic agents; fetal paralysis by the intravascular or intramuscular administration of a variety of neuromuscular agents; ultrasound guidance by a radiologist and needle insertion by an obstetrician; the use of an ultrasound transducer with needle guides; or needles of varying length (8–15 cm) or gauge (20–27).

In our centre the woman's husband is encouraged to be present during the procedure, and counselling of the parents includes discussing the risks of an affected pregnancy and the available options for further management. The procedure of cordocentesis, including its limitations and potential complications, is explained and a detailed ultrasound examination is performed for fetal biometry and the diagnosis of fetal malformations.

In cases where the placenta is anterior or lateral, the needle is introduced transplacentally into the umbilical cord. When the placenta is posterior, the needle is introduced transamniotically and the cord is punctured close to its placental insertion. The umbilical cord vessel sampled can be identified as artery or vein by the turbulence seen ultrasonically when sterile saline (200–400 µl) is injected into the vessel through the sampling needle.

The volume of fetal blood removed will depend on the gestation and the indication for sampling, but in general is 1–4 ml. In addition to prenatal diagnosis, standard investigations include blood film and Kleihauer–Betke testing, blood gases and karyotyping.

Risks

Maternal complications should be negligible, although there is one reported case of amnionitis and life-threatening adult respiratory distress developing after two attempts at cordocentesis involving a total of eight needle insertions [7]. Risk of fetal death after cordocentesis depends on the indication for blood sampling and the experience of the operator. In a series of 928 cases sampled in our unit for prenatal diagnosis of genetic diseases, or for karyotyping in cases of minor fetal malformation, there were 10 fetal deaths or spontaneous abortions within two weeks of cordocentesis. In addition there were nine deaths 4–20 weeks after cordocentesis; these losses are unlikely to be the result of the procedure. The risks are higher when the mother is obese, the placenta is posterior and the gestation at sampling 16–19 weeks rather than later.

Indications

The main indications for cordocentesis are prenatal diagnosis of inherited blood or metabolic disorders, detection of fetal infection, karyotyping of malformed fetuses, karyotyping and determination of the acid–base status of small-for-gestational-age (SGA) fetuses, and assessment and treatment of red cell isoimmunised and thrombocytopenic pregnancies.

Prenatal Diagnosis of Blood Disorders

In the 1970s and early 1980s the main indications for fetal blood sampling were prenatal diagnosis of the haemoglobinopathies and genetic defects affecting haemostasis [8,9]. Recently the application of recombinant DNA techniques to

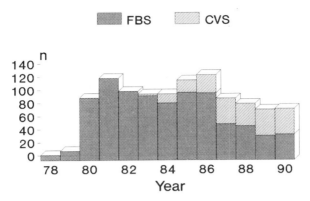

Fig. 15.1. Prenatal diagnosis of hereditary blood disorders at King's College Hospital (1978–90). An increasing number of cases is now referred in early pregnancy for chorionic villus sampling (CVS) and DNA diagnosis. However, fetal blood sampling (FBS) for phenotype diagnosis is still necessary for those patients in whom DNA analysis is not feasible either due to lack of informative DNA probes or late clinical referral.

the analysis of placental biopsy material has made it possible to diagnose many of these conditions in the first trimester of pregnancy. However, fetal blood sampling for phenotype diagnosis is still necessary for those patients requiring confirmation where normality is based on a linked probe, those who lack key relatives, those who are not informative for any of the available probes, and those in whom DNA analysis is not feasible because of late clinical referral (Fig. 15.1).

Prenatal Diagnosis of Metabolic Disorders

There are almost 200 inherited metabolic conditions for which specific enzyme deficiencies have been defined, and for which accurate diagnostic biochemical assays have been developed [10]. For many of these conditions there is no effective therapy and they result in death in early childhood or serious disability.

Prenatal diagnosis of over 100 of these conditions is now possible by analysis of amniotic fluid, placental tissue or fetal blood. From fetal blood the diagnosis can be made within a few hours of sampling and cordocentesis is, therefore, useful when the gestational age is close to the limit for elective abortion, as with late prenatal care or after failed chorionic villus or amniotic fluid techniques.

Prenatal Diagnosis and Treatment of Fetal Infection

Rubella

Primary rubella infection in the first trimester of pregnancy leads to fetal infection and congenital abnormalities in 80%–90% of cases. If the infection occurs in the fourth month the risk of infection is 50%–60% and the risk of congenital abnormalities is 20%–25%. After the fourth month fetal infection is common but seldom causes any harm. Many women with the infection find these risks unacceptably high, and more than 90% of them in the first trimester and 60% in the fourth month undergo elective abortion [11]. Prenatal diagnosis from fetal blood is by measurement of rubella-specific IgM using a sensitive radioimmunoassay [12,13]. However, fetal blood sampling should be delayed until 22 weeks' gestation when the fetal humoral immune response to infection becomes consistently detectable.

Toxoplasmosis

When primary toxoplasmosis occurs during pregnancy the risk of fetal infection increases from approximately 15% in the first trimester to 60% in the third. However, the consequences of infection in the first trimester are more serious and include fetal death or overt clinical disease with chorioretinitis, intracerebral calcification, hydrocephaly, hepatosplenomegaly and growth retardation. In contrast, more than 90% of fetal infections acquired in the third trimester are asymptomatic. Diagnosis of acute maternal infection is made by detection of specific IgM. Fetal infection can be diagnosed by fetal blood sampling after 20 weeks' gestation and a combination of tests, including demonstration of specific IgM and IgG, abnormal levels of liver enzymes, thrombocytopenia and isolation of the parasite by inoculation into mice [14,15].

Cytomegalovirus

Cytomegalovirus is the commonest cause of intrauterine viral infection. Fetal infection may occur after either primary or recurrent maternal infection, although the former is generally considered as being more serious for the fetus. The birth incidence is 0.5%–2.5%, and 5%–10% of infected infants develop major neurological sequelae [16,17]. Fetal infection can be diagnosed by fetal blood sampling after 20 weeks' gestation and demonstration of specific IgM, positive blood cultures and elevated γ-glutamic pyruvic transaminase [18,19].

Varicella-Zoster Virus

Intrauterine infection can occur after either primary or secondary viraemia in the pregnant woman. The risk of congenital varicella syndrome in mothers who contract chickenpox during the first half of pregnancy is approximately 4%, whereas the risk for women with zoster is negligible [20]. The various defects in congenital infection have been attributed to damage to neuronal cells with consequent denervation of developing organs and extremities [21]. Thus viral damage of the developing optic tracts leads to optic atrophy, chorioretinitis and microphthalmia, damage to the cervical and lumbosacral plexi causes hypoplasia of the extremities and dysfunction of the urethral and anal sphincters, whereas damage to the brain can cause microcephaly or hydrocephaly. Prenatal diagnosis is based on the ultrasonographic demonstration of the associated malformations. Recently, cordocentesis at 32 weeks' gestation in a mother who had chickenpox at

20 weeks demonstrated specific IgM in fetal blood [22]. Two weeks later there was a marked decline in fetal IgM suggesting that the fetal response may be short-lived and this may explain previous unsuccessful attempts to detect anti-varicella IgM in the serum of congenitally infected newborns.

Parvovirus

Parvovirus B19 infection is associated with inhibition of fetal erythropoiesis, which results in severe fetal anaemia and hydrops [23]. Fetal infection is diagnosed by the demonstration of viral DNA in fetal blood, because specific IgM is often negative. The prognosis for infected fetuses is poor, as all reported cases of untreated hydrops caused by parvovirus have resulted in fetal or neonatal death. Successful reversal of hydrops can be achieved by intravascular fetal blood transfusion [24] and in five such cases treated in our unit the liveborn infants were normal. However, there is concern as to the possible teratogenic effects of the virus, well documented in animals and one aborted human fetus with multiple ocular abnormalities [25,26]. Anderson et al. [27] reported no congenital defects in over 130 liveborn infants of B19 infected mothers. However, it is possible that none of these infants was infected in utero and that the majority of infected fetuses do develop hydrops and die. Therefore, the potential teratogenicity of the virus will only become apparent following survival of transfused fetuses in which antenatal infection has been proven.

Fetal Karyotyping

Fig. 15.2 shows the relative importance of the various indications for fetal blood karyotyping in our centre.

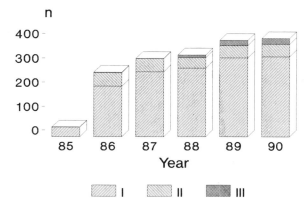

Fig. 15.2. Fetal blood karyotyping at King's College Hospital (1985–90). The main indication is ultrasonographic detection of fetal malformations or growth retardation (I). Other indications include CVS and amniocentesis failures or mosaicism, late booking and diagnosis of fragile-X (II). Recently, an increasing number of patients are referred because of abnormal maternal serum biochemical results (III).

Late Booking / Failed Amniocentesis

Karyotyping by cytogenetic analysis of fetal lymphocytes can potentially be available within 3 days of sampling [28] compared with 3–5 weeks for amniotic fluid. This rapid karyotyping reduces the duration of parental anxiety while awaiting results, and in the case of late booking or when amniocentesis culture has failed, may identify an abnormality at a gestation when termination of pregnancy is still possible. As an alternative, placental biopsy may be performed throughout pregnancy, and direct preparations can provide a result within a few hours [29,30]. However, the problems associated with direct preparations are as yet unresolved (*see* Chapter 12).

Mosaicism

When more than one cell line with different karyotypes is found in the same amniotic fluid or placental biopsy culture, it is necessary to distinguish true fetal mosaicism, with risk of physical and mental handicap, from pseudomosaicism, which results from maternal contamination or culture of a colony of cells derived from extra-embryonic tissue such as trophoblast or fetal membranes arising either in vivo or in vitro. While various culture techniques and strategies have been developed to distinguish cases of true fetal mosaicism from pseudomosaicism, the most satisfactory resolution of the problem is to determine the fetal blood karyotype [31]. In a series of 24 cases of amniotic fluid culture mosaicism investigated by fetal blood sampling, true mosaicism, with autosomal trisomy or sex-chromosome mosaicism, was found in only 1 (6%).

Fragile-X-Linked Mental Retardation

Fragile-X syndrome, occurring in 1 per 1000 live male births, is second only to Down's syndrome as a cytogenetic cause of mental retardation. Prenatal diagnosis from fetal blood and placental tissue is reviewed in Chapter 13.

Abnormal Maternal Serum Biochemistry

Despite the widespread introduction of amniocentesis for fetal karyotyping of women aged 35 years or over, in Britain the birth prevalence of babies with trisomy 21 has not fallen over the last 15 years. This is because more than 70% of chromosomally abnormal babies are born to younger women and because the uptake of prenatal diagnosis by the women at risk is poor. The latter is mainly a consequence of inadequate counselling but is also due to economic constraints on cytogenetic laboratories. For example, increased uptake is often followed by cytogenetic laboratories increasing the age limit for testing. The recent introduction of maternal serum biochemical screening (*see* Chapter 4) can potentially identify 60% of all fetuses with trisomy 21. However, it is very unlikely that the uptake of testing will increase unless funding of biochemical and cytogenetic laboratories is substantially increased. Furthermore, younger women who are suddenly confronted with the possibility of a chromosomally abnormal child and who, unlike older mothers, did not have the 'luxury' of long-term rational appreciation of the statistical risks, will find the current 3–5 weeks' delay in obtaining results from amniocentesis unacceptable. There will therefore be an increase in fetal karyotyping by cordocentesis for this indication.

Fetal Malformation

An ultrasound diagnosis of fetal malformation may require fetal karyotyping because of the common association with chromosomal abnormalities [32]. Fetal karyotyping should be considered not only in the second trimester, to allow termination of abnormal pregnancies, but also in the third trimester, because knowledge of a serious chromosomal defect may alter the management of labour and delivery. Karyotyping should also be performed for conditions, such as hydrops fetalis or severe early onset growth retardation, in which the risk of intrauterine death is high. Autolysis may render postmortem chromosomal studies, and therefore accurate genetic counselling for future pregnancies, impossible.

The incidence of chromosomal defects among fetuses with ultrasonically detectable abnormalities is much higher than the incidence reported in screening studies based on advanced maternal age or abnormal maternal serum biochemistry. In our series of 1366 fetuses with malformations or growth retardation who had antenatal karyotyping, 234 (17%) were found to be chromosomally abnormal. Some of the common malformations and the associated chromosomal abnormalities are shown in Table 15.1. The ultrasound diagnosis of a marker for a specific chromosomal defect should stimulate the search for other associated malformations and when these additional abnormalities are found the probability that the fetus is chromosomally abnormal is dramatically increased. Thus, chromosomal defects were found in 61 of our 953 (6%) cases with an isolated defect and in 173 of 413 (42%) with more than one malformation.

Table 15.1. Fetal malformations and chromosomal defects

	Total	Abnormal		Trisomies			Triploidy	Turner's	Other
	n	n	%	21	18	13			
Ventriculomegaly	74	13	18	1	1	1	6	–	4
Holoprosencephaly	40	11	28	–	1	8	–	–	2
Choroid plexus cysts	58	25	43	2	19	1	1	–	2
Nuchal oedema	74	33	45	18	1	5	2	3	4
Cystic hygromas	41	29	71	1	1	–	–	27	–
Facial cleft	44	26	59	1	10	12	1	–	2
Heart defects	115	66	57	14	23	7	4	12	6
Diaphragmatic hernia	51	13	25	–	7	3	1	–	2
Duodenal atresia	19	7	37	7	–	–	–	–	–
Exomphalos	75	29	39	–	22	4	1	–	2
Renal defects	536	79	15	16	20	16	5	4	18
Small for gestation	348	62	18	5	16	7	23	–	11

Detailed fetal ultrasound examination should be offered to all pregnant women at 20 weeks' gestation. This should detect major malformations as well as more subtle ones which may be markers of chromosomal defects. More than 90% of fetuses with trisomy 13 or 18, triploidy and Turners' syndrome have associated malformations [33] that should be easily detectable by diligent ultrasonographic examination. However, the common malformations in fetuses with trisomy 21 are more subtle (brachycephaly, relative shortening of the long bones, nuchal oedema, midphalanx hypoplasia or clinodactyly in the fifty finger, sandal gap, cardiac defects, or mild hydronephrosis), and therefore the false-positive rate for each individual feature may be unacceptably high [34–38]. Nevertheless, it should be emphasised that a single ultrasound marker, such as nuchal oedema, as a risk factor for fetal karyotyping can identify 40% of fetuses with trisomy 21, which compares favourably with the potential 25% detection rate if the risk factor is advanced maternal age.

Assessment of Small for Gestation Fetuses

Cytogenetic, biochemical, haematological and metabolic data derived from the study of blood samples obtained by cordocentesis have recently been used for evaluation of the SGA fetus [39]. Thus, severe early-onset fetal growth retardation is commonly associated with chromosomal abnormalities. Furthermore, some small fetuses are hypoxaemic, hypercapnic, hyperlacticaemic and acidaemic; they are erythoblastaemic, leukopenic and thrombocytopenic; they have disturbed carbohydrate, lipid and protein metabolism, and they demonstrate features of adrenal hyperplasia and pancreatic and thyroid hypoplasia [40–49]. The contribution of each of these abnormalities to perinatal death and long-term morbidity in this group of fetuses remains to be elucidated.

Cordocentesis and measurement of fetal blood gases have also provided an end point for evaluation and non-invasive tests aimed at predicting fetal asphyxia. Thus, in hypoxaemic growth retardation Doppler studies have demonstrated increased impedance to flow in the uterine and umbilical arteries [50,51]. Furthermore, Doppler studies of fetal vessels (reduced impedance to flow in the common carotid and middle cerebral arteries and increased impedance in the descending thoracic aorta and renal arteries) have provided evidence that in fetal hypoxaemia there is redistribution of the fetal circulation in favour of the brain at the expense of the viscera [52–54]. Similarly, fetal heart rate (FHR) monitoring has demonstrated an association between fetal acidaemia and the development of pathological FHR patterns [55]. However, with both Doppler studies and FHR monitoring there is a large scatter of results around the regression lines describing the associations with fetal hypoxaemia/acidaemia (Fig. 15.3). Therefore, if antepartum monitoring is aimed at predicting the degree of fetal acidaemia (a well-accepted aim of intrapartum monitoring), then cordocentesis and measurement of fetal blood gases is an essential adjunct in the management of high-risk pregnancies.

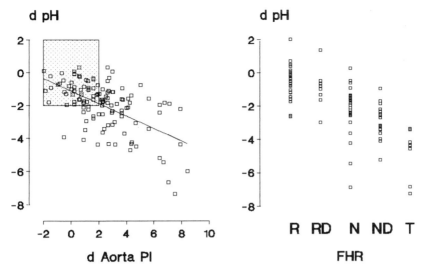

Fig. 15.3. Prediction of the degree of fetal acidaemia (d pH) by Doppler ultrasound measurement of impedance to flow (d PI) in the descending thoracic aorta (shaded area represents normal pH and PI), and fetal heart rate (FHR) patterns (R, reactive; RD, reactive with decelerations; N, non-reactive; ND, non-reactive with decelerations; T, terminal). d values represent deviations from the normal mean for gestation in SDs.

Assessment and Treatment of Fetal Anaemia

During the last 50 years, several therapeutic approaches have been undertaken in the management of red cell isoimmunised pregnancies with the aim of ameliorating the severity of the condition and preventing intrauterine fetal death. In the last five years, fetal blood sampling has made it possible to gain a better understanding of the pathophysiology of the disease and has provided an improved method for the assessment and treatment of fetal anaemia [56–61].

The severity of fetal haemolysis can be predicted from (a) the history of previously affected pregnancies, (b) the level of maternal haemolytic antibodies, (c) the amniotic fluid bilirubin concentration, (d) the altered morphometry of fetus and placenta, (e) the presence of pathological FHR patterns, and (f) changes in the flow velocity waveforms obtained by Doppler studies of the fetal circulation. However, there is a wide scatter of values around the regression lines describing the associations between the degree of fetal anaemia and the data obtained from these indirect methods of assessment [62–65].

The only accurate method for determining the severity of the disease is blood sampling by cordocentesis and measurement of the fetal haemoglobin concentration (Fig. 15.4). However, the indication for, and the timing of fetal blood sampling in the context of this disease have not yet been defined adequately [66,67]. Furthermore, transplacental cordocentesis may result in fetomaternal haemorrhage [68], an increase in the maternal haemolytic antibody concentration [69] and consequent worsening of the disease. Nevertheless, it could be argued that cordocentesis should be performed for all patients with a history of severe disease and those with high haemolytic antibody levels (>10 iu/ml), pathological

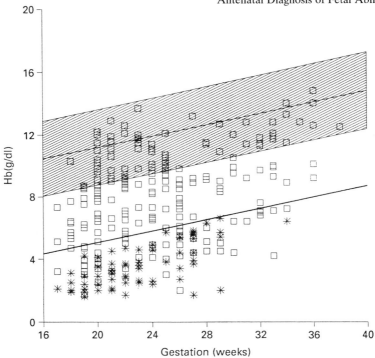

Fig. 15.4. Fetal haemoglobin concentration from red cell isoimmunised pregnancies at the time of the first fetal blood sampling are plotted on the reference range (mean, 5th and 95th centiles; shaded area) for gestation. Hydrops fetalis (*) is associated with severe anaemia.

FHR patterns or abnormal flow velocity waveforms. A fetal blood sample is first obtained, the haemoglobin concentration is then measured and an intravascular blood transfusion is given as necessary. Subsequent transfusions are given at 1–3 weekly intervals until 34–36 weeks' gestation and their timing is based on the findings of non-invasive tests, such as Doppler studies and FHR monitoring, and the knowledge that following a fetal blood transfusion the mean rate of decrease in fetal haemoglobin is approximately 0.3 g/dl per day [58].

Currently the survival rate of red cell isoimmunised pregnancies treated with cordocentesis is more than 75% [66,70]. During the last 6 years in our unit cordocentesis was performed in 249 fetuses from red cell isoimmunised pregnancies; 208 were Coomb's positive and 41 negative. A total of 931 cordocenteses were performed and on 656 occasions intravascular fetal blood transfusions were given. There were 15 intrauterine, six neonatal and two infant deaths. Furthermore, in two cases the pregnancies were terminated because the fetuses developed atrophic hydrocephalus 4–6 weeks after transfusions complicated by prolonged fetal bradycardias. Of the 249 pregnancies, infants have survived and the remaining pregnancies are continuing.

Diagnosis and Treatment of Congenital Thrombocytopenia

Congenital thrombocytopenia, resulting from decreased production or increased consumption of platelets, is a relatively common haemostatic abnormality observed among newborns admitted to neonatal intensive care units. However, in the majority of cases the condition is self-limiting and the underlying pathology is related to complications of pregnancy or delivery, such as pre-eclampsia, intrauterine fetal growth retardation and infection or intrapartum hypoxia and traumatic haemorrhage. In such cases the management and prognosis for the infant will primarily depend on the other manifestations or complications of the initiating insult.

Thrombocytopenia in the newborn is also observed in association with the transplacental passage of maternal IgG antiplatelet auto- or allo-antibodies. *Autoimmune thrombocytopenia* is secondary to maternal immune thrombocytopenic disorders, including idiopathic thrombocytopenic purpura, systemic lupus erythematosus and lymphoproliferative disorders. Platelet-associated IgG and thrombocytopenia are found in both the mother and her fetus. By contrast, *alloimmune thrombocytopenia*, with an incidence of 2–5 per 10 000 live births, is analogous to red cell isoimmunisation except that fetal platelets rather than erythrocytes are the targets of the allo-antibody. Platelet antigen-negative mothers develop IgG antibodies against the fetal-platelet specific platelet antigen, which is inherited from the platelet antigen-positive father. Although the exact incidence of intracranial haemorrhage among affected infants is not known, it is thought to be 10%–30% [71].

Current arguments on the role of cordocentesis in the management of fetal thrombocytopenia revolve around the need for blood sampling from 20 weeks and various therapeutic regimens including maternal administration of steroids or gamma-globulin and repeated fetal platelet infusions. The rationale is that fetal intracranial haemorrhage can occur antenatally [72–74]; indeed the majority of fetal deaths may occur early in pregnancy [75] and therefore focusing on the perinatal events may be inappropriate. In autoimmune thrombocytopenia there is no correlation between fetal platelet count and either the level of maternal serum-associated IgG or the degree of maternal thrombocytopenia [76–79]. Similarly in alloimmune thrombocytopenia the level of antiplatelet antigen antibodies is not helpful in the prediction of the severity of the disease [79]. Therefore, in both auto- and allo-immune thrombocytopenia cordocentesis has been advocated from as early as 20 weeks. In autoimmune thrombocytopenia the value of cordocentesis would be to provide an additional criterion (fetal thrombocytopenia) for maternal treatment with steroids or gamma-globulin and also to monitor the effectiveness of such treatment by documentation of improvement in the fetal platelet count. In alloimmune thrombocytopenia cordocentesis allows determination of the fetal platelet phenotype and count. If the latter is reduced, intravascular fetal platelet transfusions can be given at weekly intervals [80,81]. Alternatively, a second cordocentesis may be performed and a platelet transfusion given at 37–38 weeks' gestation [82]. The choice between these two suggested approaches is determined by the balance of the risk of fetal intracranial haemorrhage and the risks of cordocentesis. A further method of treatment is the maternal administration of steroids and IgG. Daffos et al. [76] reported an apparent increase in fetal platelet count after maternal treatment with prednisolone (10 mg) daily from week 23 of pregnancy until delivery. Bussel et al. [71] reported seven cases in which they documented elevation of the fetal platelet count by weekly maternal administration of high dose intravenous IgG with or without dexamethasone.

Our management of patients with severe alloimmune thrombocytopenia

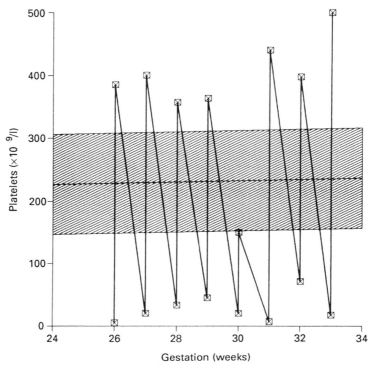

Fig. 15.5. Treatment of alloimmune thrombocytopenia in a mother who had five previous "unexplained" intrauterine deaths at 17–24 weeks' gestation. During her last pregnancy, ultrasonographic examination at 20 weeks had demonstrated fetal hydrocephalus and hydrops and at cordocentesis the fetus had severe anaemia and thrombocytopenia. In her present pregnancy she was treated with prednisolone and immunoglobulins from 16 weeks' gestation and by weekly cordocenteses and fetal platelet transfusions after 26 weeks. She had a healthy livebirth at 34 weeks.

includes maternal treatment with prednisolone and IgG from 16–20 weeks' gestation until 26 weeks, when cordocentesis is performed [80]. If the fetal platelet count is above $50 \times 10^9/l$, treatment of the mother is continued until 38 weeks when a further cordocentesis is performed to help decide on the mode of delivery. If the platelet count is less than $50 \times 10^9/l$, intravascular fetal platelet transfusions are given at weekly intervals (Fig. 15.5).

References

1. Freda VJ, Adamsons KJ. Exchange transfusion in utero. Am J Obstet Gynecol 1964; 89:817–21.
2. Hobbins JC, Mahoney MJ. In utero diagnosis of haemoglobinopathies. Technique for obtaining fetal blood. N Engl J Med 1974; 290:1065–7.
3. Rodeck CH, Campbell S. Umbilical cord insertion as source of pure fetal blood for prenatal diagnosis. Lancet 1979; i:1244–5.
4. Bang J, Bock JE, Trolle D. Ultrasound-guided fetal intravenous transfusion for severe rhesus haemolytic disease. Br Med J 1982; 284:373–4.
5. Daffos F, Cappella-Pavlovsky M, Forestier F. Fetal blood sampling via the umbilical cord using a needle guided by ultrasound. Report of 66 cases. Prenat Diagn 1983; 3:271–7.

6. Nicolaides KH, Soothill PW, Rodeck CH et al. Ultrasound guided sampling of umbilical cord and placental blood to assess fetal wellbeing. Lancet 1986; i:1065–7.
7. Wilkins I, Mezrow G, Lynch L et al. Amnionitis and life-threatening respiratory distress after percutaneous umbilical blood sampling. Am J Obstet Gynecol 1989; 160:427–8.
8. Modell B. Haemoglobinopathies-diagnosis by fetal blood sampling. In: Rodeck CH, Nicolaides KH, eds. Prenatal diagnosis. Proceedings of the XIth Study Group of the Royal College of Obstetricians and Gynaecologists. Chichester: Wiley, 1984; 93–8.
9. Mibashan RS, Rodeck CH. Haemophilia and other genetic defects of haemostasis. In: Rodeck CH, Nicolaides KH, eds. Prenatal diagnosis. Proceedings of the XIth Study Group of the Royal College of Obstetricians and Gynaecologists. Chichester: Wiley, 1984; 179–94.
10. Patrick AD. Prenatal diagnosis of inherited metabolic diseases. In: Rodeck CH, Nicolaides KH, eds. Prenatal diagnosis. Proceedings of the XIth Study Group of the Royal College of Obstetricians and Gynaecologists. Chichester: Wiley, 1984; 121–32.
11. Miller E, Cradock-Watson JA, Pollock TM. Consequences of confirmed maternal rubella at successive stages of pregnancy. Lancet 1982; ii:781–4.
12. Daffos F, Forestier F, Grangeot-Keros L et al. Prenatal diagnosis of congenital rubella. Lancet 1984; ii:1–3.
13. Morgan-Capner P, Rodeck CH, Nicolaides KH et al. Prenatal detection of rubella specific IgM in fetal sera. Prenat Diagn 1985; 5:21–3.
14. Desmonds G, Daffos F, Forestier F et al. Prenatal diagnosis of congenital toxoplasmosis. Lancet 1985; i:500–3.
15. Daffos F, Forestier F, Capella-Pavlovsky M et al. Prenatal management of 746 pregnancies at risk of congenital toxoplasmosis. N Engl J Med 1988; 318:271–5.
16. Pass RF, Stango S, Meyers GJ et al. Outcome of symptomatic congenital cytomegalovirus infection: results of long term follow up. Pediatrics 1980; 66:758–62.
17. Stango S, Pass RF, Dworsky ME et al. Congenital cytomegalovirus infection: the relative importance of primary and recurrent maternal infection. N Engl J Med 1982; 306:945–9.
18. Lange I, Rodeck CH, Morgan-Capner P et al. Prenatal serological diagnosis of intrauterine cytomegalovirus infection. Br Med J 1982; 284:1673–4.
19. Meisel RL, Alvarez M, Lynch L et al. Fetal cytomegalovirus infection: a case report. Am J Obstet Gynecol 1990; 162:663–4.
20. Enders G. Management of varicella-zoster contact and infection in pregnancy using a standardized varicella-zoster ELISA test. Postgrad Med J 1985; 61:23–30.
21. Grose C, Itani O. Pathogenesis of congenital infection with three diverse viruses: varicella-zoster virus, human parvovirus, and human immunodeficiency virus. Semin Perinatol 1990; 13:278–93.
22. Cuthberson G, Weiner CO, Giller RH et al. Prenatal diagnosis of second trimester congenital varicella syndrome by virus-specific IgM. J Pediatr 1987; 111:592–5.
23. Anand A, Gray ES, Brown T. Human parvovirus infection in pregnancy and hydrops fetalis. N Engl J Med 1987; 316:183–6.
24. Peters MT, Nicolaides KH. Cordocentesis for the diagnosis and treatment of human fetal parvovirus infection. Obstet Gynecol 1990; 75:501–4.
25. Kilham L, Margonis G. Problems of human concern arising from animal models of intrauterine and neonatal infections due to viruses: a review. I. Introduction and virologic studies. Prog Med Virol 1975; 20:113–43.
26. Weilland HT, Vermeer-Keers C, Salimans MM et al. Parvovirus B19 associated with fetal abnormality. Lancet 1987; i:521–2.
27. Anderson LJ, Hurwitz ES. Human parvovirus B19 and pregnancy. Clin Perinatol 1988; 15:273–86.
28. Gosden C, Rodeck CH, Nicolaides KH et al. Fetal blood chromosome analysis: some new indications for prenatal karyotyping. Br J Obstet Gynaecol 1985; 92:915–20.
29. Nicolaides KH, Soothill PW, Rodeck CH. Why confine chorionic villus (placental) biopsy to the first trimester? Lancet 1986; i:543–4.
30. Holzgreve W, Miny P, Gerlach B. Benefits for rapid fetal karyotyping in the second and third trimester (late chorionic villus sampling) in high risk pregnancies. Am J Obstet Gynecol 1990; 162:1188–92.
31. Gosden C, Nicolaides KH, Rodeck C. Fetal blood sampling in investigation of chromosome mosaicism in amniotic fluid cell culture. Lancet 1988; i:613–16.
32. Nicolaides KH, Rodeck CH, Gosden CM. Rapid karyotyping in non-lethal fetal malformations. Lancet 1986; i:283–6.
33. Jones KL. Smith's recognizable patterns of human malformation, 4th edn. London: WB Saunders, 1988.

34. Benacerraf BR, Frigoletto FD, Laboda LA. Sonographic diagnosis of Down syndrome in the second trimester. Am J Obstet Gynecol 1985; 153:49–52.
35. Benacerraf BR, Osathanodh R, Frigoletto FD. Sonographic demonstration of hypoplasia of the middle phalanx of the fifth digit: a finding associated with Down syndrome. Am J Obstet Gynecol 1988; 159:181–3.
36. Benacerraf BR, Mandell J, Estroff JA et al. Fetal pyelectasis: a possible association with Down syndrome. Obstet Gynecol 1990; 76:58–60.
37. Nyberg DA, Resta RG, Hickok DE et al. Femur length shortening in the detection of Down syndrome: is prenatal screening feasible? Am J Obstet Gynecol 1990; 162:1247–52.
38. Toi A, Simpson GF, Filly RA. Ultrasonically evident fetal nuchal skin thickening: is it specific for Down syndrome? Am J Obstet Gynecol 1987; 156:150–3.
39. Nicolaides KH, Economides D, Thorpe-Beeston G. Treatment of fetal growth retardation. In: Sharp F, Fraser RB, Milner RDG, eds. Fetal growth. Proceedings of the XXth Study Group of the Royal College of Obstetricians and Gynaecologists, 1989; 333.
40. Soothill PW, Nicolaides KH, Campbell S. Prenatal asphyxia, hyperlactaemia, hypoglycaemia and erythroblastosis in growth retarded fetuses. Br Med J 1987; 294:1051–3.
41. Pardi G, Buscaglia M, Ferrazzi E et al. Cord sampling for the evaluation of oxygenation and acid–base balance in growth-retarded human fetuses. Am J Obstet Gynecol 1987; 157:1221–8.
42. Nicolaides KH, Economides KH, Soothill PW. Blood gases, pH and lactate in appropriate and small for gestational age fetuses. Am J Obstet Gynecol 1989; 161:996–1001.
43. Economides D, Nicolaides KH, Linton EA et al. Plasma cortisol and adrenocorticotropin in appropriate and small for gestational age fetuses. Fetal Ther 1988; 3:158–64.
44. Economides DL, Nicolaides KH. Blood glucose and oxygen tension in small for gestational age fetuses. Am J Obstet Gynecol 1989; 160:385–9.
45. Economides D, Nicolaides KH, Gahl WA, Bernardini I, Evans MI. Plasma amino acids in appropriate and small for gestational age fetuses. Am J Obstet Gynecol 1989; 161:1219–27.
46. Economides DL, Proudler A, Nicolaides KH. Plasma insulin in appropriate and small for gestational age fetuses. Am J Obstet Gynecol 1989; 160:1091–4.
47. Economides DL, Crook D, Nicolaides KH. Hypertriglyceridaemia and hypoxaemia in small for gestational age fetuses. Am J Obstet Gynecol 1990; 162:382–6.
48. Thorpe-Beeston JG, Nicolaides KH, Snijders RJM et al. Thyroid fuction in small for gestational age fetuses. Obstet Gynecol (in press).
49. Van den Hof M, Nicolaides KH. Platelet count in normal, small and anemic fetuses. Am J Obstet Gynecol 1990; 162:735–9.
50. Soothill PW, Nicolaides KH, Bilardo C, Hackett G, Campbell S. Utero-placental blood velocity resistance index and umbilical venous pO_2, pCO_2, pH, lactate and erythroblast count in growth retarded fetuses. Fetal Ther 1986; 1:176–9.
51. Nicolaides KH, Bilardo CM, Soothill PW et al. Absence of end diastolic frequencies in the umbilical artery: A sign of fetal hypoxia and acidosis. Br Med J 1988; 297:1026–7.
52. Bilardo CM, Nicolaides KH, Campbell S. Doppler measurements of fetal and uteroplacental circulations: relationship with umbilical venous blood gases measured at cordocentesis. Am J Obstet Gynecol 1990; 162:115–20.
53. Vyas S, Nicolaides KH, Campbell S. Renal artery flow velocity waveforms in normal and hypoxemic fetuses. Am J Obstet Gynecol 1989; 161:168–72.
54. Vyas S, Nicolaides KH, Bower S, Campbell S. Middle cerebral artery flow velocity waveforms in fetal hypoxaemia. Br J Obstet Gynaecol 1990; 97:797–803.
55. Visser GHA, Sadovsky G, Nicolaides KH. Antepartum heart rate patterns in small for gestational age third trimester fetuses: correlations with blood gases obtained at cordocentesis. Am J Obstet Gynecol 1990; 162:698–703.
56. Berkowitz RL, Chitkara U, Goldberg JD et al. Intrauterine transfusion in utero: the percutaneous approach. Am J Obstet Gynecol 1986; 154:622–7.
57. Grannum PAT, Copel JA, Plaxe SC et al. In utero exchange transfusion by direct intravascular injection in severe erythroblastosis fetalis. N Engl J Med 1986; 314:1431–4.
58. Nicolaides KH, Rodeck CH, Kemp J et al. Have Liley charts outlived their usefulness? Am J Obstet Gynecol 1986; 155:90–4.
59. Nicolaides KH. Studies on fetal physiology and pathophysiology in Rhesus disease. Semin Perinatol 1989; 13:328–37.
60. Soothill PW, Lestas AN, Nicolaides KH et al. 2,3-Diphosphoglycerate in normal, anaemic and transfused human fetus. Clin Sci 1988; 74:527–8.
61. Panos MZ, Nicolaides KH, Anderson JV et al. Plasma atrial natriuretic peptide in human fetus: response to intravascular blood transfusion. Am J Obstet Gynecol 1989; 161:357–61.

62. Nicolaides KH, Soothill PW, Rodeck CH et al. Rh disease: intravascular fetal blood transfusion by cordocentesis. Fetal Ther 1986; 1:185–92.
63. Nicolaides KH, Sadovsky G, Cetin E. Fetal heart rate patterns in red blood cell isoimmunized pregnancies. Am J Obstet Gynecol 1989; 161:351–6.
64. MacKenzie IZ, Bowell PJ, Castle BM et al. Serial fetal blood sampling for the management of pregnancies complicated by severe rhesus (D) isoimmunization. Br J Obstet Gynaecol 1988; 95:753–8.
65. Sadovsky G, Visser GHA, Nicolaides KH. Heart rate patterns in fetal anemia. Fetal Ther 1988; 3:216–23.
66. Berkowitz RL, Chitkara U, Wilkins IA et al. Intravascular monitoring and management of erythroblastosis fetalis. Am J Obstet Gynecol 1988; 158:783–95.
67. Parer JT. Severe Rh isoimmunization – current methods of in utero diagnosis and treatment. Am J Obstet Gynecol 1988; 158:1323–9.
68. Nicolini U, Kochenour NK, Greco P. Consequences of fetomaternal haemorrhage after intrauterine transfusion. Br Med J 1988; 297:1379–80.
69. Bowell PJ, Wainscoat JS, Peto TEA et al. Maternal anti-D concentrations and outcome in rhesus haemolytic disease of the newborn. Br Med J 1982; ii:327–9.
70. Grannum PAT, Copel JA. Prevention of Rh isoimmunization and treatment of the compromised fetus. Semin Perinatol 1988; 12:324–35.
71. Bussel JB, Berkowitz RL, McFarland JG et al. Antenatal treatment of neonatal alloimmune thrombocytopenia. N Engl J Med 1988; 319:1374–8.
72. Zalneraitis EL, Young RK, Krisnamoorthy KS. Intracranial haemorrhage in utero as a complication of isoimmune thrombocytopenia. J Pediatr 1979; 95:611–14.
73. Jesurun CA, Levin GS, Sullivan WR et al. Intracranial haemorrhage in utero and thrombocytopenia. J Pediatr 1980; 97:695–6.
74. Herman JH, Jumbelic MI, Ancona RJ et al. In utero cerebral hemorrhage in alloimmune thrombocytopenia. Am J Pediatr Hematol Oncol 1986; 8:312–17.
75. Kelton JG, Inwood MJ, Barr RM et al. The prenatal prediction of thrombocytopenia in infants of mothers with clinically diagnosed immune thrombocytopenia. Am J Obstet Gynecol 1982; 144:449–54.
76. Daffos F, Forestier F, Kaplan C, Cox W. Prenatal diagnosis of bleeding disorders with fetal blood sampling. Am J Obstet Gynecol 1988; 158:939–46.
77. Moise KJ, Carpenter RJ, Cotton DB et al. Percutaneous umbilical cord blood sampling in the evaluation of fetal platelet counts in pregnant patients with autoimmune thrombocytopenic purpura. Obstet Gynecol 1988; 72:346–50.
78. Scioscia AL, Grannum PAT, Copel JA et al. The use of percutaneous blood sampling in immune thrombocytopenic purpura. Am J Obstet Gynecol 1988; 159:1066–8.
79. Reznikoff-Etievant MF, Muller JY, Kaplan C et al. Immunization against the ZWa (PLA1) platelet antigen: group at risk, prevention of complications. Pathol Biol (Paris) 1986; 34:783–7.
80. Nicolini U, Rodeck CH, Kochenour NK et al. In utero platelet transfusion for alloimmune thrombocytopenia. Lancet 1988; ii:506–7.
81. Murphy MF, Pullon HWH, Metclalfe P et al. Management of fetal alloimmune thrombocytopenia by weekly in utero platelet transfusions. Vox Sang 1990; 58:45–9.
82. Kaplan C, Daffos F, Forestier F et al. Management of alloimmune thrombocytopenia; antenatal diagnosis and in utero transfusion of maternal platelets. Blood 1988; 72:340–3.

Chapter 16

Intrauterine Therapy

C. H. Rodeck and N. M. Fisk

Introduction

Most conditions diagnosed in the fetus are managed either by postnatal treatment or by termination of pregnancy. Improved access to the fetus, both non-invasive and invasive, has, however, led to the intrauterine treatment of an increasing number of disorders, and also to the more rational management of others.

Tachyarrhythmias

Supraventricular tachycardia (SVT) is the most common fetal tachyarrhythmia, although atrial flutter and fibrillation also occur. As SVT is often intermittent, treatment is indicated only when SVT is sustained or associated with hydrops [1]. Therapy in utero seems preferable to delivery and neonatal treatment, as the fetus tolerates haemodynamic compromise better in utero, where gas exchange is not hindered by pulmonary oedema. Transplacental treatment by digitalising the mother leads to cardioversion in only about 25% of cases [2,3]. These poor results partly reflect difficulties in achieving therapeutic maternal levels due to the increased intravascular volume and glomerular filtration rate of pregnancy. Whether digoxin actually crosses the placenta has been called into doubt by a recent report of similar concentrations of digoxin-like immunoreactive substances in fetuses of treated and untreated mothers [4]. The addition of second-line drugs, such as propranolol, procainamide, verapamil, amiodarone and flecainide results in eventual cardioversion in just over 50% of cases [2,3]. Because waiting weeks for successful transplacental therapy may be hazardous in the presence of hydrops, there is a need for more rapid cardioversion. In two fetuses, Gembruch et al. [5] injected antiarrhythmic drugs daily into the fetal peritoneal cavity, leading to prolonged episodes of sinus rhythm. Weiner and Thompson [6] tried three 8-hourly intramuscular injections of digoxin in a fetus at 26 weeks, using a neonatal loading dose plus 25% for placental binding [7]. Fetal blood sampling (FBS) 8 hours later revealed subtherapeutic levels, and repeat loading was required to achieve therapeutic levels and sinus rhythm. The calculated elimination half-life of 15.9 h was substantially less than the 50 h previously reported in low birth weight (LBW) infants. This is in agreement with our general experience of fetal pharmacology, in which mugh higher doses are required per kilogram than in the neonate, presumably due to placental passage into the maternal circulation, and to the fetus' greater blood volume. Thus the frequency of invasive procedures renders direct fetal injection impractical for maintenance therapy. Our current approach is to commence maternal therapy at the same time as direct fetal cardioversion, administering intravascularly a quarter of the neonatal loading dose adjusted for fetal pharmacodynamics (50 µg/kg of digoxin), and repeating this if no response is observed within 30 minutes. The remainder is administered intraperitoneally prior to removing the needle. Further daily intraperitoneal injections may be required until a steady state is achieved.

Red Cell Alloimmunisation

Despite a dramatic decline in incidence, maternal sensitisation has not disappeared for a variety of reasons including antenatal sensitisation, prophylaxis failure and antibodies other than anti-D. Untreated, 45%–50% of affected infants will have no or mild anaemia, and 25%–30% will have moderate anaemia posing neonatal problems only. The remaining 20%–25% develop hydrops and usually die in utero or neonatally; in half, the hydrops develops prior to 30 weeks [8]. The aim of antenatal management is to identify severely affected fetuses, to correct their anaemia by transfusion, and then deliver them at the optimal time. At each gestational age, the risks of invasive monitoring are weighed against those of conservative management and of delivery.

Assessment of Severity

Regular monitoring is indicated from 18 weeks when the maternal antibody concentration exceeds 4 iu/ml [9]. Above this threshold, antibody levels have a limited role as they correlate poorly with the degree of fetal anaemia [10], although a rising level suggests an increase in severity. Non-invasive methods such as sonographic measurement of placental thickness, umbilical vein diameter and intraperitoneal volume [11,12], or Doppler assessment of velocities in the descending aorta [13,14] are unreliable in predicting the degree of anaemia. Demonstration of fetal ascites, however, indicates severe anaemia (haematocrit (Hct) <15%, haemoglobin concentration (Hb) <4 g/dl) in the mid-trimester, although this threshold for ascites rises with advancing gestation [12,15]. Weekly sonographic surveillance for ascites is thus an integral part of monitoring in Rh disease, but cannot be relied on as the sole method for two reasons. First, ascites is only found in two-thirds of fetuses with Hb<4 g/dl, and second, anaemia of this degree may be associated with hypoxaemia [16].

Invasive monitoring necessitates amniocentesis or FBS. Severity has traditionally been assessed by spectrophotometric measurement of the deviation in absorbance at 450 nm (ΔA_{450}) due to bilirubin in amniotic fluid [17]. Serial readings of the ΔA_{450} are plotted on Liley's charts to give an indirect index of fetal red cell destruction, with delivery or transfusion indicated when Whitfield's line is reached [18]. While this is reasonably reliable in the third trimester, amniocentesis has proved inaccurate in the mid-trimester, especially under 25 weeks [19]. FBS on the other hand allows direct assessment of fetal Hct and Hb, permits transfusion to be performed at the same procedure if anaemia is detected, and yields additional information on fetal rhesus (Rh) and blood gas status. As FBS has a slightly greater loss rate, and provokes fetomaternal haemorrhage in 70% of procedures in which the placenta is transgressed, thereby increasing antibody levels [20], amniocentesis remains the mainstay for monitoring mild/moderate disease in the late second and third trimesters [21]. The ΔA_{450} is unreliable in alloimmunisation due to Kell antibodies [22], which seem to suppress erythropoiesis rather than cause haemolysis.

Another diagnostic option, fetal blood grouping of erythrocytes eluted from first trimester chorion villi [23], may be appropriate for women with a severe history and a heterozygous partner, but again chorionic villus sampling (CVS) can provoke fetomaternal haemorrhage [24].

Fetal Blood Transfusion

Intravascular transfusion (IVT) was first administered fetoscopically [25], but is now given by ultrasound guided FBS [26–28]. The decision to administer an intravascular transfusion is based on the Hct or Hb or FBS. The needle tip is kept within the umbilical vein and fresh Rh-negative packed cells compatible with the mother are infused at 10–15 ml/min. The fetal heart rate and flow of infused blood are monitored on ultrasound to guard against inadvertent needle dislodgement and cord tamponade. Some centres advocate temporary immobilisation of the fetus by intramuscular or intravascular curare or pancuronium [29]. The volume transfused is determined by consideration of the estimated fetoplacental blood volume and the fetal and donor Hct [28] or Hb [15] levels, according to published nomograms. The Hct is rechecked after transfusion, and a further increment given if it is less than the desired 40%–45%. Exchange transfusion has been recommended to minimise circulatory overload [30], but is not widely used, as the more popular and quicker 'top-up' transfusion increases fetoplacental blood volume by up to 100% without adverse effects and with only minor changes in blood gases [31].

The second transfusion is always performed two weeks after the first because the rate of fall in Hct in each fetus at that stage is unpredictable [32]. Subsequent transfusions are timed at 2–4 week intervals when the Hct is estimated to have fallen to 20%–25%, based on each fetus's calculated rate of fall in Hct, which varies between 0.2%–3.0% per day [28]. Kleihaur testing of fetal samples indicates that erythropoiesis is usually completely suppressed after two to three transfusions [32]. As the donor blood in the fetal circulation is not susceptible to immune destruction, the rate of fall in Hct declines with increasing transfusion and thus the interval between procedures can be increased. The same principles are used in scheduling delivery between 36 and 38 weeks.

The direct intravascular route for transfusion is favoured over the older intraperitoneal route, as it yields direct information on the severity of the disease, and corrects anaemia more physiologically. Intraperitoneal transfusion (IPT) is usually ineffective in hydrops due to impaired absorption, and is associated with complications in 20% of procedures [33]. Large rises in intraperitoneal pressure have been implicated in sudden death at the time of IPT [34]. However, a small IPT may be given in combination with IVT to increase the total volume administered while avoiding polycythaemia, due to the slower absorption of the intraperitoneal blood. Thus the interval between transfusions may be prolonged to 4–5 weeks and the total number of invasive procedures per pregnancy reduced [32].

Survival

With serial intravascular transfusions, survival rates of 78%–95% have been achieved in severely affected fetuses [28,35]. In the last 4 years, we (five operators) have performed 324 transfusions in 99 fetuses, with a survival rate of 80%, similar in hydropic and non-hydropic fetuses. Fetal mortality was correlated inversely with gestational age at the first procedure, and operator experience is undoubtedly also of importance.

Fetal Alloimmune Thrombocytopenia

Perinatal alloimmune thrombocytopenia (PAIT) complicates 1:5000 births, with intracranial haemorrhage (ICH) affecting 20%–35% of cases of PAIT [36]. Maternal antiplatelet antibodies cross the placenta, in a situation analogous to Rh disease. The consequent fetal thrombocytopenia may be profound, and in PAIT there have been increasing reports of spontaneous ICH in utero in the third trimester [37–39]. The initial approach, of FBS for platelet-specific antigen (PLA) typing at 20–22 weeks with a repeat procedure at 37–38 weeks for platelet transfusion prior to delivery [40], did not address the risk of ICH in utero. Instead, from 26 weeks we have performed, weekly in utero transfusion of PLA-negative platelets in women with a previously affected pregnancy, with delivery once fetal lung maturity was attained [41,42]. As in Rh disease, this approach carries a cumulative procedure-related risk, but this appears lower than that of ICH. In this respect, a recent report that maternal high-dose immunoglobulin improves neonatal platelet counts in PAIT appeared promising. Bussel et al. [43], using a weekly dose of 1 g/kg, found that platelet counts at birth were significantly higher than in untreated siblings. However, the effect seemed mild, as five of the seven treated fetuses had platelet counts $<100 \times 10^9/l$ at birth, and two had counts of $<50 \times 10^9/l$ at birth. In addition, several case reports indicate that fetal thrombocytopenia may persist despite this treatment [42,44,45]. If high dose maternal immunoglobulin is used, it is important that the effect on fetal platelet count be monitored by FBS, with transfusion indicated in the presence of persistent thrombocytopenia after 26 weeks [42].

Fetofetal Transfusion Syndrome

Although vascular communications are found in the placentas of almost all monochorial twin pregnancies, clinical signs of fetofetal transfusion syndrome (FFTS) occur in only 4%–26% [46,47]. Certain ultrasound appearances are highly suggestive but not diagnostic of FFTS, principally discrepant growth and amniotic fluid volume in monochorial diamnionic twins [48,49]. Our study using FBS to investigate FFTS, has shown that the classic postnatal difference in Hb of 5 g/dl does not apply in the mid-trimester [50]. FBS may nevertheless have a role in confirming the diagnosis, if adult red cells are transfused into the donor twin and their passage into the other twin's circulation is then demonstrated by Kleihaur testing of blood aspirated from the recipient [50].

With FFTS in the mid-trimester, the recipient develops acute polyhydramnios and may become hydropic, while the donor twin is growth retarded and has severe oligohydramnios. As perinatal mortality in FFTS diagnosed in the mid-trimester is 80%–100%, aggressive therapeutic measures may be warranted. Selective feticide of the donor has been used to allow survival of the recipient in two cases, although this drastic procedure was performed primarily for prolongation of pregnancies threatened by gross polyhydramnios in the absence of hydrops or other fetal compromise [51,52]. Fetoscopic laser ablation of vascular anastomoses is possible when the placenta is posterior, and has been used successfully to control polyhydramnios in three cases [53]. Simpler therapies for polyhydramnios deserve consideration first, but it has been suggested that oligohydramnios in one twin contraindicates the use of indomethacin [54]. Vetter and Schneider [55] noted resolution of FFTS in two pregnancies after iatrogenic puncture of a chorionic plate vessel at amniocentesis, with consequent intra-amniotic haemorrhage. Repeated phlebotomy of the recipient and transfusion of the donor, has been suggested [52], but the degree of shunting calculated to occur in FFTS [50] indicates that this would be unlikely to be of long-term benefit. Given that the hydropic recipient seems more at risk than its donor, venepuncture of the recipient alone warrants consideration in FFTS complicated by hydrops.

Fetal Surgery

In several congenital malformations, a satisfactory outcome is often achieved with surgical correction after birth. In some, however, the uncorrected malformation results in progressive organ damage in utero, jeopardising survival. In such conditions there may be a role for intervention in utero. Attempts at fetal surgery have taken one of two forms: open surgical correction at hysterotomy, or bypassing obstructive lesions by ultrasound guided insertion of catheter shunts. These techniques should be contemplated only in a few centres with expertise, and for conditions in which animal models have demonstrated benefit from correction in utero. It is axiomatic that chromosomal and other structural malformations be first excluded. Reliable antenatal predictors are needed in selecting cases for intervention, so that fetal surgery is withheld from those which would otherwise have a satisfactory outcome, and from those in which the pathology is irreversible.

Intrathoracic Lesions

Diaphragmatic Hernia

The mortality rate from diaphragmatic hernia diagnosed in utero remains high at 75% despite optimal postnatal care [56,57]. The main determinant of outcome is the degree of pulmonary hypoplasia resulting from in utero lung compression, dependent on the timing and volume of visceral herniation through the diaphragm. Adverse factors are polyhydramnios, mediastinal shift, and a large volume of viscera within the chest. Studies in an animal model demonstrated that surgical correction in utero reverses the effects on fetal lung growth and allows survival at birth [58]. Next, Harrison's San Francisco group demonstrated in non-human primates the safety and feasibility of fetal surgery [59]. After a decade developing appropriate techniques, they have now operated in the late mid-trimester on eight human fetuses with adverse factors [57,60]. Technical problems were encountered in the first four, when herniated liver could not be returned to the abdomen without kinking the umbilical vein. Successful repairs were, however, accomplished in the next four. One died at three weeks of age from intestinal complications, one died from an unrelated accident, and there were two postneonatal survivors. Thus, successful surgical correction in utero of highly selected cases is feasible, at least in one very specialised centre.

Fetal Hydrothorax

Perinatal mortality in fetal hydrothorax exceeds 50%, and is higher when associated with hydrops than in isolation. Compression of the lung during its canalicular phase (at 17–24 weeks) produces pulmonary hypoplasia. Large effusions cause polhydramnios by impairing swallowing and cause hydrops by vena caval obstruction and cardiac compression [61,62]. Chylothorax, the commonest cause in neonates, is diagnosed after alimentation by demonstrating chylomicrons in pleural fluid. Although some claim to have made this diagnosis in the fetus by showing high mononuclear cell counts in aspirated pleural fluid, chylothorax was not confirmed postnatally, given that these effusions often resolve soon after birth. We, however, have been unable to detect any difference in lymphocyte counts or lipoprotein electrophoresis between fetuses confirmed neonatally to have chylothorax and those with hydrothoraces from other causes (unpublished observations). As hydrops often remains unexplained despite detailed investigation, congenital chylothorax has been suggested as the primary cause in many cases [63].

Ultrasound-guided aspiration of fetal hydrothoraces facilitates neonatal resuscitation if performed immediately before delivery [64]. Because the fluid reaccumulates within 6–48 hours, long-term drainage is required. This is achieved by a plastic pleuroamniotic shunt, inserted under ultrasound guidance. In our series of eight fetuses which underwent pleuroamniotic shunting, there were six survivors; polyhydramnios resolved in all six, and hydrops in three of five [65]. Two died of pulmonary hypoplasia, and in these the lungs did not re-expand and signs of hydrops did not diminish, presumably because the lungs were already hypoplastic. Thus, although pleuroamniotic shunting seems effective in reversing polydramnios and hydrops, this intervention has not been performed early enough in gestation to demonstrate any effect on lung development. A single aspiration is performed one week beforehand, since the effusion does not always reaccumulate, especially with small or unilateral effusions. In fetal hydrothoraces unassociated with other abnormalities, shunting is indicated in the presence of hydrops or polyhydramnios, or if detected in the mid-trimester. The catheters must be clamped immediately at delivery to prevent pneumothorax.

Obstructive Uropathy

Posterior Urethral Valves

In fetuses with posterior urethral valves (PUV) unassociated with other anomalies, the two main factors determining perinatal outcome are pulmonary hypoplasia and renal dysplasia [66], which seem related to the duration and severity of obstruction. Lack of a urinary contribution to amniotic fluid in the mid-trimester leads to pulmonary hypoplasia. Although its pathogenesis is not understood, numerous animal and human studies indicate that lung hypoplasia is a consequence of oligohydramnios. Urethral obstruction has also been considered responsible for renal dysplasia, presumably mediated by raised urinary pressure [67]. However, intravesical pressure seems only marginally raised in fetuses with low obstructive uropathies [68]. There is an alternative embryological theory, which suggests that renal dysplasia is not secondary to PUV, but results from the same insult – abnormally lateral orgin of the ureteric bud from the Wolffian duct [69].

As the surgical correction of PUV is relatively simple after birth, a hypothesis has emerged that bypassing the obstruction in utero would minimise renal dysplasia and restore amniotic fluid, thus preventing pulmonary hypoplasia and allowing survival at birth [67]. The basis for such intervention has been rigorously tested in animals. Urinary obstruction in fetal lambs produced both renal dysplasia and lung hypoplasia [70], whereas early decompression prevents these sequelae [71].

The presence of normal amniotic fluid volume indicates that lung hypoplasia will not occur, that the obstruction is incomplete and that the kidneys are producing adequate amounts of urine. Any benefit from bypass procedures would therefore be restricted to those fetuses with severe oligohydramnios, in which irreversible renal damage had not yet developed. Vesicoamniotic shunting of a fetus with severe renal dysplasia would not only fail to prevent neonatal death from renal failure, but would also fail to prevent pulmonary hypoplasia if the kidneys in utero were incapable of restoring amniotic fluid volume. Accordingly, accurate prediction of fetal renal function is important in selecting potential cases for intrauterine surgery. Although renal cysts are visualised in only 44% of dysplastic kidneys, their presence strongly suggests dysplasia [72]. Hyperechogenicity of the renal parenchyma predicts dysplasia with a sensitivity of 73% and a specificity of 80% [72]. In view of the inaccuracy of ultrasound in predicting renal function, biochemical analysis of urine obtained from within the fetal bladder [73] or the renal pelves [74] has become the accepted predictive parameter [75]. In 20 fetuses with PUV, Glick et al. [73] found that urinary $Na^+>100$ mEq/l, $Cl^->90$ mEq/l and osmolality>210 mosmol/l were indicative of irreversible renal damage. These cut-offs take no account of changes in urinary electrolytes with gestational age [76], and have not always been found to be accurate [77]. Using a reference range established in normal fetuses, a report on 24 fetuses with PUV [75] has shown an 80% sensitivity and specificity of a raised Na^+ for histologically confirmed renal dysplasia.

The standard method of vesicoamniotic shunting is by ultrasound-guided insertion of an indwelling plastic double pigtail catheter [66], as for pleuroamniotic shunting. As some shunts are prone to blockage or displacement [78], one group currently advocates hysterotomy and open decompression [79]. However, these technical problems are less likely with the catheters used in the UK (KCH Bladder Catheter, Rocket of London).

The results of 73 cases of shunting procedures for low obstructive uropathy reported to the International Registry were not encouraging, with only a 41% perinatal survival rate [80]. This, however, included fetuses with pathologies other than PUV, with chromosomal and other abnormalities, and with normal amniotic fluid volume, whereas 70% of the contributing centres had had experience of only one case. Most of the scepticism about this procedure can thus be attributed to poor case selection. Until the value of shunting procedures is established, selection criteria should be rigorously applied to restrict shunting to those few fetuses in whom it may be of benefit: i.e. otherwise normal male fetuses of normal karyotype with PUV, severe oligohydramnios, and biochemical evidence of adequate renal function. Accordingly only 2.5% of 200 fetuses referred with bilateral hydronephrosis in a recent series, were considered suitable for shunting [81].

References

1. Kleinman CS, Copel JA, Weinstein EM, Santulli TV, Hobbins JC. In utero diagnosis and treatment of fetal supraventricular tachycardia. Semin Perinatol 1985; 9:113–29.

2. Stewart PA, Wladimiroff JW. Cardiac tachyarrhythmia in the fetus: diagnosis, treatment and prognosis. Fetal Ther 1987; 2:7–16.
3. Maxwell DJ, Crawford DC, Curry PV, Tynan MJ, Allan LD. Obstetric importance, diagnosis and management of fetal tachycardias. Br Med J 1988; 297:107–10.
4. Weiner CP. Diagnosis and treatment of twin to twin transfusion in the mid-second trimester of pregnancy. Fetal Ther 1987; 2:71–4.
5. Gembruch U, Hansmann M, Redel DA, Bald R. Intrauterine therapy of fetal tachyarrhythmias: intraperitoneal administration of antiarrhythmic drugs to the fetus in fetal tachyarrhythmias with severe hydrops fetalis. J Perinatol Med 1988; 16:39–44.
6. Weiner CP, Thompson MI. Direct treatment of fetal supraventricular tachycardia after failed transplacental therapy. Am J Obstet Gynecol 1988; 158:570–3.
7. Saarikowski S. Placental transfer and fetal uptake of ^{3}H digoxin in humans. Br J Obstet Gynaecol 1976; 83:879–84.
8. Bowman JM, Pollack J. Amniotic fluid spectrophotometry and early delivery in the management of erythroblastosis fetalis. Pediatrics 1965; 35:815–35.
9. Bowell P, Wainscoat JS, Peto TE et al. Maternal anti-D concentrations and outcome in rhesus haemolytic disease of the newborn. Br Med J 1982; 285:327–9.
10. Nicolaides KH, Rodeck CH, Mibashan RS. Obstetric management and diagnosis of haematological disease in the fetus. In: Letsky EA, ed. Haematological disorders in pregnancy. Clinics in haematology, vol. 14. Philadelphia: WB Saunders, 1985; 775–805.
11. Nicolaides KH, Fontanarosa M, Gabbe SG, Rodeck CH. Failure of ultrasonographic parameters to predict the severity of fetal anemia in rhesus isoimmunization. Am J Obstet Gynecol 1988; 158:920–6.
12. Chitkara U, Wilkins I, Lynch L, Mehalek K, Berkowitz RL. The role of sonography in assessing severity of fetal anemia in Rh- and Kell-isoimmunized pregnancies. Obstet Gynecol 1988; 71:393–8.
13. Copel JA, Grannum PA, Belanger K, Green J, Hobbins JC. Pulsed Doppler flow-velocity waveforms before and after intrauterine intravascular transfusion for severe erythroblastosis fetalis. Am J Obstetr Gynecol 1988; 158:768–74.
14. Nicolaides KH, Bilardo CM, Campbell S. Prediction of fetal anemia by measurement of the mean blood velocity in the fetal aorta. Am J Obstet Gynecol 1990; 162:209–12.
15. Nicolaides KH, Clewell WH, Mibashan RS, Soothill PW, Rodeck CH, Campbell S. Fetal haemoglobin measurment in the assessment of red cell isoimmunisation. Lancet 1987; i:1073–5.
16. Soothill PW, Nicolaides KH, Rodeck CH. The effect of anaemia of fetal acid/base status. Br J Obstet Gynaecol 1987; 94:880–3.
17. Liley AW. Liquor amnii analysis in the management of the pregnancy complicated by rhesus sensitisation. Am J Obstet Gynecol 1961; 82:1359–70.
18. Whitfield CR. A three year assessment of an action line method of timing intervention in rhesus isoimmunisation. Am J Obstet Gynecol 1970; 108:1239–44.
19. Nicolaides KH, Rodeck CH, Mibashan RS, Kemp JR. Have Liley charts outlived their usefulness? Am J Obstet Gynecol 1986; 155:90–4.
20. Nicolini U, Kochenour NK, Greco P et al. Consequences of fetomaternal haemorrhage after intrauterine transfusion. Br Med J 1988; 297:1379–81.
21. Rodeck CH, Letsky EA. How the management of erythroblastosis fetalis has changed. Br J Obstet Gynaecol 1989; 96:759–63.
22. Berkowitz RL, Beyth Y, Sadovsky E. Death in utero due to Kell sensitisation without excessive elevation of the delta OD450 value in amniotic fluid. Obstet Gynecol 1982; 60:746–9.
23. Kanhai HH, Gravenshorts JB, Gemke RJ, Overbeeke MA, Berrini LF, Beverstock G. Fetal blood group determination in first trimester pregnancy for the management of severe immunization. Am J Obstet Gynecol 1987; 156:120–3.
24. Warren RC, Butler J, Morsman JM, McKenzie C, Rodeck CH. Does chorionic villus sampling cause feto-maternal haemorrhage? Lancet 1985; i:691.
25. Rodeck CH, Holman CA, Karnicki J, Kemp J, Whitmore DN, Austin MA. Direct intravascular fetal blood transfusion by fetoscopy in severe rhesus isoimmunisation. Lancet 1981; i:625–7.
26. Bang J, Bock JE, Trolle D. Ultrasound guided fetal intravenous transfusion for severe rhesus haemolytic disease. Br Med J 1982; 284:373–4.
27. Berkowitz RL, Chitkara U, Goldberg JD, Wilkins I, Chervenak FA. Intrauterine intravascular transfusion for severe red blood cell isoimmunization; ultrasound guided percutaneous approach. Am J Obstet Gynecol 1986; 60:746–9.
28. Nicolaides KH, Soothill PW, Rodeck CH, Clewell W. Rh disease: intravascular fetal blood transfusion by cordocentesis. Fetal Ther 1986; 1:185–92.

29. de Crespigny LC, Robinson HP. Amniocentesis: a comparison of 'monitored' versus 'blind' needle insertion technique. Aust NZ J Obstet Gynaecol 1986; 26:124–8.
30. Grannum PA, Copel J, Plaxe SC, Scioscia AL, Hobbins JC. In utero exchange transfusion by direct intravascular injection in severe erythroblastosis fetalis. N Engl J Med 1986; 314:1431–4.
31. Nicolini U, Santolaya J, Fisk NM et al. Changes in fetal acid/base status during intravascular transfusion. Arch Dis Child 1988; 63:710–14.
32. Nicolini U, Kochenour NK, Greco P, Letsky E, Rodeck CH. When to perform the next intrauterine transfusion in patients with Rh alloimmunisation: combined intravascular and intraperitoneal transfusion allows longer intervals. Fetal Ther 1989; 4:14–20.
33. Robertson EG, Brown A, Ellis MI, Walker W. Intrauterine transfusion in the management of severe rhesus isoimmunization. Br J Obstet Gynecol 1976; 83:694–7.
34. Nicolini U, Talbert DG, Fisk NM, Rodeck CH. Pathophysiology of pressure changes during intrauterine transfusion. Am J Obstet Gynecol 1989; 160:1139–45.
35. Poissonier M-H, Brossard Y, Demedeiros N et al. Two hundred intrauterine exchange transfusions in severe blood incompatibilities. Am J Obstet Gynecol 1989; 161:709–13.
36. Mennuti M, Schwarz RH, Gill F. Obstetric management of isoimmune thrombocytopenia. Am J Obstet Gynecol 1974; 118:565–6.
37. Friedman JM, Aster RH. Neonatal alloimmune thrombocytopenia purpura and congenital porencephaly in two siblings associated with a 'new' maternal antiplatelet antibody. Blood 1985; 65:1412–15.
38. Morales WJ, Stroup M. Intracranial hemorrhage in utero due to isoimmune neonatal thrombocytopenia. Obstet Gynecol 1985; 65:20S–21S.
39. de Vries L, Connell J, Bydder JM et al. Recurrent intracranial haemorrhages in utero in an infant with alloimmune thrombocytopenia. Br J Obstet Gynaecol 1988; 95:299–302.
40. Daffos F, Forestier F, Kaplan C et al. Prenatal treatment of neonatal alloimmune thrombocytopenia. Lancet 1984; ii:632.
41. Nicolini U, Rodeck CH, Kochenour NK et al. In-utero platelet transfusion for alloimmune thrombocytopenia. Lancet 1988; ii:506.
42. Nicolini U, Tannirandorn Y, Gonzalez P et al. Continuing controversy in alloimmune thrombocytopenia: fetal hyperimmunoglobulin fails to prevent thrombocytopenia. Am J Obstet Gynecol, in press.
43. Bussel JP, Berkowitz RL, McFarland JG et al. Antenatal treatment of neonatal alloimmune thrombocytopenia. N Engl J Med 1988; 319:1374–8.
44. Kaplan C, Daffos F, Forestier F et al. Management of alloimmune thrombocytopenis: antenatal diagnosis and in utero transfusion of maternal platelets. Blood 1988; 72:340–3.
45. Mir N, Samson D, House MJ, Kovar IZ. Failure of high-dose immunoglobulin to improve fetal platelet count in neonatal alloimmune thrombocytopenia. Vox Sang 1988; 55:188–9.
46. Galea P, Scott JM, Goel KM. Feto-fetal transfusion syndrome. Arch Dis Child 1982; 57:781–3.
47. Robertson EG, Neer KJ. Placental injection studies in twin gestation. Am J Obstet Gynecol 1983; 147:170–4.
48. Brennan JN, Diwan RJ, Rosen MG, Bellon EM. Fetofetal transfusion syndrome: prenatal ultrasonographic diagnosis. Radiology 1982; 143:535–6.
49. Brown DG, Benson CB, Driscoll SG, Doubilet PM. Twin–twin transfusion syndrome: sonographic findings. Radiology 1989; 170:61–3.
50. Fisk NM, Borrell A, Hubinont C, Tsannirandorn Y, Nicolini U, Rodeck CH. Fetofetal transfusion syndrome: do the neonatal criteria apply in utero? Arch Dis Child 1990; 65:657–61.
51. Wittmann BK, Farquharson DF, Thomas WDS, Baldwin VJ, Wadsworth LD. The role of feticide in the management of severe twin transfusion syndrome. Am J Obstet Gynecol 1986; 155:1023–6.
52. Weiner CP. Diagnosis and treatment of twin to twin transfusion in the mid-second trimester of pregnancy. Fetal Ther 1987; 2:71–4.
53. de Lia JE, Cruickshank DP, Keye WR. Fetoscopic neodymium: YAG laser occlusion of placental vessels in severe twin–twin transfusion syndrome. Obstet Gynecol 1990; 76:1046–53.
54. Lange IR, Harman CR, Ash KM, Manning FA, Menticoglou S. Twins with hydramnios: treating premature labour at source. Am J Obstet Gynecol 1989; 160:552–7.
55. Vetter K, Schneider KT. Iatrogenous remission of twin transfusion syndrome. Am J Obstet Gynecol 1988; 158:221.
56. Adzick NS, Harrison MR, Glick PL, Nakayama DK, Manning FA. Diaphragmatic hernia in the fetus: prenatal diagnosis and outcome in 94 cases. J Pediatr Surg 1985; 20:357–61.
57. Harrison MR, Langer JC, Adzick NS et al. Correction of congenital diaphragmatic hernia in utero. V: Initial clinical experience. J Pediatr Surg 1990; 25:47–57.

58. Harrison MR, Bressack MA, Churg AM, de Lorimer AA. Correction of congenital diaphragmatic hernia in utero. II: Simulated correction permits fetal lung growth with survival at birth. Surgery 1980; 88:260–8.
59. Adzick NS, Harrison MR, Glick PL et al. Fetal surgery in the primate. III: Maternal outcome after fetal surgery. J Pediatr Surg 1986; 21:477–80.
60. Harrison MR, Adzick NS, Longaker MT et al. Successful repair in utero of a fetal diaphragmatic hernia after removal of herniated viscera from the left thorax. N Engl J Med 1990; 322:1582–4.
61. Petres RE, Redwine FO, Cruickshank DP. Congenital bilateral hydrothorax: antepartum diagnosis and successful intrauterine surgical management. JAMA 1982; 248:1360–1.
62. Benacerraf BR, Frigoletto FD. Mid-trimester fetal thoracentesis. JCU 1985; 13:202–4.
63. Roberts AB, Clarkson NS, Pattison MG, Mok PM. Fetal hydrothorax in the second trimester of pregnancy: successful intrauterine treatment at 24 weeks gestation. Fetal Ther 1986; 1:203–9.
64. Schmidt W, Harms E, Wolf D. Successful prenatal treatment of non-immune hydrops fetalis due to congenital chylothorax. Br J Obstet Gynaecol 1985; 92:671–9.
65. Rodeck CH, Fisk NM, Fraser DI, Nicolini U. Long-term in utero drainage of fetal hydrothorax. N Engl J Med 1988; 319:1135–8.
66. Harrison MR, Golbus MS, Filly RA et al. Management of the fetus with congenital hydronephrosis. J Pediatr Surg 1982; 17:728–42.
67. Harrison MR, Filly RA, Parer JT, Faer MJ, Jacobson JB, de Lorimer AA. Management of the fetus with a urinary tract malformation. JAMA 1981; 246:635–9.
68. Nicolini U, Tannirandorn Y, Vaughan J, Fisk NM, Nicolaidis P, Rodeck CH. One day interval urine sampling and bladder pressure measurement for the prediction of renal dysplasia in fetuses with lower urinary tract obstruction. Prenat Diagn, in press.
69. Stephens FD. Congenital malformations of the urinary tract. New York: Praeger, 1983; 433–62.
70. Glick PL, Harrison MR, Noall RA, Villa RL. Correction of congenital hydronephrosis in utero. II: early mid-trimester ureteral obstruction produces renal dysplasia. J Pediatr Surg 1983; 18:681–7.
71. Glick PL, Harrison MR, Adzick NS, Noall RA, Villa RL. Correction of congenital hydronephrosis in utero. IV: In utero decompression prevents renal dysplasia. J Pediatr Surg 1983; 19:649–57.
72. Mahoney BA, Filly RA, Callen PW, Hricak H, Golbus MA, Harrison MR. Fetal renal dysplasia: sonographic evaluation. Radiology 1984; 152:143–6.
73. Glick PL, Harrison MR, Golbus MS et al. Management of the fetus with congenital hydronephrosis. II: Prognostic criteria and selection for treatment. J Pediatr Surg 1985; 20:376–87.
74. Nicolini U, Rodeck CH, Fisk NM. Shunt treatment for fetal obstructive uropathy. Lancet 1987; ii:1338–9.
75. Nicolini U, Fisk NM, Beacham J, Rodeck CH. Fetal urine biochemistry: an index of renal maturation and dysfunction. Br J Obstet Gynaecol, in press.
76. Rodeck CH, Nicolini U. Physiology of the mid-trimester fetus. In: Whitelaw A, Cooke RWI, eds. The very immature infant less than 28 weeks' gestation. Edinburgh: Churchill Livingstone, Br Med Bull 1988; 44:826–49.
77. Wilkins IA, Chitkara U, Lynch L, Goldberg JD, Mehalek KE, Berkowitz RL. The nonpredictive value of fetal urinary electrolytes: preliminary report of outcomes and correlations with pathological diagnoses. Am J Obstet Gynecol 1987; 157:694–8.
78. Golbus MS, Harrison MR, Filly RA, Callen PW, Katz M. In utero treatment of urinary tract obstruction. Am J Obstet Gynecol 1982; 142:383–8.
79. Harrison MR, Golbus MS, Filly RA et al. Correction of congenital diaphragmatic hernia in utero. V: Initial clinical experience. J Pediatr Surg 1987; 25:47–57.
80. Manning FA, Harrison MR, Rodeck CH and members of the International Fetal Medicine and Surgery Society. Catheter shunts for fetal hydronephrosis and hydrocephalus. N Engl J Med 1986; 315:336–40.
81. Longaker MT, Adzick NS, Harrison MR. Fetal obstructive uropathy. Br Med J 1989; 299:325–6.

Chapter 17

Magnetic Resonance Imaging (MRI) Scanning

I. R. Johnson

The principle of magnetic resonance imaging (MRI) was demonstrated by Block [1] and Purcell [2] in the 1940s. It was not until the early 1970s [3,4] that imaging was attempted, but since that time tremendous advances have been made in evaluating the usefulness of magnetic resonance imaging in clinical and research practice. MRI does not use ionising radiation and, so far, appears harmless. Spatial resolution is excellent, comparable with that of computed tomography (CT). MRI images of tissues can be varied in their nature by altering the weight attached to the two relaxation times, T1 and T2. This enables the research worker or clinician to alter the effects on the image of characteristics such as fat content or hydration.

Although MRI is now being used extensively by neurologists and orthopaedic surgeons in clinical practice it has not yet found a regular place in obstetrics. The procedure itself does not usually trouble the patient greatly. Suitable positioning of the patient avoids the problem of vena caval compression and only a few patients complain of claustrophobia. Fetal safety is, of course paramount. MRI has never been demonstrated to cause any harm to the fetus, or indeed any adult patient. It is, however, difficult to be sure that MRI will not cause some subtle abnormality in the fetus. MRI systems with magnetic field strengths of less than 1.5 T have been shown to have no effect on the development of mouse embryos [5,6] or on embryonic development in amphibians at 4 T [7]. Long-term follow-up of patients submitted to MRI in pregnancy is clearly required, but for the moment the National Radiological Protection Board has published guidelines limiting imaging of the fetus in utero to the second and third trimesters with field strengths of less than 2.5 T [8]. These guidelines have been accepted by local ethical committees allowing research and investigations to take place.

Imaging the Fetus

Early Work

Fetal imaging was first performed in 1983 [9,10]. Patients were imaged in the first and second trimesters of pregnancy by two groups using resistive magnets at 0.04 T and 0.15 T. These patients were about to undergo termination of pregnancy. In all cases the placenta was seen in great detail and easily localised, but because of the relatively long scanning times of 2 minutes the fetus was not seen clearly. Attempts were made to measure biparietal diameters, crown–rump length and fetal circumference. Because of difficulties with motional artefact and plane acquisition true measurements could not be made.

After the publication of the National Radiological Protection Board's revised guidelines on acceptable limits of exposure during nuclear magnetic clinical imaging [8] ethical committees began to give permission for imaging of patients to take place in the second and third trimesters of pregnancies when the pregnancy was to continue to term. Johnson et al. [11] reported on images obtained from 15 patients between 30 and 40 weeks' gestation using a 0.15 T resistive magnet. Images took 2 minutes to produce with a section thickness of 1 cm. All the patients went on to term and delivered normal infants. With increasing gestation the fetus became larger and more detail became apparent because of reduction in mobility, but even so most of the internal structures of the fetus were seen very poorly. Further studies in Aberdeen [12] were performed to compare biparietal diameter measurements between ultrasound and magnetic resonance imaging. These appeared to show a good correlation, although because of the restriction of plane selection to sagittal, coronal, and transverse maternal planes, measurement of the actual biparietal diameter could not be guaranteed. Crown–rump length was measured in 9 out of 15 cases with a passable correlation between the two techniques.

More powerful superconducting magnets were used in American studies [13] to investigate the internal structure of fetal organs. Claims were made suggesting that fetal movements did not significantly degrade images. Other groups [14] were less optimistic, concluding that fetal motion would result in blurring of fine detail at the relatively long imaging times employed, characterstically between 4 and 17 minutes. In fetuses near term, or in the presence of oligohydramnios, magnetic resonance was of much more use as fetal movement was decreased. The fetal heart could be resolved into left and right sides and in some patients the aorta and vena cava could be identified. Fetal lungs were usually seen clearly as was the liver, but images of the genital tract were disappointing [15]. The fetal brain was relatively uninteresting because of the lack of myelination until late in pregnancy, although myelination in the basal ganglia could be seen after 34 weeks [16] and in the thalamus and posterior internal capsule areas of the fetal brain after 36 weeks [17]. Fig. 17.1 shows the head of a 38-week fetus engaged in the maternal pelvis. In this T1 weighted image the lighter areas in the central part of the fetal brain represent myelination [18].

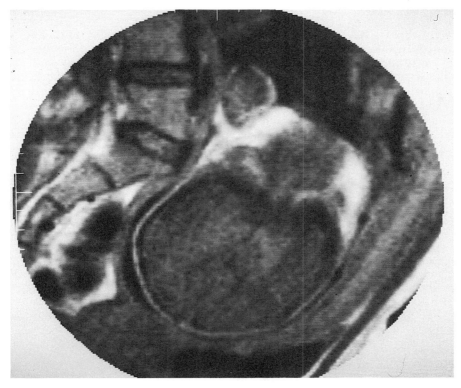

Fig. 17.1. A T1 weighted image of a 38 week fetus showing the fetal head engaged within the maternal pelvis. The lighter areas within the fetal brain represent areas of myelination. (From Powell [18].)

Fetal Immobilisation

It was clear that the major drawback to accurate imaging of fetal organs was the motional artefact caused by random fetal movements at all stages of pregnancy. Although as term approached fetal movements became less, particularly in those few patients with oligohydramnios, image degradation was almost always so great as to make accurate delineation of fetal organs impossible. Attempts were made to sedate the fetus through the mother [19] by giving mothers 5–10 mg of diazepam intravenously. These attempts were partially successful, as with sedation fetal structures were frequently identified in the third trimester, although fetal motion still degraded images in most cases. Although hearts and livers were seen, kidneys were not seen in the majority of cases.

Complete immobilisation of the fetus was achieved using neuromuscular blocking agents injected into an umbilical vein under ultrasound guidance [20]. Fetal movements disappeared within a few seconds, returning approximately 2 hours later with no apparent compromise to the fetus. This technique considerably improved differentiation between fetal tissues and some abnormalities of the fetal brain could be distinguished from one another.

Fetal Abnormality

Several reports have been made of cases of fetal abnormality evaluated by magnetic resonance imaging. In almost all these cases the abnormality had previously been diagnosed, at least partially, by ultrasound. McCarthy et al. [21] demonstrated fetuses late in pregnancy with ventricular asymmetry and hydrocephalus. In another case with urethral atresia severe oligohydramnios and pulmonary hypoplasia were demonstrated but the kidneys could not be seen. Using fetal sedation with diazepam (given to the mother) conditions such as the Dandy–Walker syndrome and meningomyelocele were reported [22] although fetal movement, and the fact that imaging planes were maternally orientated, caused considerable difficulties in establishing the diagnosis. Neuromuscular blockade using pancuronium bromide to immobilise the fetus via the umbilical vessels allowed diagnosis of other abnormalities [23,24]. Even using these invasive techniques MRI rarely added very much to the findings already discovered with ultrasound.

The conclusion reached was that MRI gave excellent soft tissue definition, but this was reduced by fetal movement and movement of the maternal structures. Unlike ultrasound, magnetic resonance was not hindered by oligohydramnios, and indeed because of the reduced fetal movement the images were superior. MRI was also superior to ultrasound when imaging a full-term abdominal pregnancy [25] when the lack of fluid around the baby made visualisation by ultrasound difficult, and in two cases of conjoined twins [26] where the relationship between the fetuses was demonstrated more clearly than with ultrasound.

Intrauterine Growth Retardation

Many workers have speculated about the possibility of measuring fetal fat content in order to estimate degrees of intrauterine growth retardation (IUGR). Stark et al. [27] estimated fetal fat volumes in a series of fetuses and correlated these with clinical outcome. The magnetic resonance tissue characteristics of fat were shown to be quite specific and selection of appropriate imaging techniques allowed clear delineation of fetal fat from other tissues and from the amniotic fluid. Volumetric analysis in slices allowed the fat content of the fetus to be expressed as a percentage of total body volume. This was spoilt only by the ever-present problem of fetal motion. Qualitative differences in fat content have also been noted in cases of twins subsequently shown to have significantly different birth weights [28].

Limitations of Fetal Imaging

The two major problems in all cases reported using conventional MRI are motional artefact due to the length of acquisition time and difficulty in selecting specific planes.

As the fetus approaches term it becomes less mobile and easier to image but this severely limits the usefulness of the technique in anything but the late third trimester. In severe oligohydramnios the fetus moves very little, but this is not a common condition. Although the fetus can be immobilised by neuromuscular blockade this is an invasive technique which would be appropriate only when the fetus was known to have a specific problem for investigation. Motional artefact can be overcome only by using faster techniques which are not generally available to conventional MRI. The low flip angle technique, in which a radio frequency is used which does not deflect the proton precession as much as more conventional techniques, has been experimented with, but as yet clear fetal images have not been published.

Plane selection is progressively becoming less of a problem. Initially it was possible to image the fetus only in the sagittal, coronal and transverse planes in relation to the mother. With improving computer software it is becoming increasingly possible to select any plane. Difficulty is still experienced because of long image acquisition time so that there is no guarantee that the fetus will remain in the same plane between two image acquisitions. This clearly further complicates plane selection.

Placental Localisation

Imaging the placenta is much more straightforward than imaging the fetus. MRI has been found to be the equal of ultrasound in the localisation of the placental site and superior to ultrasound in definition of the degree of placenta praevia [29]. More importantly in a large proportion of the patients the management was altered because of the MRI result. The technique is particularly useful in assessing the relationship between the lower edge of the placenta and the cervical os especially when the placenta is attached to the posterior surface of the uterus. Fig. 17.2 shows a fetus presenting by the breach at term. The fetal heart can be seen as a dark area in the chest, with the fetal liver immediately below it. Further into the abdomen loops of bowel can be seen. The legs are extended and the fetus is facing posteriorly. The placenta is seen as a grey area extending down the posterior wall of the uterus against the maternal spine. The vagina and cervix are immediately below the fetal buttocks. It can be seen that although the placenta dips into the pelvis the breech has reached a position below its lowest edge.

Although it should be possible to demonstrate functioning and non-functioning areas within placentas and particularly to demonstrate retroplacental clots, no reports of these have yet appeared.

Maternal Anatomy in Pregnancy

Powell et al. [30] studied 154 women in pregnancy and were able to demonstrate the usefulness of MRI for the assessment of pelvic size. Cervical architecture was studied in considerable detail and it was suggested that MRI might offer some insight into cervical dystocia and incompetence. Hydatidiform moles have been examined [31]. Although ultrasound is superior in reaching a diagnosis, MRI

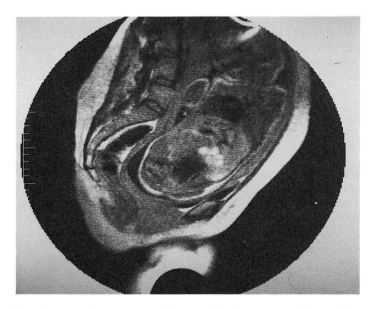

Fig. 17.2. A breech presentation near term with a posterior placenta extending into the pelvis but not amounting to a significant degree of placenta praevia. (From Powell [18].)

tissue differentiation is potentially of considerable importance in demonstrating myometrial invasion. Pelvic masses present coincidentally during pregnancy have been identified [30,32] as have Mullerian abnormalities [33,34].

Echo Planar Imaging

Echo planar imaging (EPI) is a type of magnetic resonance imaging in which a complete image is obtained in 64 or 128 ms [35,36]. Using a 0.5 T superconductive magnet two-dimensional images are obtained as a snapshot. The resolution varies between 1.5 mm and 3 mm. Using this technique fetal motional image degradation is reduced to a minimum. Currently images have been produced only in sagittal and transverse planes, but further developments will allow acquisition in any plane. Because of the speed of image acquisition plane selection will not be subject to the same problems as in conventional MRI with fetal movement between each snapshot. Using this technique pictures have been acquired of fetuses at all stages of the second and third trimesters showing considerable fetal detail [37].

Eight examples of the images obtained using EPI are shown in Fig. 17.3 In each image the mother is lying on her back, the white spot in the middle of the lower part of the picture representing cerebrospinal fluid in the maternal spinal canal. The uterus lies anteriorly with the placenta appearing as a mid-grey region in the anterior portion of the uterus. The amniotic fluid carries a high signal in this

Magnetic Resonance Imaging (MRI) Scanning

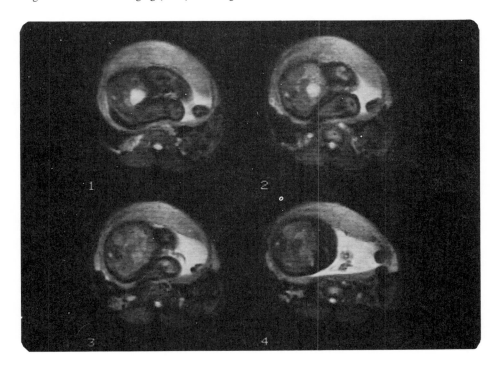

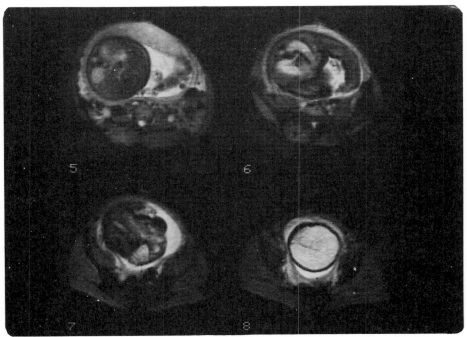

Fig. 17.3. Eight transverse slices through a 36-week fetus obtained using echo planar imaging.

particular spin sequence and is starkly white, outlining the fetus, placenta and uterine borders. The sequence begins with image 1 which is a transverse section through the pelvis in this 36-week fetus. The high signal area in the pelvis represents fluid in the fetal bladder. The legs are seen extending from the trunk and the low signal area of bone can be made out. In image 2, the transverse slice is at a slightly higher level through the pelvis still demonstrating the bladder but by image 3, the bladder has disappeared and loops of bowel can be seen within the fetal abdomen. In image 4, the fetal trunk is outlined more clearly within the amniotic fluid as is a transection of the umbilical cord showing vessels within it. In image 5, parts of the umbilical cord can be seen more clearly. The fetal abdomen is shown with the kidneys on either side of the fetal spinal canal. In front of the high signal fetal cerebrospinal fluid the vertebral body (very low signal) and the aorta and vena cava can be made out. In image 6, the transverse slice is taken through the chest. The arms are seen extending from the trunk, and in the chest itself, the fluid-containing fetal lungs are seen to either side of the darker fetal heart. Characteristically the lungs are white and fan shaped. Image 7 is more difficult to interpret as it includes multiple structures in the fetal neck and jaw. In image 8, the fetal head is shown within the maternal pelvis, the brain being homogeneous with little detail within it because of the lack of myelination.

By reducing the image degradation caused by fetal motion whilst allowing a full range of spin sequences to be used, echo planar imaging has considerably increased the potential of MRI to study fetal physiology and pathology. Experiments are under way to study growth of fetal organs during pregnancy as well as more accurate evaluation of congenital abnormalities.

Conclusions

During the last 15 years ultrasound imaging of the fetus has been demonstrated to be convenient, relatively inexpensive and effective in altering clinical practice. A large body of experience has been built up in using and interpreting ultrasound images. Despite the advent of colour Doppler enabling us to look at some aspects of blood flow there are still considerable gaps in our knowledge in the physiology and the pathology of the fetus during pregnancy. What has to be considered is what can magnetic resonance imaging do that is better than or different from ultrasound. MRI is significantly better than ultrasound in tissue differentiation and in gaining some measure of functional activity.

When viewing fetuses with congenital abnormalities MRI has the ability to view the fetus slice by slice and is of particular value when multiple abnormalities are present or have to be excluded. It is also of value in the precise definition of the extent of an abnormality, for instance, when the extent of movement of abdominal contents into the chest needs to be estimated in cases of diaphragmatic hernia. MRI techniques have considerable potential in improving our understanding of intrauterine growth retardation. Fetal fat volume estimations have already been attempted with conventional MRI and are in the process of being evaluated with EPI. Changes in the volume of the fetal liver are similarly being estimated.

With increasing size of magnets, spectroscopy becomes possible. All conventional MRI and EPI performed so far has depended on the magnetic moment of protons. With increasing magnetic fields it is possible to look at magnetic moments of the nuclei of other elements such as phosphorus. This has already been used to evaluate high phosphate compounds and inorganic phosphate in neonatal brains and in a relatively short while it will also be possible to do this in the fetus. Experiments are also under way to evaluate changes in fetal brain lipids in late pregnancy and to evaluate changes in the glycogen content of the liver.

Rapid acquisition of images using EPI particularly lends itself to evaluating blood flow in the placenta. In addition to this, spectroscopy of the placenta, looking for high energy compounds and evidence of ischaemic damage, will be available within a short while.

The use of magnetic resonance imaging in obstetrics has lagged behind its use in other specialties because of the peculiar problems of fetal motion. The advent of the echo planar technique has removed that problem. As the software programs improve enabling acquisition of pictures in any plane with EPI, and as larger magnetic fields allow spectroscopy to be performed more readily, our understanding of patho-physiology of the fetus in pregnancy will improve. With echoplanar imaging obstetric MRI has come of age.

References

1. Block F, Hansen WW, Packard M. Nuclear induction. Phys Rev 1946; 69:127.
2. Purcell EM, Torrey HC, Pound RV. Resonance absorption by nuclear magnetic movements in a solid. Phys Rev 1946; 69:37.
3. Lauterbur PC. Image formation by induced local interactions: examples of employing nuclear magnetic resonance. Nature 1973; 242:190–1.
4. Garroway AN, Grannell PK, Mansfield P. Image formation in NMR by a selective irradiation process. J Phys Chem 1974; 7:757.
5. Heinrichs LW, Fong P, Flannery M et al. Midgestational exposure of pregnant BALB/c mice to magnetic resonance imaging conditions. Magn Reson Imaging 1988; 6:305–13.
6. McRobbie D, Foster MA. Pulsed magnetic field exposure during pregnancy and implications for NMR fetal imaging: a study with mice. Magn Reson Imaging 1985; 3:231–4.
7. Prasad N, Wright DA, Ford JJ et al. Safety of 4-T MR imaging: Study of effects on developing frog embryos. Radiology 1990: 174:251–3.
8. The National Radiological Protection Board ad hoc advisory group on nuclear magnetic resonance clinical imaging. Revised guidance on acceptable limits of exposure during nuclear magnetic resonance clinical imaging. Br J Radiol 1983; 56:974.
9. Smith FW, Adam AH, Philips WDP. NMR imaging in pregnancy. Lancet 1983; i:61–2.
10. Worthington BS, Kean DM, Johnson IR et al. Nuclear magnetic resonance imaging of the pregnant uterus. Radiology 1983; 149:145.
11. Johnson IR, Symonds EM, Kean DM et al. Imaging the pregnant human uterus with nuclear magnetic resonance. Am J Obstet Gynecol 1984; 148:1136–9.
12. Smith FW, MacLennan F, Abramovich DR et al. NMR imaging in human pregnancy: a preliminary study. Magn Reson Imaging 1984; 2:57–64.
13. McCarthy S, Stark DD, Higgins CB. Demonstration of the fetal cardiovascular system by MR imaging. J Comput Assist Tomogr 1984; 8:1168–9.
14. Weinreb JC, Lowe T, Cohen JM et al. Human fetal anatomy: MR imaging. Radiology 1985; 157:715–20.
15. McCarthy SM, Filly RA, Stark DD et al. Obstetrical magnetic resonance imaging: fetal anatomy. Radiology 1985; 154:427–32.
16. Smith FW, Kent C, Abramovich DR et al. Nuclear magnetic resonance imaging – a new look at the fetus. Br J Obstet Gynaecol 1985; 92:1024–33.
17. Powell MC, Worthington BS, Buckley JM et al. Magnetic resonance imaging (MRI) in obstetrics. II Fetal anatomy. Br J Obstet Gynaecol 1988; 95:38–46.

18. Powell MC. A clinical evaluation of magnetic resonance imaging (MRI) in obstetrics and gynaecology. Nottingham: DM thesis, 1990.
19. Weinreb JC, Lowe TW, Santos-Ramos R et al. Magnetic resonance imaging in obstetric diagnosis. Radiology 1985; 154:157–61.
20. Daffos F, Forestier F, Macaleese J et al. Fetal curarisation for prenatal magnetic resonance imaging. Prenat Diagn 1988; 8:311–14.
21. McCarthy SM, Filly RA, Stark DD et al. Magnetic resonance imaging of fetal anomalies in utero: early experience. AJR 1985; 145:677–82.
22. Lowe TW, Weinreb JC, Santos-Ramos R et al. Magnetic resonance imaging in human pregnancy. Obstet Gynecol 1985; 626:629–33.
23. Persutte WH, Lenke RR, Kurczynski TW et al. Antenatal diagnosis of Pena–Shokeir syndrome (1) with ultrasonography and magnetic resonance imaging. Obstet Gynecol 1988; 72:472–5.
24. Williamson RA, Weiner CP, Yuh WTC, Abu-Yousef MM. Magnetic resonance imaging of anomalous fetuses. Obstet Gynecol 1989; 73:952–6.
25. Cohen JM, Weinreb JC, Lowe TW et al. MR imaging of a viable full term abdominal pregnancy. AJR 1985; 145:407–8.
26. Turner RJ, Hankins GDV, Weinreb JC et al. Magnetic resonance imaging and ultrasonography in the antenatal evaluation of conjoined twins. Am J Obstet Gynecol 1986; 155:645–9.
27. Stark DD, McCarthy SM, Filly RA et al. Intrauterine growth retardation: evaluation by magnetic resonance. Radiology 1985; 155:425–7.
28. Brown CEL, Weinreb JC. Magnetic resonance imaging appearance of a growth retardation in a twin pregnancy. Obstet Gynecol 1988; 71:987–8.
29. Powell MC, Buckley J, Price H et al. Magnetic resonance imaging and placenta praevia. Am J Obstet Gynecol 1986; 154:565–9.
30. Powell MC, Worthington BS, Buckley JM et al. Magnetic resonance imaging (MRI) in obstetrics. I. Maternal anatomy. Br J Obstet Gynaecol 1988: 95:31–7.
31. Powell MC, Buckley JM, Worthington BS et al. Magnetic resonance imaging and hydatidiform mole. Br J Radiol 1986; 59:561–4.
32. Lubbers PR, Izuno C, Goff WUB et al. Magnetic resonance imaging of a pelvic mass complicating pregnancy. J Am Osteopath Assoc 1988; 88:1010–14.
33. Yuh WTC, DeMarino GB, Ludwig WD et al. MR imaging of pregnancy in bicornuate uterus. J Comput Assist Tomogr 1988; 12:162–5.
34. Kelley JL, Edwards RP, Wozney P et al. Magnetic resonance imaging to diagnose a Mullerian anomaly during pregnancy. Obstet Gynecol 1990; 75:521–3.
35. Mansfield P. Multi-planar image formation using NMR spin echoes. J Phys C 1977; 19:L55–L58.
36. Mansfield P, Morris PG. NMR imaging in biomedicine. New York: Academic Press, 1982.
37. Johnson IR, Stehling MK, Blamire AM et al. Study of internal structure in the human fetus in utero by echo planar magnetic resonance imaging. Am J Obstet Gynecol 1990; 163:601–7.

Discussion

Nevin: Professor Rodeck mentioned that one of the primary purposes of prenatal diagnosis was to influence management of the patient and the fetus. In relation to NTDs would he comment on the possibility of a relationship between the ultrasonography behaviour of the fetus and the severity of the NTD?

Second, is there any evidence, and I know there are anecdotal accounts, that caesarean section as opposed to normal vaginal delivery will leave a fetus with less neurological defect if it has a neural tube defect?

Rodeck: I am not the best person here to comment on that. My own feeling is that there is no good evidence that one can assess the severity of a neural tube defect by looking at fetal limb movements, for example.

Again, regarding the question of route of delivery there is a lot of anecdotal experience around that babies with spina bifida that are delivered vaginally do as well as those that are delivered by caesarean.

Campbell: We certainly have not made any progress in assessing the severity of the neurological deficit that the child will have. Clearly a large lesion will give a more severe neurological handicap than a small one. A sacral lesion is better from the prognostic point of view. Finding things like talipes clearly indicates a neurological lesion.

The size of the ventricles gives no idea of the neurological deficit because some of the smallest lesions have the biggest ventriculomegaly.

Allan: Have loss rates from fetal blood sampling been analysed according to either gestation or age, or the position of the placenta, or the difficulty of the procedure?

Nicolaides: Yes. It is much higher in early gestation, so much so that I now delay blood sampling until 19 or 20 weeks. It is also higher with posterior than with anterior placentas.

Allan: These are all very important points to put into the equation when one is faced with an individual patient, as to when or whether to offer fetal blood sampling.

Whittle: There is another point which this sort of group should be giving guidance on. Mr Nicolaides said that there were about ten centres in the UK doing cordocentesis. Is that about the right number?

Nicolaides: I think that will be resolved in the next ten years through subspecialisation. With sub-specialisation we shall create the centres and will be able to respond to new realities. For example, it would have been impossible in 1980 to predict what the main indications for fetal blood sampling would be and there has been a dramatic change. In 1985 I was advocating fetal blood sampling as a routine, I now believe that that is impossible. I believe we shall have many centres, perhaps 50, and that that will come with the creation of specialist centres and subspecialisation.

Campbell: Professor Rodeck discussed the various therapeutic procedures and I do not think we have made much progress in the last seven years. Apart from transfusion and the treatment of arrhythmias, it is still totally uncertain whether or not shunting is effective.

He ended in the area which may perhaps be exciting, marrow cell transplantation. Could he give us some idea of where we are in this field?

Rodeck: We have worked on this for some years and Mr Nicolaides has been involved with others in a monkey model utilising the father as the donor of the erythropoietic stem cells and in some cases T-cell depleting the marrow and in others not. Basically the results were very disappointing. The only evidence of the graft taking in some of those fetuses was the finding that a number of them got hydrops and died in utero.

Mike Harrison's group in San Francisco have used fetal liver stem cells and have achieved survival of erythropoietic stem cells, up to four months, which is very encouraging.

Campbell: Has transplantation been performed too late in gestation?

Rodeck: I do not think so. In our work, and also in Harrison's, intervention was at about 60 days of gestation when term is 160, and so it is probably equivalent to 14 weeks in the human pregnancy. All the evidence available suggests that human fetal immunocompetence does not develop until later.

Allan: Can I mention balloon valvuloplasty in the fetus, which was not mentioned at all? We have only done it on two fetuses and it is technically feasible. It is not difficult and it would be logical to think about it in those fetuses where otherwise the cardiac muscle will be irreparably damaged by the obstruction.

Lilford: Conditions that are amenable to surgery seem to be desperately rare.

Allan: Aortic stenosis is not a rare cardiac abnormality but in the majority of cases when it is seen in early pregnancy the prognosis is so poor the mother will not choose to continue with the pregnancy.

Section VI
Counselling, Economics and Ethical Issues

Chapter 18

Psychological Implications of Prenatal Diagnosis

T. M. Marteau

Alongside the benefits from diagnosing severe abnormalities in the fetus are some potential psychological costs. These include anxiety, loss of confidence about the pregnancy and negative attitudes towards the baby. Some of these are inevitable consequences of the process of diagnosing fetal abnormality; others could be avoided or reduced with changes in the way clinical services are organised and delivered. Careful consideration of the psychological implications of prenatal diagnosis is necessary to establish that the benefits of this new technology outweigh the costs. There are separate implications for all prospective parents, health professionals and society in general.

All Prospective Parents

The availability of prenatal screening and diagnostic testing has changed the experience of pregnancy [1,2]. Before the development of prenatal testing for fetal abnormality, the fetus was assumed to be healthy, unless there was evidence to the contrary. The presence of prenatal testing and monitoring shifts the balance towards having to prove the health or normality of a fetus. Prenatal testing provides a much wanted choice for some prospective parents, particularly those at known risk of having a child with a severe abnormality. Such parents will only embark on pregnancies knowing that testing and termination of affected fetuses are available. Yet for other parents this is a choice they would prefer not to confront.

Uptake of Prenatal Testing

Whether or not women undergo prenatal testing in pregnancy depends on three main factors: whether the test is available; the knowledge and attitudes of the health professionals they consult; and the knowledge and attitudes of the women themselves.

There is relatively little routine information on the availability of prenatal screening and diagnostic tests. Ultrasound scanning for dating and routine measurements is almost universally used in UK hospitals at the present time. Maternal serum alphafetoprotein (MS-AFP) screening is available in the majority of the 15 Health Regions in England and Wales, although it is not always offered in every hospital in a Region [3].

Amniocentesis is readily available in all Health Districts in England and Wales, although it may not be routinely performed in all hospitals. Currently, chorionic villus sampling is neither routinely available nor routinely offered for prenatal diagnosis.

Several studies have documented the importance of doctors' knowledge and attitudes in influencing the use of prenatal testing. Lippman-Hand and Piper [4] reported that professional hesitation and under-referral are a more important cause of the limited use of prenatal diagnosis than women's knowledge and attitudes. Younger physicians are more in favour of both prenatal testing [5] and termination [6]. Such age-related differences may stem from differences in knowledge about or attitudes towards the process and techniques of prenatal diagnosis.

Health professionals' knowledge and attitudes will in turn influence women's knowledge and attitudes, and hence their decisions about whether to undergo testing. Surveys of women's attitudes towards MS-AFP screening show that over 95% are in favour of such screening [7,8], although fewer actually use it.

Where MS-AFP testing is available, the main factors influencing whether or not a woman undergoes the test include:

1. Knowledge of the test [9]
2. Attitude to termination [8-10] (Fig. 18.1)
3. Perceived reliability of the test [11]

For amniocentesis, the factors influencing uptake are slightly different and include:

1. Perceived risk of having an affected child [12] (Fig. 18.2)
2. Attitude to termination [9,13] (Fig. 18.1)
3. Fear of miscarriage [9]

Women undergoing routine prenatal screening are generally under-informed about the tests they are being offered and may subsequently undergo. For example, 39% of women who had recently undergone MS-AFP screening for open neural tube defects were unaware that they had even had the test [14]. For parents to be active participants in the decision about what prenatal testing, if any, to undergo, they need to be more informed about the tests available to them.

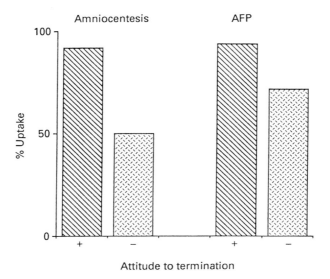

Fig. 18.1. Relationship between uptake of amniocentesis and AFP screening, and attitudes towards termination of pregnancy (+ positive attitude; − negative attitude).

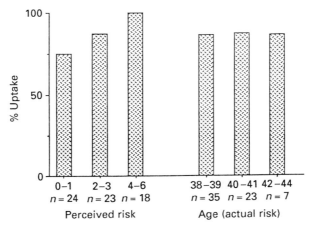

Fig. 18.2. Relationship between uptake of amniocentesis, perceived risk of having an affected child and actual (age-related) risk. Perceived risk: 0 = not at all likely; 7 = extremely likely. (From Marteau et al. [12] with permission.)

Recommendations to Increase Informed Uptake of Prenatal Testing

Health professionals should provide those who are eligible with detailed information, both oral and written, explaining:

1. Purpose of testing; i.e. to promote informed reproductive decisionmaking by offering the opportunity for diagnosing disorders in the fetus [15]

2. Conditon(s) screened for
3. Likelihood that an abnormality will be detected
4. The test procedure, including any risks
5. Meaning of test results, both negative and positive
6. Possible actions following a positive result, including: termination, planning for the birth of a handicapped child, or adoption.

Impact of Prenatal Testing

Women's experiences of undergoing prenatal testing will be influenced by: their reasons for undergoing the tests; their understanding of the test results; and the care they receive from the health professionals providing the services.

The majority of women undergoing prenatal testing will receive a negative result. It would appear, however, that this confers little if any significant benefit to women in terms of reassurance [16]. This is the case for both ultrasound and MS-AFP testing. It is possible that if women were better informed about such tests, receipt of a negative result could be a more reassuring event. In the short term, undergoing amniocentesis is associated with marked levels of anxiety, both about the procedure and about the possibility of an adverse result. It is also associated with more negative attitudes towards the baby and uncertainty about the baby's health. After the receipt of a negative result, anxiety levels drop to pre-test levels, as do negative attitudes towards the baby and certainty about the baby's health [17–19] (unpublished observations) (Fig. 18.3)

Amniocentesis is reassuring for women who choose to undergo it, although this finding cannot be generalised beyond this self-selected sample.

Overall, there would appear to be little if any difference between the experience of undergoing chorionic villus sampling and that of undergoing amniocentesis in terms of attachment to the baby and maternal anxiety [20,21].

Inherent in all screening tests is the possibility of a false positive result. The routine use of ultrasound may result in the detection of symptomless minor anomalies, the incidence and natural history of which are unknown. Although they are not indications for a termination, their detection means that women face the rest of their pregnancies with the knowledge that their child has an abnormality, the implications of which are unknown. Although single case studies must be treated with caution, there is a published case of a woman who rejected her baby at birth after prenatal ultrasound diagnosis of gastroschisis [22].

Receipt of a positive result on MS-AFP screening causes much distress both for the woman and her partner [23–26]. For most of these women, subsequent testing will reveal no abnormality. Yet, for some women, despite no abnormality being revealed by further testing, their anxiety levels will remain raised [27] (Fig. 18.4).

These results are consistent with those from other screening programmes [28–31]: negative results following earlier positive results do not necessarily provide complete reassurance that all is well.

Routine screening tests do not detect all cases. MS-AFP detects about 80% of cases of spina bifida. Although smaller, there is a fase negative rate from both CVS and amniocentesis. Some women will therefore give birth to an affected child although no abnormality was detected on prenatal testing. There have been

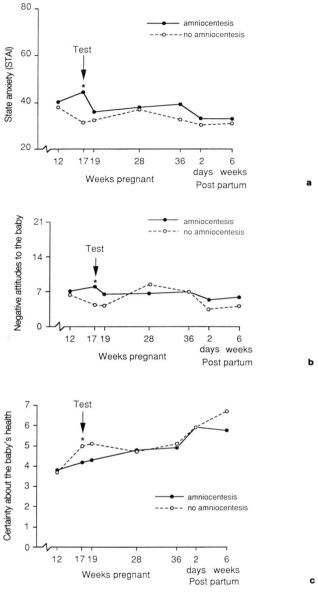

Fig. 18.3. Comparisons across pregnancy between women undergoing amniocentesis and those not for: **a**, anxiety; **b**, negative attitudes to the baby; **c**, certainty about the baby's health.

no studies to date documenting how this experience affects adjustment to the birth of a handicapped baby.

A small proportion of those undergoing prenatal diagnostic testing will receive a positive result, indicating a fetal abnormality. For severe abnormalities, this

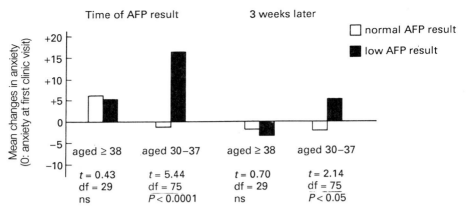

Fig. 18.4. Anxiety at the time of the test result and three weeks later (when further tests show no abnormality) in women receiving a normal AFP result and those receiving an abnormally low result. (From Marteau et al. [2] with permission.)

news is invariably greeted with shock and distress [32,33]. Women in this situation may choose to terminate the pregnancy, the option chosen by the majority, or they may choose to continue the pregnancy.

Terminating a pregnancy for fetal abnormality causes acute grief as well as some relief. While there is much variation in the extent and duration of distress experienced by women, there is some evidence to suggest that better coping is associated with the presence of more social support from partner and friends, termination earlier in pregnancy, and with terminations for conditions that are incompatible with life, such as anencephaly [34,35].

Some women choose to continue their pregnancies following a diagnosis of severe fetal abnormality. They may do so with one of several outcomes in mind: first, with the knowledge that the child will die at or shortly after the birth; second, with the intention of having the child adopted; or third, with the intention of keeping the child. There have been few studies of women who choose not to have a termination following diagnosis of a severe abnormality. One account by a general practitioner of parents who made such a decision following prenatal diagnosis of anencephaly raises the possibility that continuing with a pregnancy when the baby will not survive, may be a preferable option for some couples [36].

Recommendations to Reduce the Adverse Psychological Consequences of Prenatal Testing

1. Health professionals should provide full details about a test before testing.
2. Negative as well as positive results should be relayed to women, and their implications explained.
3. For women receiving positive results:
    a) A decison to terminate the pregnancy is not to be assumed; other options should also be discussed

b) Full and accurate information should be given about the condition diagnosed
c) For those choosing to terminate the pregnancy, routine follow-up should be given by a midwife, and a genetic counsellor.

Health Professionals

Health professionals' clinical practice both affects and is affected by the presence of routine prenatal testing for fetal abnormalities.

The quality and quantity of information given and the manner in which it is presented will influence whether or not women undergo a test. There is little, if any, routine training for staff in how to present prenatal testing to women, to ensure maximum informed uptake and minimal distress. Staff are not always fully informed themselves about the tests they are offering. For example, in a study of midwives' knowledge of MS-AFP screening, Sanden [37] found that although they were involved in providing information to patients, 45% of them lacked even basic knowledge about the test. Interviews with parents who had given birth to children with thalassaemia after a local screening programme had commenced, revealed both failure of doctors to recommend carrier testing, and quite frequent misinformation by inadequately informed health professionals [38].

Because of the serious consequences of a positive result from prenatal testing, it is recommended that (contrary to the usual procedure on routine testing of blood during pregnancy) the decision on whether or not to undergo prenatal screening or diagnostic testing should be made by the prospective parents [16,39]. In a survey of over 300 obstetricians [23], 25% stated that their policy was to present MS-AFP screening routinely without offering women any explanation of its purpose or any choice about participation in the screening programme. These attitudes appear to be reflected in actual behaviour in the consultation. In 102 tape-recorded consultations, MS-AFP testing was presented as one of a series of routine tests, with little attendant explanation in over half the consultations (unpublished observation).

Farrant [23] found that at all stages of the screening process, counselling by obstetricians was systematically biased towards encouraging women to undergo tests (MS-AFP and amniocentesis) and have a termination if any abnormality was detected, rather than towards providing women with information and the support required to make an informed choice.

As well as providing health professionals with the task of informing women about the available tests, prenatal testing also presents them with some new and at times difficult problems. Whelton [40] writes about the need for staff to respect parents' choices, including the choice not to abort a fetus with an abnormality incompatible with life. Both Whelton [40] and Farrant [23] describe the problems for parents who decide not to terminate a pregnancy with a diagnosed fetal abnormality, problems which include not only the distress of coping with a handicapped child but also the need to defend their decision to health professionals.

The existence of prenatal testing for fetal abnormality may affect how obstetricians and other health professionals manage pregnancy. Obstetricians are perhaps increasingly reluctant to accept uncertainty concerning the outcome of pregnancy. Enkin and Chalmers [41] suggest that the marked improvements in pregnancy outcomes since the Second World War have encouraged expectations to rise even faster than achievements. The complexity and pervasiveness of the screening procedures that characterise antenatal care today, they argue, are manifestations of an unwillingness to tolerate these uncertainties.

Even when all available tests have been done, there is no guarantee of a perfect child. Although prenatal testing can be used to diagnose in utero several severe birth abnormalities, only a minority of all birth abnormalities can be tested for. So while prenatal screening and diagnosis reduce some of the uncertainty surrounding birth outcomes, they are perhaps presented as reducing more of the uncertainty than they actually do.

Recommendations to Improve Staff Delivery of Prenatal Testing

1. Designation of one individual to monitor the running of prenatal testing services in each centre, and to run training and refresher courses for staff.
2. All health professionals involved in delivering antenatal care require brief training in the following aspects of prenatal testing:
    a) Education about the tests they are offering
    b) Training in what information to present, and when and how best to do this.

Society

It has been argued that the availability of prenatal screening and diagnosis, together with the termination of affected pregnancies, both reflect and reinforce the negative attitudes of our society towards those with handicaps [42,43]. We hold a mixture of attitudes towards handicap. On the one hand, there is an effort to create an environment in which people with a handicap are accepted into society and seen as having a worthwhile life, as depicted in a recent poster from the Down's Syndrome Association (Fig. 18.5).

Such an effort is also reflected in the change of policy to educate children with handicaps within mainstream schooling, and the spread of equal opportunity policies and practices for people with handicaps. A further area in which the promotion of the rights of people with handicaps is evident is in the language used to describe them. It is no longer acceptable to refer to such individuals as cripples or subnormals; instead, they are referred to as being people with disabilities or handicaps, or people with special needs.

At the same time as encouraging a more positive environment for people with severe handicaps, resources are spent on preventing their births (Fig. 18.6).

Given the option of prenatal diagnosis and abortion of affected fetuses, some parents may feel that to produce a child with a potentially diagnosable handicap is to be blame-worthy for that child's birth. Indeed in the US, some disabled

Fig. 18.5. Advertising poster presenting a positive image of a child with Down's syndrome. (Reproduced with kind permission of the Down's Syndrome Association.)

children, including some with Down's syndrome, have been awarded damages for wrongful life, on the basis that their disability should have been diagnosed before their birth and they should have been aborted [43]. Women undergoing prenatal testing describe these pressures as influencing their use of the new technology, pressures that they experience from health professionals, friends and relatives, as well as from society more generally [23,44]. When asked what effect, if any, prenatal diagnosis is likely to have on attitudes towards those with a disability, 46% of parents who had recently undergone prenatal diagnosis for Down's syndrome felt that it made such attitudes more negative [44].

Recommendation

Research is needed to document the effects of prenatal testing and termination of affected fetuses upon society's attitudes towards and care of those who, despite this technology, are born with severe abnormalities.

Fig. 18.6. Bubble gum dispenser in the United States of America, sponsored by a charity funding research on prenatal diagnosis.

Conclusions

Prenatal testing for fetal abnormality is now an integral part of obstetric care in Britain. To improve its implementation in practice, attention needs to be focused on providing more information to those eligible for these tests both prenatally and ideally, preconceptually. To achieve this, all health professionals involved in prenatal diagnostic testing need to be trained in the medical and technical aspects of the tests, as well as in what information to give to those eligible for testing, when to give this information and how best to do so.

While acknowledging the evident benefits of prenatal screening and diagnosis to many prospective parents, this new technology brings new issues and problems in its wake. To ensure that we are not doing more harm than good in providing prenatal testing for fetal abnormality, it is essential that as much effort is put into planning, delivering and evaluating services as is put in to the development of the tests themselves.

Acknowledgements. The work described in this chapter conducted by the research group at the Royal Free Hospital School of Medicine was supported by grants from the Medical Research Council, Birthright, and the Marie Stopes Foundation.

References

1. Rothman BK. The tentative pregnancy. New York: Viking, 1986.
2. Lilford RJ. In my day we just had babies. J Reprod Infant Psychol 1989; 7:187–91.
3. Cuckle HS, Wald NJ, Cuckle PM. Prenatal screening and diagnosis of neural tube defects in England and Wales in 1985. Prenat Diagn 1989; 9:393–400.
4. Lipmann-Hand A, Piper M. Influence of obstetricians' attitudes on their use of prenatal diagnosis for the detection of Down's syndrome. Can Med Assoc J 1981; 122:1381–6.
5. Margolin CR. Attitudes toward control and elimination of genetic defects. Soc Biol 1978; 25:33–7.
6. Julian C, Huard P, Gouvernet JF, Mattei JF, Ayme S. Physicians' acceptability of termination of pregnancy after prenatal diagnosis in Southern France. Prenat Diagn 1989; 9:77–89.
7. Bennett MJ, Gau GS, Gau DW. Women's attitudes to screening for neural-tube defects. Br J Obstet Gynaecol 1980; 87:370–1.
8. Kyle D, Cummins C, Evans S. Factors affecting the uptake of screening for neural tube defects. Br J Obstet Gynaecol 1988; 95:560–4.
9. Marteau TM, Johnston M, Shaw RW, Slack J. Factors influencing the uptake of screening for open-neural tube defects and amniocentesis to detect Down's syndrome. Br J Obstet Gynaecol 1989; 96:739–48.
10. Faden RR, Chwalow J, Quaid J et al. Prenatal screening and pregnant women's attitudes towards the abortion of defective fetuses. Am J Public Health 1987; 77:288–90.
11. Berne-Fromell K, Josefson G, Kjessler B. Who declines from antenatal alpha-fetoprotein screening – and why? Acta Obstet Gynecol Scand 1984; 63:687–91.
12. Marteau TM, Kidd J, Cook R, Michie S, Johnston M, Slack J, Shaw RW. Perceived risk not actual risk predicts uptake of amniocentesis. Br J Obstet Gynecol, in press.
13. Bundy S. Attitudes of 40 year old college graduates towards amniocentesis. Br Med J 1978; ii:1475–7.
14. Marteau TM, Johnston M, Plenicar M, Shaw RW, Slack J. Development of a self-administered questionnaire to measure women's knowledge of prenatal screening and diagnostic tests. J Psychosom Res 1988; 32:403–8.
15. Royal College of Physicians. Prenatal diagnosis and genetic screening. Community and service implications. London: The Royal College of Physicians, 1989.
16. Marteau TM, Slack J. Psychological implications of prenatal diagnosis for patients and health professionals. In: Brock D, Rodeck C, Ferguson-Smith MA, eds. Prenatal diagnosis and screening. London: Churchill Livingstone (in press).
17. Marteau TM, Johnston M, Shaw RW, Michie S, Kidd J, New M. The impact of prenatal screening and diagnostic testing upon the cognitions, emotions and behaviour of pregnant women. J Psychosom Res 1989; 33:7–16.
18. Astbury J, Walters WAW. Amniocentesis in the early second trimester of pregnancy and maternal anxiety. Aust Fam Physician 1979; 8:595–9.
19. Beeson D, Golbus MS. Anxiety engendered by amniocentesis. Birth Defects Original Articles Series. 1979; 15:191–7.
20. Sjogren B, Uddenberg N. Prenatal diagnosis and maternal attachment to the child-to-be. J. Psychosom Obstet Gynaecol 1988; 9:73–87.
21. Spencer JW, Cox DN. A comparison of chorionic villi sampling and amniocentesis acceptability of procedure and maternal attachment to pregnancy. Obstet Gynecol 1988; 72:714–18.
22. Griffiths MD, Gough MH. Dilemmas after ultrasonic diagnosis of fetal abnormality. Lancet 1985; i:623–4.
23. Farrant, W. Who's for amniocentesis? The politics of prenatal screening. In: Homans H, ed. Sexual politics of reproduction. London: Gower, 1985; 96–177.
24. Robinson J, Hibbard BM, Laurence KM. Anxiety during a crisis: emotional effects of screening for neural tube defects. J Psychosom Res 1984; 28:163–9.
25. Burton BK, Dillard RG, Clark EN. The psychological impact of false positive elevations of maternal serum alpha-fetoprotein. Am J Obstet Gynecol 1985; 15:77–82.
26. Evans MI, Bottoms SF, Carlucci T et al. Determinants of altered anxiety after abnormal maternal serum alpha-fetoprotein. Am J Obstet Gynecol 1988; 159:1501–4.

27. Marteau TM, Kidd J, Cook R et al. Screening for Down's syndrome. Br Med J 1988; 297:1469.
28. Rothenberg MN, Sills EM. Iatrogenesis: the PKU anxiety syndrome. J Am Acad Child Psychol 1968; 7:689–92.
29. Bodegard G, Tyro K, Larsson A. Psychological reactions in 102 families with a newborn who has a falsely positive screening test for congenital hypothyroidism. Acta Paediatr Scand [Suppl] 1983; 304:3–21.
30. Tymstra T. False positive results in screening tests: experiences of parents of children screened for congenital hypothyroidism. Fam Pract 1986; 3:92–6.
31. Bloom JR, Monterossa S. Hypertension labelling and sense of well-being. Am J Public Health 1981; 71:1228–32.
32. Statham H. Cold comfort. Guardian 1987; 24 March.
33. Brown J. The choice: a piece of my mind. JAMA 1989; 262:2735.
34. Iles S. The loss of early pregnancy. In: Oates MB, ed. Psychological aspects of obstetrics and gynaecology. Balliere's Clin Obstet Gynaecol 1989; 3:769–90.
35. Black RB. A 1 and 6 month follow-up of prenatal diagnosis patients who lost pregnancies. Prenat Diagn 1989; 9:795–804.
36. Watkins D. An alternative to termination of pregnancy. Practitioner 1989; 203:990–1.
37. Sanden M-L. Midwives' knowledge of the alpha-fetoprotein test. J Psychosom Obstet Gynecol 1985; 4:23–30.
38. Modell B. Social implicaidies of fetal diagnosis. In: Rodeck CH, Nicolaidies KH, eds. Prenatal diagnosis. Proceedings of the eleventh study group of the Royal College of Obstetricians and Gynaecologists. Chichester: Wiley, 1984.
39. Black Report. Report of the working group on the screening for neural tube defects. London: Department of Health and Social Security, 1979.
40. Whelton JM. Sharing the dilemmas: midwives' role in prenatal diagnosis and fetal medicine. Profess Nurse 1990; July: 514–18.
41. Enkin M, Chalmers I. Effectiveness and satisfaction in antenatal care. In: Clinics in developmental medicine. nos. 81/82. London: Spastics International Medical Publications, 1982; 266–90.
42. Stacey M. Manipulation of the birth process: a sociologists view. Paper presented to the European Advisory Committee on Health Research; Fourteen Session, Copenhagen, 1988.
43. Hayes M. "The defective baby test": the social implications of antenatal diagnosis. Paper presented to the British Association for Science. Swansea, 1990.
44. Sjogren B, Uddenberg N. Attitudes towards disabled persons and the possible effects of prenatal diagnosis. An interview study among 53 women participating in prenatal diagnosis and 20 of their husbands. J Psychosom Obstet Gynecol 1987; 6:187–96.

Chapter 19

Counselling After Prenatal Diagnosis

D. Donnai and L. Kerzin-Storrar

As the range of prenatal tests becomes wider due to technological advances, and the number of women offered tests increases, so does the need for appropriate counselling provision. The ascertainment and counselling of those at high risk, as well as counselling of all pregnant women offered prenatal screening tests, must be addressed. Care for couples undergoing prenatal diagnosis hinges on communication and much of this, especially for those at relatively high risk, may be provided through the process of genetic counselling.

Genetic Counselling

An ad hoc committee of the American Society of Human Genetics [1] defined genetic counselling as follows:

> Genetic counselling is a communication process which deals with the human problems associated with the occurrence, or risk of occurrence, of a genetic disorder in a family. This process involves an attempt by one or more appropriately trained persons to help the individual or the family:
> 1. Comprehend the medical facts, including the diagnosis, probable course of the disorder, and the available management.
> 2. Appreciate the way heredity contributes to the disorder, and the risk of recurrence in specified relatives.

3. Understand the options for dealing with the risk of recurrence.
4. Choose the course of action which seems appropriate to them in view of their risk, and the family goals, and act in accordance with that decision.
5. Make the best possible adjustment to the disorder in an affected member and/or to the risk of recurrence of that disorder.

In 1989 the Royal College of Physicians of London [2] outlined the main requirements for genetic counselling. These were the correct diagnosis of the propositus, estimation of genetic risk, communication of genetic risks and the options for avoiding them, assistance to the consultant to assimilate the information and reach an appropriate decision, and accessibility for long-term contact with genetic services where appropriate. Both the definition and the requirements emphasise the comprehensive nature of the process of genetic counselling, which cannot effectively be provided in a single consultation nor necessarily by a single person. Genetic counselling is provided to many different age groups and in many circumstances, but is particularly important at the time of reproductive planning and during pregnancy.

Groups Presenting for Prenatal Diagnosis

Couples presenting for prenatal diagnosis fall primarily into three groups; those who have experienced previous problems or have a family history of genetic disorder, those with an increased risk but without family history, and those from the general population identified through screening tests in pregnancy. Counselling after prenatal diagnosis depends very much on previous information given to the couple and their previously perceived level of risk. There are many factors contributing to the decision to have the prenatal tests. These include the individual's or couple's prior and present view of the magnitude of the risk, the way the risk was conveyed to them in counselling, their personal experience of the condition for which tests are being performed, and the family's personal circumstances, religious and moral views [3].

The characteristics of each group presenting for prenatal diagnosis will be outlined below, followed by discussion of counselling at the times of the test, result giving, and outcome as it applies to the three groups.

Positive Family History

Many couples falling in this group will have a high risk of a problem occurring. Genetic counselling may have been provided beforehand and the couple be very familiar with their risk and with the available tests. This may have been reinforced by the provision of detailed summaries after their counselling appointment. In many regions now, genetic registers have been set up with the objective of ascertaining those at risk and completing counselling and preliminary investigations before a pregnancy occurs [4]. The couples may have made a positive decision to embark on prenatal diagnosis before a pregnancy and some may have avoided pregnancy until prenatal diagnostic tests became technically feasible. A number of studies have shown that this is particularly the case for groups with a high risk (greater than 15%), for whom the availability of prenatal diagnostic tests has induced changes in reproductive planning [3,5].

Increased Risk Without Family History

Increased public awareness of the availability of prenatal testing for Down's syndrome means that many couples in older age groups, although their risk tends to be low (5% or less) will be at least partly informed before pregnancy. This group represents a large proportion of prenatal diagnoses and unlike the first group will not routinely receive formal genetic counselling. Their counselling needs impose considerable demands on obstetric services.

Heterozygote screening is available for a number of recessive disorders common in specific ethnic groups, and may soon be introduced for cystic fibrosis for the whole population. Many women identified as carriers through screening programmes have partners who are not carriers and thus they can be reassured. However, those whose partners also carry the genetic trait fall into a high risk group, probably without previous personal experience of the condition, and should have detailed counselling.

Screened Pregnant Population

Screening the whole pregnant population, which collectively has a low risk of a particular abnormality, is aimed at identifying those at higher risk so that more specific tests can be applied. It is a major problem adequately to inform the pregnant population of the nature of the abnormalities being sought, the risks of such problems being found, the interpretation of the results of screening tests and the options for further investigations. Information is also needed about risks and limitations of such investigations. The Royal College of Physicians' report [2] discusses dissemination of information about available screening tests and the organisation of prenatal diagnosis through specialist clinics and at a community level. There is a considerable literature on the ethical and counselling issues raised when introducing prenatal screening to women not previously aware of a risk. Sjogren and Uddenberg [6] discuss factors involved in decisionmaking during the prenatal diagnostic procedure and found that although most women felt that prenatal diagnosis was voluntary, many reported that it was difficult to decline tests when they were offered. Many of the participants had decided on an abortion if the test indicated an abnormality, but a significant number expressed ambivalence and reported differences between their own attitudes and those of their partners.

In general, couples undergoing prenatal diagnosis divide into two main risk groups: those at high (greater than 15%) risk and those at lower risk. Many in the high-risk group will have had prepregnancy counselling and will have been through a long decisionmaking process, and they are prepared for the possibility of an abnormality. The group at lower risk may not have had such individual counselling, and on the whole are less informed and less prepared for problems. These two groups are not always equally distributed in the patients presenting for the various test procedures. For example, where the availability of chorionic villus biopsy is limited, the high-risk group will be over-represented and the amniocentesis group will have a large proportion of those at lower risk. Those attending for ultrasound scan diagnosis may include many in the high-risk group in regional centres, whereas it may be a screening procedure with a largely low-risk group in district general hospitals.

Counselling at the Time of Testing

Counselling about a couple's risk and the risk of the procedures, and the decision whether or not to have the test, should have been completed before the actual day of the test. Pre-test counselling is considered in detail in Chapter 18. Once the decision has been made that a couple wish to undergo prenatal diagnostic tests, every effort should be made to ensure that the tests are performed as soon as technically possible. At the test appointment, mainly organisational and supportive aspects need consideration. It is important that a woman understands possible sequelae – physical and emotional – following the test procedure and that she has a contact point should there by any concerns. Precise arrangements should be made for the giving of results, after discussion with the couple as to whether they wish the results to be communicated by telephone or in person. A specific date should be arranged and if this needs to be altered later, the couple should be informed immediately. It is important to remember that couples at high risk are aware at the time of the test that the pregnancy may well be terminated. They often choose at least to attempt to delay emotional attachment to the baby until prenatal tests are complete. This is particularly so for those undergoing first trimester prenatal diagnosis, during which many couples prefer not to view the ultrasound screen. We have found it reassuring to inform couples that if the tests are positive and they choose termination this will be available as speedily as they feel appropriate. This was added into our counselling following anxieties expressed by many women that they "would have to go on a waiting list".

Sometimes a problem is identified at the time of the test, especially with ultrasound scanning. Those involved in these investigations need to have training in communication in this difficult area and to know from whom further help may be obtained. The problems identified may include a missed abortion, in which case it is appropriate for the obstetrician to be the first person of referral. Multiple pregnancy may be diagnosed and again the obstetrician will need to discuss whether previously planned tests are still feasible or whether modifications in the test plan need to be made. When an anomaly is detected on ultrasound scan the available information should be explained to the couple. Many centres have arrangements for immediate counselling of such families so that if the full extent of the problems appears clear a decision can be made, or if further investigations are indicated these can be discussed and speedily instituted.

Counselling After Prenatal Diagnosis

In most cases, the results of prenatal diagnostic tests are normal, and this provides immense relief to the couple. Phipps and Zinn [7] suggested that any amniocentesis-related psychological disturbances are transient and are outweighed by the receipt of normal results which enhance emotional adaptation to pregnancy. The same enhanced adaptation is observed after ultrasound scanning.

Abnormal Results, Easy to Interpret

These include trisomy 21, other autosomal trisomies, the unbalanced product of a parental chromosome translocation, DNA diagnosis of single gene disorders, specific metabolic defects or a multiple malformation syndrome in a family where there has been a previously affected child. This group also includes neural tube defects clearly visualised on ultrasound scan. In many of these situations, because of a previously recognised risk, the couple have elected to have the tests with the intention of termination of pregnancy. There are a number of issues in decisionmaking at this point, and regardless of the couple's previously stated intention to terminate if an abnormality were found, there is bound to be some ambivalence when a problem is actually identified [8]. If a couple already have a living affected child or family member, there are problems in reconciling the decision to terminate with positive regard for the affected person. Where an abnormality is detected by screening, as opposed to previously recognised risk, the couple will probably need more time and counselling before making their decision. For all couples with an abnormal result there may be moral or religious objections or social pressures about termination, there may be disagreements between the couple as to the correct course of action and it is not always possible to give the couple a clear idea of the degree of disability of that particular fetus. There is also the extremely important aspect of feeling responsible for the loss of a wanted child, which many couples describe as guilt. For a number of single gene disorders, prenatal diagnosis is based on DNA linkage studies where the risk given to the fetus will be less than 100% certain. Clearly in these situations the decisionmaking is more difficult, and for those couples who choose to terminate a pregnancy at less than 100% risk of the fetus being affected, there will be lingering thoughts over whether the baby may have been normal. Our experience in Duchenne muscular dystrophy has shown that two couples who have fetuses at the same risk may make different decisions about termination, and they must be supported whichever choice they make.

When the results are available, they should be transmitted as soon as possible to the woman and her partner in the manner decided with them at the time of the test. Time should be allowed for the inevitable questions and immediate grief and an appropriate room should be found for this discussion away from the routine clinic. It is helpful if a member of the prenatal team with particular experience in supportive counselling becomes involved at this stage. Some patients may wish for immediate hospital admission for termination of pregnancy, and others may wish to have a little time to consider their decision and to make domestic arrangements. Great sensitivity is required for these women undergoing termination of wanted pregnancies. All medical and nursing staff working in this area may need special training and support. As in the case of stillbirth, some couples may wish to see and hold the fetus whereas others would choose not to take up this option. Some couples may elect not to terminate the pregnancy. Their subsequent antenatal management will require great sensitivity, and perhaps they will wish to consult with a neonatologist or paediatric surgeon who will be involved with the subsequent care of the child.

Abnormal Results, Difficult to Interpret

There are a number of groups of disorders where either the significance of the test findings is unclear, or the abnormality identified has a very variable outcome. Examples of these problems are listed below.

Sex Chromosome Aneuploidies

Most women in whom a fetal sex chromosome abnormality is identified have had their investigation because of raised maternal age. Our experience over a ten-year period has been of 58 fetuses shown to have a sex chromosome abnormality on amniocentesis [9]. Our practice upon detection of a sex chromosome abnormality is for the cytogeneticist to inform a member of the clinical genetics team. The obstetrician concerned is contacted and we offer to counsel the parents. The offer is usually accepted but if the obstetrician wishes to be the only person speaking to the parents, we can supply relevant information. In some cases the couple are jointly seen by both obstetrician and geneticist, and a genetic fieldworker is often also involved. The amniocentesis findings are explained, often with the aid of diagrams and karyotypes. We give information about the condition, based on the findings of long-term prospective studies of children with sex chromosome abnormalities detected by screening at birth [10,11]. This information is less biased than much information in the standard textbooks, which is based on observations in children who were diagnosed after birth because of the occurrence of problems. Many couples ask to see relevant literature and photographs, and some are shown photographs of patients already known to us (parental permission having been obtained). These photographs show the physical features of the conditions involved, but present a more realistic picture than the extreme examples of the different phenotypes usually found in textbooks. The main parental anxieties concern congenital abnormalities, intelligence quotient, sexual development and orientation, short or tall stature, psychological problems and fertility. We aim to cover all these topics in the counselling session. Turner's syndrome is the only condition consistently associated with structural abnormalities [12] and an ultrasound scan is arranged to check for associated malformations. The parents are told that there is no "right" decision to be made and we make it clear that whatever their decision we will support it. Table 19.1 gives the decisions of the 58 families in our study and the type of anomaly detected. Among the 21 couples who elected to abort the pregnancy, eight had an abnormal scan, most commonly hydrops associated with a 45,X karyotype. Other reasons given for termination included the concern of genital ambiguity in those with two cell lines, one with and one without a Y chromosome.

Tab. 19.1. Sex chromosome aneuploidies detected at amniocentesis 1980–90

	No.	Continue	Abort
X	11	2	9[a]
XXX	15	11	4
XX/X/XXX	6	6	0
X/XdelX	1	0	1
XXY	14	10	4
XYY	4	2	2[b]
XY/XXY	1	1	0
XX/XXY	1	0	1
XX/XX/XY	1	1	0
Y rearrangements	4	4	0
Total	58	37	21

[a] Seven had abnormal scan.
[b] One had abnormal scan.

The initial reaction of most couples on first hearing that they had an abnormal amniocentesis result was to opt for abortion. In counselling we felt it reasonable to be optimistic about the population of children diagnosed to have a sex chromosome abnormality at amniocentesis. These children are screened by ultrasound scan for major structural abnormalities, and the paediatric surveillance they receive should lead to earlier detection of associated problems and institution of appropriate therapy such as speech therapy or hormone treatment. In our experience, appropriate counselling before birth has a positive influence on the parent/child relationship [9].

De Novo, Apparently Balanced Chromosome Rearrangements

This is a particularly difficult area for counselling because of lack of data. Warburton [13] presented frequency data from 76952 prenatal diagnoses from the USA of de novo structural rearrangements and markers and added other cases from around the world to estimate outcome. Even from such a large group, the number of anomalies was small and it was concluded that further studies and improved follow-up were required. In total, 66 rearrangements were reported, of which 14 were terminated. In the remainder there was only one known living abnormal child, but of the other 51 live births only four were followed-up beyond the first year of life. The Association of Clinical Cytogeneticists in the UK has recently established a working party on de novo structural rearrangements and markers detected prenatally, with the aim of combining experience from centres in the UK and collecting sufficient information to be helpful for families in decisionmaking. A control has been selected for each case identified, and both cases and controls will be examined by clinical geneticists. A detailed medical and developmental history will be taken and an asessment of behaviour and development made by an adaptive behaviour scale [14]. It is hoped that at least 85 cases will be available for this follow-up study. At the present time, a risk of physical or developmental abnormality of 10% is quoted based on the Warburton study, but this may represent an overestimate.

A Mosaic Chromosome Pattern

A mosaic pattern describes the situation where two cell lines are found with differing chromosome complements. The difficulty is to ascertain whether this represents the pattern in the fetus, is a reflection of placental or membrane mosaicism, or whether it is an in vitro artefact.

Chromosomal mosaicism may be fairly common in placental tissue whereas the fetus has a normal karyotype. The Association of Clinical Cytogeneticists' collaborative study of chorionic villus biopsy [15] indicated that there was an aneuploid cell line in the tiny sample taken in 1% of pregnancies. It would be reasonable to speculate that if more areas of the placenta were analysed, mosaicism would be detected even more frequently. In the majority of cases in the study, the resulting baby was cytogenetically normal with no clinical abnormalities and no evidence on cord blood analysis of the mosaicism. There are anecdotal reports of infants with intrauterine growth retardation who were cytogenetically normal but whose placenta had a large proportion of trisomic cells [16]. A converse situation may also apply and a recent publication [17] reported cytogenetic analyses of 14 placentas from liveborn infants, or from terminated pregnancies, with trisomy 13 or 18 and revealed that all were mosaic, the mosaicism consisting of trisomic and normal cell lines confined to the cytotrophoblast. The percentage of cells with normal karyotype varied from 12% to 100%. The authors interpreted these findings as suggesting that a postzygotic loss of a trisomic chromosome in a progenitor cell of trophoectoderm facilitated the intrauterine survival of trisomy 13 or 18 conceptuses. If a mosaic pattern is detected on chorion villus biopsy (CVB) examination, then it is reasonable to proceed to amniocentesis at the appropriate time. It may also be reasonable to consider the place of fetal blood sampling after CVB as well as it being the usual course of action after mosaicisms detected on amniocentesis samples.

Fetal Examination After Termination

Part of the care of a family after detection of abnormality at prenatal diagnosis is the careful examination of terminated fetuses. This is important for quality control of the tests employed and for accurate genetic counselling of the families. In a prospective study over 5 years [18], we changed or modified the pretermination diagnosis in 53 of 133 fetuses aborted after scan diagnosis of abnormality and in three of 115 fetuses terminated after amniocentesis diagnosis. In the scan group, the modified diagnosis led to an increased risk of recurrence in 25, a decreased risk of recurrence in 24 and no change in recurrence risk in four. We believe that examination of mid-trimester fetuses by a clinical geneticist with experience in dysmorphology is a worthwhile service and improves the accuracy of diagnosis, which in turn benefits the diagnostic team's and parental counselling. The collaboration of cytogeneticists and paediatric pathologists is important in defining the full extent of the anomalies in order to arrive at the final post-termination diagnosis, on which the parents will base their future reproductive decisions. Becker et al. [19] have also stressed the importance for families of receiving autopsy results. It is vital that confirmatory tests are done since doubts are bound to arise in families as to whether the fetus was indeed abnormal. A recent study found that of 166 women who had had a termination of pregnancy for neural tube defect, one quarter did not have, and were not invited for, a post-termination appointment, and thus did not have the opportunity to ask questions or to discuss the findings in the fetus [20].

Support for the Family After Termination and the Next Pregnancy

Two retrospective studies [21,22] have demonstrated the adverse psychological sequelae after termination for fetal abnormality and have stressed the need for co-ordination of follow-up and counselling services. Every effort must be made to avoid unnecessarily distressing events such as reminders to attend antenatal clinic appointments or parentcraft classes, because of poor communication. The primary care team should be involved in the post-termination period and in many regions genetic associates or specialist health visitors associated with the regional genetics service can now offer post-termination support. At the very least, a visit within the first fortnight and again 2–3 months later should be made. Some voluntary organisations are beginning to set up support networks for couples undergoing termination for fetal abnormality, and obstetric staff should be aware if such help is available locally.

A recent retrospective study asked women who had experienced a second trimester termination of pregnancy for a neural tube defect, their views of the care they had received [20]. The majority felt satisfied with the care received during screening, prenatal diagnosis and during termination. Patients were less satisfied with the post-termination care in hospital, and post-termination sequelae were mentioned to only a small number, which left over 80% disturbed by the post-partum reactions of their bodies and by their strong emotions. The authors suggest that further prospective and retrospective research is needed to determine the long-term medical and psychosocial sequelae of termination and the possible vulnerable groups.

When a couple have undergone a termination for fetal abnormality, any subsequent pregnancy is bound to cause a great deal of anxiety. Some obstetric units hold preconception clinics where plans for care and investigations in a pregnancy can be made before the pregnancy occurs. Often the risks and possible investigations will have been fully discussed and a summary provided by the genetic service, and a genetic field worker may be able to provide support during the pregnancies following the one which ended in termination. The time before prenatal tests can be stressful, and another period of increased stress is the last few weeks before the birth. After delivery, even when the baby is normal, anxieties are common and the expected relief following the birth of a normal child does not always occur. Sympathetic support and explanation to the woman and her partner about their feelings can be helpful and prevent long-term sequelae.

References

1. Ad Hoc Committee on Genetic Counselling. Genetic counselling. Am J Hum Genet 1975; 27:240–2.
2. A Report of the Royal College of Physicians. Prenatal diagnosis and genetic screening. Community and service implications. The Royal College of Physicians of London, 1989.
3. Frets PG, Duivenvoorden HJ, Niermeijer MF, van de Berge SMM, Galjaard H. Factors influencing the reproductive decision after genetic counselling. Am J Med Genet 1990; 35:496–502.
4. Read AP, Kerzin-Storrar L, Mountford RC, Elles RG, Harris R. A register based system for gene tracking in Duchenne muscular dystrophy. J Med Genet 1986; 23:581–6.
5. Evers-Kiebooms G, Denayer L, van den Berghe H. A child with cystic fibrosis: II. Subsequent family planning decisions, reproduction and use of prenatal diagnosis. Clin Genet 1990; 37:207–15.
6. Sjogren B, Uddenberg N. Decision making during the prenatal diagnostic procedure. A questionnaire and interview study of 211 women participating in prenatal diagnosis. Prenat Diagn 1988; 8:263–73.
7. Phipps S, Zinn AB. Psychological response to amniocentesis: I. Mood state and adaptation to pregnancy. Am J Med Genet 1986; 25:131–42.
8. Drugan A, Greb A, Johnson MP et al. Determinants of parental decisions to abort for chromosome abnormalities. Prenat Diagn 1990; 10:483–90.
9. Clayton-Smith J, Andrews T, Donnai D. Genetic counselling and parental decision following antenatal diagnosis of sex chromosome aneuploidies. J Obstet Gynecol 1989; 10:5–7.
10. Leonard MF, Sparrow S. Prospective study of development of children with sex chromosome anomalies: New Haven Newborn Study IV. Adolescence. Birth Defects 1986; 22:221–49.
11. Ratcliffe SG, Murray L, Teague P. Edinburgh study of growth and development of children with sex chromosome abnormalities III. Birth Defects 1986; 22:73–118.
12. Connor JM. Prenatal diagnosis of the Turner's syndrome: what to tell the parents. Brit. Med J. 1986; 293:711–12.
13. Warburton D. Outcome of cases of de novo structural rearrangements diagnosed at amniocentesis. Prenat Diagn 1984; 4:69–70.
14. Donnai D. The clinical significance of de novo structural rearrangements and markers detected prenatally by amniocentesis. J Med Genet 1989; 26:545.
15. Chorionic Villi Working Party Report. Chorionic Villi 1987 and 1989. Association of Clinical Cytogeneticists, 1988.
16. Stiovi S, De Silvestris M, Molinari A, Stripparo L, Ghisoni L, Simoni G. Trisomic 22 placenta in the case of severe intrauterine growth retardation. Prenat Diagn 1989; 9:673–80.
17. Kalousek DK, Barrett IJ, McGillivray BC. Placental mosaicism and intrauterine survival of trisomies 13 and 18. Am J Hum Genet 1989; 44:338–43.
18. Clayton-Smith J, Farndon PA, McKeown C, Donnai D. Examination of fetuses after induced abortion for fetal abnormality. Br Med J 1990; 300:295–7.
19. Becker J, Glinski L, Laxova R. Long-term emotional impact of 2nd trimester pregnancy termination after detection of fetal abnormality. Am J Hum Genet 1984; 36:122.
20. White-van Mourik MCA, Connor JM, Ferguson-Smith MA. Patient care before and after termination of pregnancy for neural tube defect. Prenat Diagn 1990; 10:497–505.
21. Donnai P, Charles N, Harris R. Attitudes of patients after 'genetic' termination of pregnancy. Br Med J 1981; 282:621–2.
22. Lloyd J, Laurence KM. Sequelae and support after termination of pregnancy for fetal malformation. Br Med J 1985; 290:907–9.

Discussion

Lilford: A word about another method of genetic counselling altogether. This was originally suggested by the Paukers [1]. What they did was to ascertain their patients' values through a series of independent "gambles". They then calculated, knowing the probabilities, the best bet for a given patient, and then gave counselling around that. They claim that this is an extremely effective way of counselling, reduces patient anxiety, increases patient understanding, and they propose it as a method for clinical use. I think we should be aware of it [2].

Pembrey: How detailed is the prior analysis? My gut reaction is that this does not sound right somehow and that they would end up quantifying just those things that can be quantified and missing out all the other nuances.

Lilford: That takes one into the whole discussion of decision analysis, which is a huge subject in economics and a large subject in business. It is quite a large subject in medicine in the United States, but very new in the UK.

Williamson: I agree with Dr Marteau about the importance of information and non-directive counselling for parents, and especially for pregnant women, and of training for professionals. But there were three things she said that I was unclear whether there are any data. In the first place, she said that the experience of being given choice, that is, having the option of accepting or not accepting prenatal diagnosis, was an experience that some people would rather not confront. And then she said there are some who are made anxious by a negative result. I can see how some people can be falsely reassured by a negative result, and I can see how people can become anxious by a positive result, but I have never met anyone, in the case of CF at least, who was made anxious by a negative result.

Finally, a negative impact on handicap. The majority of oganisations representing the handicapped through the Genetic Interest Group (GIG) support the availability of prenatal diagnosis, and none of the organisations to my knowledge, oppose it.

So it seems that on those three points at least, there are no data supporting those contentions. At least from my experience.

Pembrey: What about the anxiety after what we would regard as a good news result?

Marteau: There are no data on that because they are very poor studies. What is alluded to there is that anxiety is caused, not by the negative result but by the process of having prenatal testing for women who did not realise that they were at risk of fetal abnormality. So although they might have got a negative result on testing, nonetheless the whole issue and the fact that there could be something wrong with their pregnancy has been raised.

Pembrey: It is residual anxiety in a sense.

Williamson: Dr Marteau says that there are some people who regard it as an experience they would rather not confront? That has not been my experience.

Marteau: Professor Williamson's experience, from what he has been saying, has been related to families who enter pregnancy knowing themselves to be a high genetic risk. Parents who enter pregnancy not at particularly high risk, but who are aware that there are these tests, experience the dilemma. So it is different from the cystic fibrosis group.

Professor Williamson was saying that many of the groups working for people with handicaps do not have a negative attitude. At the King's Fund Consensus Forum three years ago on Prenatal Screening/Prenatal Diagnosis there were groups there who were represented who had quite strong negative feelings towards the use of prenatal diagnosis and how it may be affecting society's attitudes. So I think it is a mixed picture.

Whittle: Part of the problem with CVS is the difficulty getting to the low-risk groups and talking to them before they get into the system. What happens in many places is that mothers come to an antenatal clinic and then the issue of their being suitable for CVS is raised at that point. There is a lot of information to be imparted to the mother at a time when she is probably already anxious and confused.

We need to devise a mechanism by which those mothers perceived at risk by, say, maternal age, are reached and counselled before they ever get to the hospital.

Donnai: We do that in two ways. One is by having a big and active genetic service with genetic registers, and so we hope to have counselled people at high risk before ever they become pregnant. The other thing that we have done within our hospital, because we have got limited resources for CVS, is to book people for the test through the genetic department and we can judge how many people we can see in a week at high risk, and whether we have spare capacity for anybody who wants it but is at relatively low risk.

Campbell: I wanted to mention the positive effect of the ultrasound scan in terms of the psychological effect on the woman. We did a randomised study some time ago in which all women were counselled on the basis of their routine scan, but only half the women saw the image on the screen. Those who saw the image on the screen were very profoundly affected in terms of loving feelings towards their fetus, reducing smoking, reducing alcohol intake, and were very much more positive.

We have also done another study in women with high maternal serum AFP who had very high anxiety scores, and with the visual impact of the scan their scores came plummeting down to normality. Obviously it has to be associated with reassuring comment.

Marteau: That first study is widely quoted, and one of the problems with it is that there was not a random allocation to no scan at all. There was low feedback and high feedback, and what we do not know with that study is whether if the mother can sit down with a health professional for ten minutes and talk about her baby, that has the same effect. So it is suggestive, but it is not conclusive.

References

1. Pauker SP, Pauker SG. The amniocentesis decision: ten years of decision analytic experience. Birth Defects 1987; 23:151–69.
2. Thornton JG. Decison analysis in prenatal diagnosis: measuring patient values. In: Lilford R, ed. Diagnosis and prognosis. London: Butterworths, 1990.

Chapter 20

Economic Aspects of Prenatal Diagnosis

J. B. Henderson

Introduction

There seems to be growing public demand for prenatal diagnosis and screening. Health service resources, however, are limited. Therefore those who work within health care have to choose the best ways of deploying the resources that are available, so as to bring about the maximum benefits. Economic evaluation can help with such unavoidable choices by organising relevant information about costs and benefits in a systematic way.

Choices

As the range of available options gets wider, the choices increase. At the regional level, health service managers with responsibility for allocating resources may have to choose which regional specialties to provide, which to expand and which to contract. Similarly, at the local level there are choices to be made about which services deserve the highest priority, which groups of people should receive them, and how extensive they should be.

Managers may wish consciously to consider which prenatal diagnostic services would be most appropriate for the populations served by their hospitals. Choices could include the provision of services such as extra ultrasound scans, maternal serum testing for identifying those at risk of fetal abnormalities, amniocentesis for diagnosis of fetal chromosomal anomalies and spina bifida, and screening for carrier status for inherited diseases such as sickle cell disease or cystic fibrosis. Choices could also include the scale of service, and whether to provide it universally or selectively to those at higher risk, such as older pregnant women for fetal chromosomal anomalies.

Resource Considerations

Resources are never sufficient to do everything that we would like to do. By resources, economists mean factors such as labour and capital, rather than money. In health care, important resources would include the time of health professionals, the equipment that they use, and the time and materials used by the staff who support them. Resources also include the time of patients.

The prevention of ill-health and suffering can be seen as an economic production process with resource inputs and health outputs [1]. Economic studies can assess the efficiency of this production process.

Every time resources are committed to the production of one commodity, those resources are unavailable to produce some other desired commodity. Thus a cost is incurred whenever resources are used in one way – the cost is the value of the output that could have been produced if they had been used in another way. Economists refer to this as "opportunity cost" – the sacrifice of the value of the alternative benefits forgone.

Efficiency

Economic efficiency studies help to bring together information about choices, benefits and limited resources in a systematic way. This assists policymakers with decisions about which programmes to implement so as to produce the maximum benefits from available resources. What, though, is meant by the term "economic efficiency"?

There are two types of economic efficiency. The first is "cost-effectiveness" or technical efficiency. "Cost-effectiveness analysis" (CEA) in health care addresses questions such as: "Could more services be produced without any net increase in resource use?" Or: "Could at least as much service be produced using fewer resources in total?" Or: "Which way of using a given extra sum of resources would produce the maximum service improvement?" Production is technically efficient when the maximum benefits are produced for a fixed cost, or, conversely, when a fixed benefit is achieved at minimum cost.

The second type is "cost-benefit" or social efficiency. "Cost-benefit analysis" (CBA) in health care addresses questions such as: "Would the social benefits of devoting more resources to preventive services outweigh the opportunity costs?" And: "If so, how much more should be spent on preventive services?"

Production is socially efficient when resources are allocated so as to gain the maximum excess of benefit over cost (where neither the benefit nor the cost is fixed in advance).

Ethics and Economics

Screening for fetal abnormalities poses special ethical problems. Prenatal diagnosis is commonly offered with the intention that if a serious abnormality is detected then the woman may have her pregnancy terminated. Abortion is a subject about which many people have deeply held and widely differing views. For some people the idea of abortion is so morally abhorrent that they could never contemplate it. Clearly consideration of other costs and benefits is, for them, irrelevant.

Other people, however, do wish to know whether their prospective offspring will have any serious abnormalities and, if so, would consider abortion, which would be permissible under existing law. Consideration of other costs and benefits may be relevant for them and for the health service decision makers whose decisions affect their access to prenatal diagnosis.

Economic evaluation claims neutrality with respect to the ethical issues surrounding abortion. Rather it stresses the generality of its approach, in the full knowledge that different people attach different values to particular factors and may thus arrive at different conclusions while using a common framework [2].

Evaluating the Efficiency of Preventive Services

Empirical economic studies of preventive health care tend to examine issues such as: the size of the gains from devoting more resources to preventive services; the implications for the health or quality of life of sufferers and their families; and the net costs for the health sector, for families, and for society in general.

Typically a programme will be deemed more cost-effective than the status quo if it:

1. Reduces net resource costs within the health service (and does not add to resource costs elsewhere), and brings health benefits that outweigh any disbenefits, i.e. "it pays for itself directly and has net benefits"
2. Increases net resource costs within the health service but brings larger savings outside the health service, and brings health benefits that outweigh any disbenefits, i.e. "it pays for itself indirectly and has net benefits"

A programme will typically be deemed more socially efficient than the status quo if it:

3. Increases net resource costs within the health service, does not bring larger savings outside the health sector, but brings net health benefits that are

sufficiently large and important to justify the extra resource costs, i.e. "it does not pay for itself, but the benefits justify the costs".

Options

To assess whether a policy or programme is efficient it must be compared against alternatives.

First comes the "status quo option" – the baseline against which the other options are to be assessed: no (extra) prenatal diagnosis. The main options to compare against this are:

1. Prenatal diagnosis for commoner conditions
2. Prenatal diagnosis for rarer conditions

There are also the alternatives of groups of women to be offered screening:

3. Women who have a general risk of having an abnormal fetus, i.e. unselected population
4. Women who have a high risk of having an abnormal fetus, i.e. selected population;

Suboptions might concern:

5. Where screening should take place:
   a) At open access clinics (primary health care centres)
   b) At general hospitals (secondary referral centres)
   c) At specialist hospitals (tertiary referral centres)
6. Which prenatal diagnostic methods to use: combinations of imaging, biochemistry, cytogenetics etc
7. How often prenatal diagnosis should take place: how many times during pregnancy.

Costs and Benefits

An evaluation should describe all the significant consequences and identify all the potential costs and benefits. Thus an economic study needs to specify, for both true positive detections and false positive detections:

The number of abnormalities of each type that would be detected under the status quo option and under the other options (clearly this will depend critically on the sensitivity and specificity of prenatal diagnosis under the different options)

The change in health care management (i.e. resource use) following each detection, as compared with the status quo option

The change in resource consumption outside the health service following each detection, as compared with the status quo option

Economic Aspects of Prenatal Diagnosis

The change in health and welfare (morbidity and mortality) for the fetus and/or the parents following each detection, as compared with the status quo option.

Costs and benefits might be measured as shown in Table 20.1 (see also references [2] and [3]). The "tangible" factors are those where it is easiest to measure the resource costs and savings. "Intangible" is a convenient shorthand term for the rest.

Table 20.1. Costs and benefits of prenatal diagnosis

Tangible costs	Tangible benefits
Identifying women at risk and informing of services available	In event of termination:
	Avoided health services expenditure
Diagnostic procedures	Avoided education services expenditure
Laboratory analysis	Avoided other public services expenditure
Counselling about diagnosis	Avoided loss of mother's job output
Women's time for above	Avoided family expenditure on child
Repeat procedures (as above)	Avoided lifetime consumption by child of other goods and services
In event of termination:	
Abortion services	
Special counselling	
Women's time for above	
Loss of child's potential future productive output	
Intangible costs	Intangible benefits
Anxiety aroused through being informed of risk	Greater information
	Wider choices
Discomfort of diagnostic procedure	Reassurance
Worry about test results	Reduction of uncertainty
Distress over miscarriage	Greater confidence in trying to conceive
Risk of fetal damage caused by diagnostic procedure	In event of termination:
	Avoided distress by not having handicapped child
Qualms about contemplating abortion	
Possible complication for woman	Greater likelihood ultimately of having a non-handicapped child
Wrong decisions caused by false results	

A money value can eventually be put on most of the tangible costs and benefits, but the intangible effects must also be considered and if possible their magnitude assessed. Foremost among these are the benefits and costs to families. One measure of such effects that has been devised is in terms of quality-adjusted life-years gained, or QALYs. QALYs attempt to move beyond simply measuring life expectancy, to take account also of the quality of life: whether lives are improved as well as extended.

Obviously the measurement of "quality of life" is fraught with difficulties, but useful attempts to grapple with the difficulties have already been made. To date considerable progress has been made in measuring QALYs, but there are still major questions over how valid and reliable the existing QALY measures are. Also they seem to be more suited to measuring physical symptoms, and poorer at measuring effects such as the impact of information, reassurance, confidence, uncertainty and regret [3].

Economic Efficiency Studies of Prenatal Diagnosis

Of the list of options set out above, some of the main options for some of the commoner congenital abnormalities have been assessed in terms of their resource consequences. However, few studies have examined the rare congenital abnormalities, the suboptions listed above, or many of the intangible consequences. Hence it is difficult to make unequivocal statements about the economic efficiency of many prenatal diagnostic programmes. Nevertheless, some results are not much in doubt and some of these are discussed below.

Open Neural Tube Defects

Hibbard et al. [4] have reported the costs of providing screening for open neural tube defects (NTDs) in South Wales. The programme that they considered consisted of routine ultrasound scanning and serum alpha fetoprotein (AFP) measurement for all women, followed by diagnostic ultrasound and amniocentesis for women with a serum AFP level above the 97th centile, if they so desired. They estimated the cost at £735 600 (1980 prices) per 100 000 women, and that this would lead to the avoidance of the birth of 340 infants affected by open NTDs – a cost per birth avoided of £2164.

If, however, open NTDs were four times rarer than for the population they studied in South Wales, i.e. 1.25/1000 instead of 5/1000, then the cost per birth avoided would be four times higher, £8656. This figure represents a more reasonable average for open NTDs for Great Britain as a whole where the "natural" birth prevalence rate would probably be about 1.25/1000 [5].

The tangible savings from avoiding the birth of one child with open NTD have been estimated at about £18 500 [6]. The saving comprised about £2000 to the NHS, about £3000 to education services, about £1000 to social services, about £4000 net on other private resources, and about £8500 in mother's extra job output (1980 prices).

Thus the savings were estimated to be well in excess of the costs, and the programme would pay for itself, albeit indirectly. Since the programme was voluntary the intangible benefits might also be assumed to be greater than the intangible costs, and the programme could be deemed efficient.

Down's Syndrome

The tangible savings from avoiding the birth of one Down's syndrome baby fall into similar categories as those for open NTDs, i.e. reduced costs to the family, to the health and education services, of institutional care, and of the mother's lost job output. These savings are thought to be fairly constant for all ages of mother.

On the other hand, it is known that the risk of having a Down's syndrome baby rises with maternal age. The NHS cost of detecting one case of Down's syndrome, e.g. through amniocentesis, will be lower in older pregnant women, where there

Economic Aspects of Prenatal Diagnosis

are more cases per 1000 pregnancies, than in younger women, where there are fewer cases per 1000 pregnancies. Threfore the cost per case detected will fall with maternal age.

When estimates of the tangible saving per birth avoided have been compared with estimates of the cost per birth avoided, it has been found that there is a "cut-off age". Above the cut-off age screening for Down's syndrome has been found to pay for itself (indirectly). Below the cut-off age, because the rate per 1000 pregnancies is lower, the costs of detecting one fetus with Down's syndrome have been found to be greater than the associated savings. Hagard and Carter [7] estimated that this cut-off age would be 35 years in Scotland. Other studies, in other countries, have tended to come to a similar conclusions [2].

It has been found that low serum AFP is also associated with high risk of Down's syndrome in the fetus. In populations where serum AFP assessment is carried out for NTD screening, the serum AFP results can be used to define another group of women at high risk of having a Down's syndrome baby. Gill et al. [8] studied the North East Thames Region in England, where amniocentesis for Down's syndrome was offered to women aged over 37 years. They found that another group of women aged 32–37 years, at equal or greater risk than those aged 38 years, could be defined on the basis of AFP results. Thus offering amniocentesis to this other group would also pay for itself (indirectly).

The concept of the "cut-off age" could be generalised to that of a "cut-off risk". Fig. 20.1 shows the idea of a cut-off risk below which net tangible costs are incurred and above which there are net tangible savings.

In the case of Down's syndrome it is tempting to conclude, from those studies that have estimated he costs and benefits of prenatal diagnosis for women at age-related risk, that the cut-off risk may be somewhere around 1 in 200. However, in the absence of firmer information about the overall costs and benefits of newer methods of prenatal diagnosis, this would be speculative [2]. Moreover, the studies that have been undertaken so far have not been able to include the intangible costs and benefits for direct comparison with the tangible resource consequences. It is not yet possible, therefore, to make firm pronouncements on the cut-off risk above which prenatal diagnosis for chromosomal abnormalities can be deemed to be efficient.

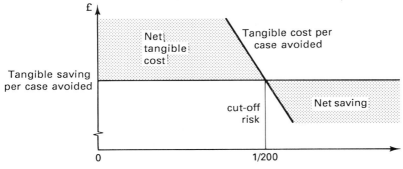

Fig. 20.1. Cut-off risk level for resource savings

How Much Screening?

The idea of the cut-off risk does, however, show how economic efficiency studies can be of help to decision makers in addressing the question of how much screening there should be. It illustrates the common economic phenomenon that, when programmes are expanded, the "returns" generally diminish. If too much of a programme is provided then the returns may become negative, i.e. the costs will outweight the benefits.

If prenatal diagnosis is extended to groups at lower and lower risk, then there is likely to come a point where the programme will be inefficient. This means that the health service would secure greater benefits by putting its resources to some other use.

Other Conditions

There have been studies of the resource consequences of screening for other conditions. Attanasio et al. [9] examined the costs and benefits of screening for thalassaemia major, on Sardinia, for women at high risk and concluded that the tangible benefits exceeded the tangible costs.

Nelson et al. [10] and Dagenais et al. [11] examined the costs and benefits of screening for Tay–Sachs disease, in Houston and Quebec respectively. Again the conclusion was that, for women at high risk, the tangible benefits exceeded the tangible costs.

A Scottish working party [12] considered the implications of screening for Duchenne muscular dystrophy. A similar, albeit tentative, conclusion was that such a programme would produce net savings if offered to known high risk groups.

Intangibles

In the future, economic efficiency studies are likely to make greater efforts to improve the comprehensiveness of evaluations. They will measure more of the intangible costs and benefits, especially those that relate to quality of life.

Bush et al. [13] assessed the benefits of neonatal diagnosis of phenylketonuria in terms of QALYs gained and found, as well as net savings, an increase of 47 QALYs per case treated. At present prenatal diagnosis of chromosomal or gene defects does not generally lead to improvements in the QALYs of the fetus. However, gene therapy for some anomalies is a prospect that is drawing nearer.

Much prenatal diagnosis is for the benefit of the parents rather than the fetus. Hence it is appropriate to try to assess their perceptions of quality of life. Some economists have already begun to try to make such assessments [14].

Conclusions

Economic evaluation attempts to measure and compare the gains and losses to society arising from the provision of various programmes. Clearly this can be both relevant and helpful for devising effective health care policy, and is essential if the best uses are to be made of limited health care resources.

The studies discussed here indicate that several prenatal diagnostic programmes more than pay for themselves. Examples include programmes for open NTDs and for chromosomal anomalies for those at high risk of having an affected fetus. The same may turn out to be true for those at high risk of passing on inherited gene defects.

Whether or not the resources available for prenatal diagnosis are increased, it would seem important, in evaluating the potential of new screening possibilities, to set priorities in the light of information about their costs and benefits. It would be prudent, if contemplating the extension of existing programmes, to measure the extra costs and extra benefits directly, rather than making projections from existing average costs, in order to allow for diminishing returns.

To make robust claims about the efficiency of prenatal diagnosis, it will be necessary to try to measure any resulting changes in health and welfare. This will involve, in particular, assessments of the intangible costs and benefits to parents. Hence there is likely to be a key role for further economic studies of the kind discussed here.

Note. Although John Henderson works as an economic adviser to the Department of Health, any opinions expressed are those of the author and do not necessarily represent the views of the Department.

References

1. Cohen DR, Henderson JB. Health, prevention and economics. Oxford: Oxford University Press, 1989.
2. Henderson JB. Economic evaluation of screening for fetal and genetic abnormality. In: Screening for fetal and genetic abnormality, King's Fund consensus development conference, programme and abstracts. London: King's Fund, 1987.
3. Mooney GH, Lange M. Economic appraisal in prenatal screening: some methodological issues. Paper presented to UK Health Economists' Study Group Meeting, Dublin. Copenhagen: University of Copenhagen, Institute of Social Medicine, 1990.
4. Hibbard BM, Roberts CJ, Elder GH, Evans KT, Laurence KM. Can we afford screening for neural tube defects? The South Wales experience. Br Med J 1985; 290:293–5.
5. Carstairs V, Cole S. Spina bifida and anencephaly in Scotland. Br Med J 1984; 289:1182–4.
6. Henderson JB. Measuring the benefits of screening for open neural tube defects. J Epidemiol Commun Health 1982; 36:214–19.
7. Hagard S, Carter FA. Preventing the birth of infants with Down's syndrome: a cost–benefit analysis. Br Med J 1976; 1:753–6.
8. Gill M, Murday V, Slack J. An economic appraisal of screening for Down's syndrome in pregnancy using maternal age and serum alpha fetoprotein concentration. Soc Sci Med 1987; 24:725–31.
9. Attanasio E, Galanello R, Rossi-Mori A. Analisi costi-benefici di un intervento preventivo per la talassemia. In: La prevenzione delle malattie microcitemiche, VI Congresso Internazionale dell' Associazione Nazionale per la lotta contro le microcitemie in Italia, Roma, 17–19 Aprile, 1980.
10. Nelson WB, Swint JM, Caskey CT. An economic evaluation of a genetic screening programme for Tay–Sachs disease. Am J Hum Genet 1978; 30:160–6.

11. Dagenais DL, Courville L, Dagenais MG. A cost–benefit analysis of the Quebec network of genetic medicine. Soc Sci Med 1985; 20:601–7.
12. National Medical Consultative Committee Working Group. Clinical Genetic Services in Scotland. Edinburgh: Scottish Home and Health Department, 1986.
13. Bush JW, Chen MM, Patrick DL. Health status index in cost effectiveness: analysis of PKU programme. In: Berg RL, ed. Health status indexes. Chicago: Hospital Research and Education Trust, 1973; 172–94.
14. Boyle MH, Torrance GW, Sinclair JC, Horwood SP. Economic evaluation of neonatal intensive care of very low birthweight infants. N Engl J Med 1983; 308:1330–7.

Chapter 21

Ethical Aspects of Prenatal Diagnosis

J. Harris

It is not news that prenatal diagnosis is becoming more and more common and more and more effective. We can now screen for all sorts of disorders from Down's syndrome and spina bifida to Huntington's chorea and AIDS. At the moment the main point of such screening is to give early warning of a disorder so that the parents can either prepare themselves for what's to come or can have time for a termination of the pregnancy.

The detection of any defect in an individual, but particularly a genetic defect, is hugely consequential for that individual, and may be equally consequential for society and for all who have dealings with that individual throughout life.

The process of screening of course has significant and by now well-rehearsed ethical dimensions. What level of fetal abnormality and what degree of likelihood of disability justify abortion? Might parents be wrong to fail to abort a disabled fetus? Are parents entitled to be screened as a matter of routine or must they be to some extent "at risk"? Are health professionals obliged or even permitted to breach patient confidentiality to protect other patients? This may happen where genetic screening reveals information which affects family members and warning one family member may well reveal to that member information about other members of the family. Do disabled children have a moral and legal right of redress against parents and health professionals who permit them to come into existence in a disabled condition? These so called "wrongful life" cases are familiar in the United States [1,2] and are even appearing in the United Kingdom [3].

All these issues have been discussed at length elsewhere [2,4,5] and on this occasion it would seem appropriate to examine two larger underlying questions.

They are: "What should we screen for?" and "What should we do about what we find?"

What Should We Screen For?

At the moment prenatal diagnosis, and screening, is done in a relatively piecemeal fashion with a whole battery of techniques and tests available, from amniocentesis and chorionic villus sampling to specific gene probes for particular genetic disorders such as Huntington's chorea. As the title of this volume suggests, such screening is at the moment largely designed to identify disease or abnormality, and in present circumstances the likeliest response to finding such abnormality is for termination of pregnancy to be considered. Where, using in vitro fertilisation (IVF) techniques, preimplantation screening is possible, the "remedy" is to decline to implant anomalous embryos. In the future we may expect genetic engineering and other methods of intervention to permit correction of at least some abnormalities, allowing the engineered or treated embryo to develop normally.

While "abnormality" is currently assumed to refer either to a disease state or to a genetic constitution which would result in some form of disability, there is already pressure to screen for (and to screen out) other identifiable features or phenotypes. The embryologist Mark Ferguson has reported that:

> It is now possible to take a single cell, extract the DNA from it and then, if one knows the region of the genome one wishes to examine, to amplify up that region to obtain enough DNA for analysis by gene probing. Currently there are some diseases which are diagnosable using the quantity of DNA present in a single cell. This technology is rapidly expanding and it seems likely that nearly all future diagnoses could be made on the basis of single cell biopsy [6].

Whether or not it will be possible to diagnose by these methods conditions like spina bifida is difficult to say. Although not regarded as a single gene disorder, it is likely that a number of genes are in fact involved in the closure of the neural tube and consequently genetic screening might pick out individuals predisposed to spina bifida. Certainly it is expected that once the genome project is completed each individual's genome might be mapped by such a single cell biopsy. When this happens a constellation of conditions and susceptibilities to disease will be revealed. These revelations will not only indicate the presence of genetic disorders but will also enable the detection of asymptomatic carriers of defective genes. These individuals can then be given counselling as to the risks to their children and other family members.

Privacy and Third Parties

The genome map will also reveal the likely susceptibility of an individual to various occupational illnesses and dangers and perhaps to some environmental hazards also. There are, not unnaturally, pros and cons to the provision of such information. On the one hand it may assist in the development of therapies and preventive strategies. It will also yield information which might interest third parties whether these are other family members, employers, the government or government agencies, and perhaps most consequential of all, insurance companies. The interest taken by these third parties will of course be their own. It will probably not coincide with the interests of the subject.

Early diagnosis of the dangers of third party access to such information will enable us to take early preventive measures if we can find the political will so to do.

Educational and Social Dimensions

We should also note the educational and social dimension to the new screening techniques. If, as is likely, genetic connections are established for things like musical ability, athleticism and so on, there may be considerable pressure to tailor the education of children earmarked for success or failure in particular areas. And the knowledge that such characteristics might be "diagnosed" will in turn create pressure for the provision of such tests. (Of course this seems remote from current clinical concerns, but we should remember that it is important to consider what it is right and wrong to do in advance of the pressure created by popular demand and clinical need. We should also be forewarned and forearmed that the pressure to tailor education towards the abilities and potentialities of particular children may well come from educationists, the State, employers and so on as well as parents.)

Gender Selection

The question of gender selection has been much canvassed. This of course has been with us for some time and there is already some considerable discussion of the particular considerations which might or might not justify such selection. In addition to the obviously respectable applications in the case of sex-linked disorders there are obvious opportunities here for chauvinism and prejudice, whether cultural, religious or personal, to result in the systematic preference for children of one gender or another. (I believe these problems are likely to be self-correcting over time, although this does not of course mean that systematic gender preference would not be highly damaging in the short term.) On a more parochial level, Baroness Warnock for example has not been slow to divide the sheep from the goats and has apparently distinguished frivolous from non-frivolous motives for gender selection. The securing of a male heir in the case of hereditary peerages is, for her, an example of a non-frivolous use of the reproductive technology [7]!

The Definition of "Abnormality"

One problem in trying to determine what should and should not be permitted is that it is simply not possible to define "abnormality" in a way which would rule out screening for genetic constitutions which might render an individual just slightly more susceptible to environmental or occupational hazards, or which might give them a statistically small (but perceived to be significant) increased risk of contracting some condition or of passing it on to their heirs. And in a strained but defensible sense, conditions like musical aptitude are also "abnormalities".

Legal Complications

Before we leave this survey we should note some obvious legal complications. Suppose a genetic counsellor knows that a patient carries a gene likely to result in a damaged child. The patient will not consent to this information going to anyone else. The patient eventually produces a damaged baby as do her two sisters. It may be that the patient's husband and the two sisters would sue the genetic counsellor for negligence (Margaret Brazier, unpublished). We have already mentioned wrongful life cases in which the damaged child sues a medical practitioner and/or the parents for the alleged wrong of being brought to birth in a damaged state. I hope that some of the principles which emerge below will help to resolve if not the legal position at least the moral rights and wrongs of situations like this.

The Issues

We will have to decide two issues. The first involves fears about what might follow from the detection of each condition in terms of developing or offering remedial therapy or in terms of termination of pregnancy. The second considers whether to offer (or whether we are entitled to decline to offer) diagnosis or screening for some conditions but not others. I will first consider the respectability of two arguments that are often produced as objections to prenatal diagnosis followed by termination of pregnancy (or of course, in the case of preimplantation screening, a decision not to implant). I will then try to identify the general and more abstract ethical principle which should govern our approach to such a question. In considering the second issue I will try again, and of necessity excessively briefly, to identify the general ethical principles which might help to determine what our attitude to the provision of diagnostic services should be.

What Should We Do About What We Find?

Aborting Beethoven

In any discussion of prenatal diagnosis when the question of screening for heritable disabilities is raised, the tantalising possibility of robbing the world of Beethoven is almost always raised. The story goes like this: "you are advising a pregnant mother and tests show that the child she is carrying is suffering from inherited syphilis and is highly likely to develop an associated deafness. Would you recommend a termination?" This story is told to a stooge who is supposed to reply "yes", to which the triumphant response is: "you have just aborted Beethoven." Even George Steiner has been seduced by the desire to produce this particular jack-in-the-box. In a recent television programme on genetic engineering he said [8]:

> It turns out that what in many cases is a hideous disease, a handicap can also be profoundly creative. Without the kind of meningital deafness which comes of inherited syphilis and alcoholism you and I would be sitting here without Ludwig Van Beethoven. Now that to me is absolutely key.
>
> Much of what has been the deepest, the most joyful, let me underline that the most joyful in human thought and creation has sprung out of very profound physical but also mental handicaps. I am not prepared to say that it would have been very much better never to engender the Muscular Dystrophy, if it was that, ... to which we owe Toulouse Lautrec.

(It is generally believed, *contra* Steiner, that Toulouse Lautrec was likely to have suffered from a skeletal defect called pyknodyostosis.)

The power of this argument lies in the idea that aborting Beethoven can only seem a good thing to do if we, or the world, or his family, or perhaps even Beethoven himself would have been better off without him. And since this seems an unlikely possibility we seem to be forced to the conclusion that Beethoven should not have been aborted and so neither should other fetuses in related circumstances.

But, to believe it right to abort a fetus is not to be necessarily committed to the view that the world would be better off without that individual, nor that the individual would eventually wish she had never been born, nor that that individual will be unhappy, nor that the individual will suffer. Nor in aborting an individual with Beethoven's syndrome are we in any sense aborting Beethoven or a potential Beethoven. In all these cases what we are aborting is an actual fetus and the rights and wrongs of that are determined by a consideration of the moral status of the fetus [4]. The fetus we abort will never become anything, and it is nothing but a fetus at the time it is aborted. It is as senseless to bemoan its loss as the loss of a Beethoven as it is to celebrate its loss as the pre-empting of a Hitler.

However, there is one other disturbing dimension to Steiner's ode to joy which we should note. It is the preoccupation with his own and humanity's pleasure at the expense of others. "You and I" he says, "would be sitting here without Ludwig van Beethoven", and this, he seems to believe, would be a tragedy even if it were very costly in terms of human pain and suffering to secure such pleasures for the rest of us. It is clear that Steiner is really committed to something like this, for the passage quoted above continues:

> ... We know very little about how much pain and suffering are positive in equations ... of human dignity, of human decency of human altruism ... somebody who has to be up all night with an incontinent relative or someone screaming in pain of Alzheimer's disease. To say to such a person we will not try and take your burden away is a hideous impudence, at the same time to say in a kind of Aldous Huxley way we are going to create a new sanitised essentially pain free, essentially beauteous set of worlds may be to alter the

balance of man's moral dignity, of man's religious questioning far out of any proportion [8].

Despite the equivocation of this passage Steiner pretty obviously is inclined to believe that it is preferable to save the art works rather than the people if the museum is on fire. He seems to believe that it is better to have a world with muscular dystrophy and Toulouse Lautrec's paintings, than a world without both.

But of course this isn't quite the possibility envisaged by the genetically engineered eradication of genetic defects nor by their pre-emption by prenatal diagnosis followed by termination of pregnancy. Toulouse Lautrec's paintings and Beethoven's Ninth Symphony will continue to give joy. Steiner is supposing that to eradicate genetic defects is somehow to cut off the wellspring of genius.

> We are playing with what you call on the Stock Exchange that haunting word "futures", and futures can stretch in this case into the centuries, perhaps into the millennia. Who are we to cut off what have been the sources often of our eminence, for a Nietzsche, for a Dostoevsky, for a Pascal ... how many, what are the units, the Benthamite Utilitarian units of sane good healths – it may be that they outweigh genius. [8]

Again it seems clear that Steiner doesn't think so, but this is not the point. It seems highly unlikely that genius is solely or even principally drawn from the ranks of people with genetic disorders, nor that genetic disorders play a causal role in the genesis of genius. There is equally no reason to suppose that in reducing human susceptibility to genetic disability we will, at the same time, be reducing our capacity for genius; that has to presuppose that geniuses are only going to be thrown up by genetic defects and there seems to be no evidence for that at all. There is then no reason to think that in reducing the sum total of human misery we will be reducing the sum total of human genius. But suppose the reverse to be true. Would we be morally justified in condemning individuals to a life of misery or disability in order to secure for ourselves a few more excellent symphonies or a painting or two (or even two dozen?)?

Advantageous Disadvantages

Another argument that purports to limit our interventions to prevent genetic disorders was produced by Germaine Greer in the course of her participation in the same television programme. Just as George Steiner wants to insist that there may be a positive side to pain and suffering which would justify our declining to eradicate it when we had the chance, so Germaine Greer believes that humankind should have the advantage of the disadvantages of particular people.

> Curing a disease is one thing, stopping it existing is another ... people also have the advantage of their disadvantages. If a trait is totally maladaptive then it doesn't survive. If Huntington's chorea has survived in ... special populations there's a reason for it and we just simply haven't found it ... When we understand more about it we might be able to deal with the deleterious activity of the gene and preserve the positive aspect of the gene [8].

Of course this might be true but we must ask at what price would we have purchased this understanding and this eventual ability to deal with the deleterious consequences? First, it is simply an impious hope on Germaine Greer's part that there is anything good about Huntington's chorea, that it does have some hidden major usefulness to humankind. I know of no reason to suppose that Greer's claim is true in the sense that all traits or all genetic defects or even perhaps all diseases are adaptive in the positive sense of being good for humankind or good for the universe or the ecosystem. They may simply for example be good for the destructive trait, organism, virus or whatever, in question.

But more important what price should we put upon the outside chance of Greer being right? Well if it costs nothing to bet on Greer and wait and see then we should certainly do so. But of course it doesn't cost nothing. It costs *her* nothing, but it costs those who have and who, *ex-hypothesi*, will continue to have Huntington's chorea and other terrible genetic disorders a great deal in terms of pain suffering and premature death.

We would surely have to be very confident indeed of a number of things before we might be justified in accepting Greer's cruel gamble. The first would be that there is indeed a positive side to Huntington's chorea, second that such a positive side is sufficiently important to be worth preserving at the terrible cost in human suffering that would be required to preserve it, and finally that such positive effects could not be, or were not likely to be, achievable another way.

In the absence of such confidence I hope that no sane, let alone moral, being would think it worth preserving disability and disease on the off-chance that some good might come of it at some unspecified and unpredictable point in the future.

The Obligation To Prevent Suffering

If we search for a principle that explains and justifies the decision to decline to bring into the world an individual who will, or will very probably be damaged it is perhaps this:

> *That we have an obligation to prevent suffering and disability, or, more abstractly, that we should try to produce a world with less rather than more suffering in it, that we should try to produce a happier world.*

Moreover, that if abortion is permissible to preserve the life and health of the mother, then it is permissible to prevent suffering and disability to others. It is not sensible to suppose that in doing so we might have "aborted Beethoven" nor should we condemn some individuals to suffering or to a restricted life on the off-chance that this might secure some benefit to third parties. Indeed there is clearly a moral obligation to provide such screening where possible so that parents can have the opportunity to choose not to bring suffering or disability into the world.

Now, in so far as we have available only specific tests which would make possible the diagnosis of specific conditions, like amniocentesis for Down's syndrome and spina bifida or using a genetic probe to detect Huntington's chorea, it is clear both that we should provide these, and that we should be prepared to act on the results to prevent disability or suffering. Equally, since being of a particular gender is not, in the absence of sex-linked abnormality, a disability nor a cause of suffering (I leave aside the issue of suffering caused by prejudice and discrimination and susceptible to social rather than medical therapy) nor is possessing or lacking musical ability or green eyes, there is not the same moral imperative to provide diagnostic tests for these conditions.

It would not, of course, be wrong to screen for features such as these, but with the sorts of exceptions we have discussed, there would be no moral obligation to do so.

On completion of the genome project, however, there may be available a single procedure, a single cell biopsy perhaps, which would generate a map of the entire genome and all of the information, both consequential and trivial that this implies.

Could we be selective in passing on this information? Could we be selective in acting on it?

Access to Information

Again, this issue of access to the information generated by diagnostic testing is going to become more and more central and increasingly troublesome as such screening becomes more routine. It is worth trying to say something about the ethical principles which should govern our approach to these issues.

First, we must consider whether there is an obligation to tell patients everything that diagnostic tests reveal about the patients themselves. Second, we must consider the extent of the confidentiality of such information.

Telling Patients the Truth

There are two good and mutually reinforcing reasons why patients should be given all remotely relevant information about them which comes into the hands of health professionals. The first has to do with the necessity of obtaining the patient's consent to any treatment, the second has to do with the nature of the relationship between patient and health professional.

Informed Consent

Any treatment a health professional would wish to give requires the consent of the patient, for any treatment given without consent would constitute a moral violation and a legal battery. No consent can be genuine if it is not fully informed, for the point of consent is that what happens to the patient should be of the patient's own choosing, she must be *minded* to have it happen to her. And to be minded to have something happen, you must know not only just what is going to happen and what its likely consequences will be but also what the alternatives are. In short, consent is necessary for treatment and full information is necessary for consent to be genuine. (The word "information" here includes not only facts about the patient but diagnoses, judgements and speculation relevant to clinical decisions which flows from the facts. I do not of course imagine that it is easy to draw parameters of relevance round the wanderings of the health professional's mind.) If health professionals wish to treat patients they must provide full information. And this is so even if the professionals believe that the information will cause harm or suffering to the patients.

To argue this point fully requires more space than we have here, but the point is that the professional cannot know how important the possession of such information would be to the patient, however traumatic the information might be. The professional may not deny the patient the opportunity for self-determination that the information gives. In short, only the patient can say whether or not the information was so traumatic that he would rather not be told, but he can only say this in knowledge of what the information is. Thus paradoxically he must have the information in order to be sure that he would have wished not to have it. This shows that wherever any treatment is involved full disclosure must be made.

In the United Kingdom the traditional legal mechanism for protecting patient autonomy has been this requirement for valid consents to treatment enforced through the torts of trespass to the person and negligence.

However, where health professionals do not wish or intend to offer any treatment this argument does not provide a compelling reason for disclosure. Nor does this legal and moral protection operate where the health professional does not need to make physical contact with the patient to acquire information. For example, if the diagnostic test requires a blood test or even a mouth swab taken by a health professional, then such an act would be unlawful without an autonomous consent. However, if the cells for biopsy are obtained by a self-administered mouthwash then the legal protections do not apply, although of course there would still be a moral obligation for full disclosure.

This is why we must also consider the nature of the patient/health professional relationship.

The Professional Relationship

The professional only acquires information about the patient because the patient freely gives it or because the patient consents to the investigations which yield the information whether or not these investigations require physical contact. Either way the patient is entitled to give the consents which generate the information *conditionally*. That is to say the patient may stipulate that his consent is only given on condition that all information which flows from it is made available to the patient.

Since there is an obligation to provide screening because it is both in the patients' and in society's interests we cannot, morally, prevent a patient receiving all the information generated by the screening process.

Moreover, the professional cannot know or sensibly predict what information, however seemingly trivial, will turn out to be important to the patient and hence what harm might be done to the patient by concealing the information.

From this it follows that if the patient wants to know anything about himself/herself of which his/her medical advisors are aware then he/she is entitled to that information. There are still problems about whether the entitlement to information is only "information on demand" or whether it should be provided undemanded. This problem will not be considered here.

Information from Third Parties

There is, of course, a small category of information about patients which comes not from the patients themselves nor from observation of those patients nor from tests done on them but from third parties, as is the case, for example, in family investigations during genetic counselling. Here there can be a conflict between the duty of confidentiality on the one hand and the obligation to make relevant information available on the other.

To resolve this we need to consider another and more abstract moral principle which falls on all of us equally whether we are health professionals or not [4].

Do No Harm

This is the obligation not to be responsible for harm to others and where we cannot help harming someone, to do the least harm possible. And of course such responsibility refers both to act and omissions. We should neither make harmful interventions nor omit interventions which would prevent harm.

In these rare cases we have to compare the harm done to patient A if confidentiality is breached coupled with the harm that the weakening respect for confidentiality that this may occasion does on the one hand, with the harm to patient B if he/she is not made aware of the information.

Conclusion

I have argued that in answer to our first question as to whether we could be selective in passing on the information generated by diagnostic tests it must be "no". We have an obligation to pass on information to patients about themselves unless doing so will cause greater harm to third parties. This leaves the problem of the sorts of requests for treatment and other services that genetic and other diagnostic screening may generate.

This is a large problem to embark upon so I will content myself with just two observations. We need to distinguish the questions: What are we obliged to provide? And what are we obliged to decline to provide?

The answer to the first question is easy to give but difficult to resource. The answer must of course be that we should provide all services necessary to secure the health and well-being of our citizens and to enable them to secure this for the children they wish to have.

The second question we should answer conservatively. We should not outlaw or ban any services unless it would be wicked to provide them or profoundly harmful to the patients themselves or to others including of course to society. But we should not take it upon ourselves to adjudicate as to which requests for services are frivolous and which are not.

A useful rule of thumb test to apply might be this. If it is not morally wrong for someone to wish something for themselves then it is not wrong to help them to grant themselves that wish. So if it is not wrong for a parent to wish for their next child to be a girl, or to be brown-eyed, it is not wrong to grant the wish. This does not of course mean that anyone is obliged to grant it, simply that we should not legislate to prevent it being granted.

Acknowledgements. This paper has greatly benefitted from the legal counselling of my colleague Margaret Brazier. I am also grateful to Dian Donnai who has saved me from a number of errors.

References

1. Steinbock B. The logical case for wrongful life. In: The Hastings Centre Report, 1986.
2. Harris J. The wrong of wrongful life. J Law Soc 1990; 17(1).
3. McKay v Essex Area Health Authority. 2 All E.R.771, 1982.
4. Harris J. The value of life. London: Routledge, 1989.
5. Harris J. Wonderwoman and superman: ethics and human biotechnology. Oxford: Oxford University Press (in press).
6. Ferguson MJW. Contemporary and future possibilities for human embryonic manipulation. In: Dyson A, Harris J, eds. Experiments of embryos. London: Routledge, 1990.
7. The Sunday Correspondent, 22 April 1990; 1.
8. BBC TV, The Heart of the Matter on genetic engineering. 22 October 1989. (For quotations I am relying on the transcript of the programme prepared by the BBC. I have added what I hope is appropriate punctuation. Since the quotations are transcriptions of "live" unscripted interviews neither George Steiner nor Germaine Greer express themselves with their characteristic elegance.)

Discussion

R. Harris: Professor (John) Harris was splendid in being so provocative. He makes people think.

Can I be devil's advocate and say that doctors and people who work with doctors are devoted to the principle of reducing the amount of misery in the world. That is based on pragmatic principles, and in genetic counselling, being specific, if one is directive or one insists upon giving every bit of information even though the patients are putting up their hands and saying, 'Stop!', then pragmatically one is probably not in a position to reduce the amount of misery because one loses the patient's support. It is all a system of checks and balances.

J. Harris: I was not suggesting that people like Professor (Rodney) Harris or Dr Donnai should be directive in their counselling. I was reserving to myself the liberty of being directive and as it were in an abstract way by saying what conclusions people ought to come to, but not what conclusions they should be told to come to by their genetic counsellors.

The other part of what he says, there is a problem about when to deliver and how to deliver information. It may be that the patient is holding up her hands and saying no more now and that she has been there for three hours and enough is enough, and that is a reasonable request. But there are ways and ways of delivering information.

Galjaard: The argument about reducing suffering and making a happier world is supported by many people, but there are also many other arguments. Parents who have integrated their suffering, having had one or more handicapped children now stand up and defend the birth and the experience of these handicapped children as having made them happier, their marriage better, and so on. They comment that I am talking about reducing suffering and question what I know about happiness. It is not a universally held view.

Williamson: It is the introduction of choice that is important here. There are families who have overcome the problems of an accidental handicap. That is really quite different from being in a position of advocating this as Professor (John) Harris has.

Allan: There are couples who make a positive experience out of making an informed choice to continue with their disastrous pregnancy and at the end of the day feel that that was the right decision. They must have the room to do that.

Wald: It is possible to have different forms of consent all of which are valid and proper. It can be explicit consent and it can be implied consent. In addition, someone can consent to something in the normal sense of the word without knowing all the implications of what they are doing, provided it is clear that they do not want to know that and they are making their decision on the basis of trust. This element of trust is often neglected in the discussions that concern the relationship between patients and doctors. We have moved to a situation where we are treating the doctor–patient relationship much more as a contractor–customer relationship, which I think is inappropriate. My own view is that the job of a good doctor or counsellor is to gauge what the patient or the individual wishes, and sometimes one has to judge that intuitively, sometimes explicitly. And there is a valid point about harming people by burdening them with knowledge that they do not want, and at the same time there is a risk of not providing information that the individual would wish. And the skill of medicine is to strike that balance, but once that balance has been struck in good faith, an individual who makes a decision is consenting, but consenting in terms of different types of consent, all of which are valid provided they fulfil the kind of considerations that I have just described.

Pembrey: I shall take the liberty of following up on that. Professor Wald's analysis begs the question of whether the genetic services we have been discussing are seen as within the province of medicine. This is one of the problems with the issues of how we look at the service, how people use the service, and this apparent discrepancy between caring for a handicapped child on the one hand and aborting a pregnancy on the other.

We really have to tackle the anxieties that the politicians and other people have about the genetic servce, that it is a state system with the aim of reducing the birth incidence to reduce the costs, that it is seen as outside the usual area of medicine where the element of trust operates. In fact I would like to see the definition of the aims of genetic services to be along the lines of "maintaining or restoring family life in the face of genetic risks or in the face of a fetal abnormality". We know perfectly well we can reduce birth incidence just by frightening people, they are left with one affected child, they are too terrified to have one again. This will achieve birth reduction but not their goal, which is to have their desired family size.

So it comes down to the fact – we cannot use the trust argument unless we can put medicine and these services into the medical domain, and many people worry that we are not doing that.

Lilford: I always feel uneasy about the trust argument in the context of prenatal diagnosis. With a subject like prenatal diagnosis where views among doctors, as among ordinary people, have such wide scatter, the obligation to give the patient more information, perhaps even more than she may initially want, is greater than it is in other aspects of medical care. What people do with that information is so much broader in prenatal diagnosis. There is such a wide range of possible human responses.

Marteau: If I might follow up on Professor Wald's argument about clinical intuition.

Clinical intuition has a poor track record in the area of doctors guessing how much information patients want. Invariably doctors tend to underestimate the amount of information that people want, and that is a danger.

Wald: I do not think I was saying that one should rely on it. I was simply recognising that in a dialogue, in the discussions that we have now, only a small part of the communication is explicit. A considerable part of the communication is the way we look, the way we behave, the way we move, eye contact and so on, and these kind of exchanges are very important in establishing an understanding and a rapport with the aim of ultimately serving the needs of that individual as that individual would wish them, which I am sure must be the aim of most good doctors, counsellors, advisers. Is it any different from that? Need we complicate it further?

Campbell: Can I get back to ultrasound?

When the patient is having a diagnosis by ultrasound, she sees the diagnostician in operation and he takes her through the diagnosis, and she will see the physical defect. She respects the diagnostician because he is an expert and she believes he knows everything about her baby by the time he has finished that scan. And so the whole of the remainder of the counselling is based on trust. Here is an expert, he has seen the baby, he can take that patient through the anatomy in detail and he can then counsel as to the long-term future. This makes ultrasound quite unique in terms of counselling.

J. Harris: Professor Campbell says she can see the defect. When I look at an ultrasound I never see anything, and it really is a question of trust that that bit is a perfectly healthy embryonic head and that bit is an unhealthy one. What she can see is what she is told. An educated observer might have disagreements with ultrasound operators about what they are seeing.

Can I come back on the happiness argument. It is very important. Rarely in philosophy is there a simple test, but we can show that Professor Galjaard's suggestion is wrong as follows. Imagine a pregnant woman who has a condition. The fetus is damaged, but there is a simple risk-free procedure which will remove the damage. She just has to imbibe orange juice and the handicap will be removed. But she says no, she does not want to do it, she does not want this therapy because the last handicapped child she had made her so happy she intends to have another. What one would feel about such a decision gives the key to the respectability of the happiness argument from other people's misery.

Nevin: A point relevant to the two legal arguments. The question of a genetic counsellor who is dealing with a patient, has that patient's confidentiality. The patient who has a balanced translocation may have similarly affected relatives and we would need to see them so as to counsel them. It was suggested that from the legal point of view we have an obligation to see those relatives, but from the confidentiality of the doctor–patient relationship we cannot break that confidence. How is that argument resolved?

J. Harris: Often such people do consent. The problem arises where they do not. I am not sure whether legal action would succeed. Where there is an absolute obligation of confidentiality one would have to ask oneself what it will cost in terms of other people's lives to keep this patient's confidentiality. One would have to weigh up the relative importance of keeping this patient's secrets on the one hand and what she will do if that confidentiality is breached with what those people will lose if they are not forewarned about their condition.

It is even clearer, for example, in AIDS counselling. Do we tell somebody that if they sleep with or exchange blood products with this person they are likely to die. If the patient does not want us to tell their partner, what then? We have to weigh up what the AIDS victim stands to lose and what their partner stands to lose, and when we compare the two we find we are not comparing like with like and we have a clear imperative to protect innocent third parties in those things.

Modle: I want to draw from Professor (John) Harris some comment about prenatal diagnosis, and possibly termination of pregnancy in relation to genetic disorders which may not present for many years after birth, possibly after the parents themselves are dead. Huntington's chorea would be a possibility.

J. Harris: Certainly. Should terminations follow where there is 20 or 30 years of happy normal life expectancy?

Modle: How much in the mother's interest can it be to have all this doubt when she may be dead before the child's illness manifests?

J. Harris: I do not think we do it because it is in the mother's interests necessarily. One is deciding whether it would be better to bring somebody with a disease like Huntington's into existence where there is a choice. The fact that the mother may be dead before she sees the distress of her child would not be part of my reasoning about that case. I think it would be better not to bring that degree of suffering, albeit postponed, into the world. Taking a decision when no person is in being is quite different from saying to a 20-year-old who has Huntington's and who will die from it that their life has not been worth having. When it is an embryo or a fetus before it has a conscious life, the calculation to be made is which action causes the least suffering, and I think termination is the answer to that question.

R. Harris: Chapters 20 and 21, considered economy and ethics, and the discussion has been dominated by ethics. However, it is the economy that we shall have to face in the next year or two, with the reform of the Health Service.

It is outcome of medical procedures which really matters, and that is something that everybody knows how to measure, and that is good clinical research. That means there must be more money for research into outcomes to see whether things do good or not; not whether they are effective or efficient in economic terms.

Second, I was delighted with the emphasis on the intangibles, and on the benefits outside the Health Service. Quite often the things that we have been discussing have benefits of those sorts which may outweigh any monetary considerations.

Finally, one thing not mentioned – and that was the question of discounting. If by spending a relatively small amount of money now, money can be saved over the next 20 or 30 years, economists always claim that by the time it is discounted the savings can be dismissed. How do we counter that argument?

Henderson: I am not sure I fully understand the question. It is suggested that by spending a small amount of money now we get a large return later.

Harris: By the time those savings have come along, the pound is not worth what it is now, and therefore the savings are far less than are predicted.

Henderson: Perhaps to simplify the example, there is a choice of two building societies, one offering 10% and the other offering 6%. Where would one put one's money?

If we are comparing two investments, we want the one with the greatest return. Whichever investment is selected, savings in 20 years are worth less than savings in 10 years. But what one tries to do is to find the investment with the greatest return overall.

Wald: I do not think one can counter that argument. Prevention inevitably means spending money now for returns a long time in the future. The discount effect is a reality that one has to simply accept. It is true with breast cancer screening and most kinds of screening.

Section VII
Service Provision

Chapter 22

Organisation of Genetic Services in the Netherlands

H. Galjaard

Introduction

In the Netherlands the clinical application of laboratory methods in cytogenetics and biochemical analysis started in departments involved in basic research and in hospital departments with a special interest in early diagnosis of congenital and genetic disorders. During the 1960s and early 1970s the various activities that later were concentrated in departments of clinical genetics were quite randomly distributed over departments of biology and biochemistry in science faculties, departments of cell biology, biochemistry, human genetics and pathology in medical faculties and in hospital departments of paediatrics, obstetrics and gynaecology, endocrinology, haematology and ophthalmology and in some institutes for the handicapped.

The main emphasis in the early days was on chromosome analysis in relation to spontaneous abortion and the diagnosis of congenital malformations, mental retardation and sexual developmental abnormalities. In addition there was an increasing interest in the search for biochemical abnormalities in children with a possible genetic metabolic disease. All activities mentioned above were performed within the organisational framework of the departments and without special funding for clinical applications.

From 1968 onward the Department of Cell Biology and Genetics, Erasmus University in Rotterdam, in collaboration with the University Hospital Department of Obstetrics and Gynaecology, started with prenatal diagnosis of chromosomal aberrations and with the application of microchemical methods in the prenatal diagnosis of inborn errors of metabolism.

In the early 1970s a substantial grant was provided by a semigovernmental foundation "Praeventiefonds" to the Rotterdam group to allow further development in prenatal diagnosis of congenital disorders. A few years later it became clear that clinical and laboratory methods aimed at early diagnosis of index patients as well as genetic counselling of couples at risk and prenatal monitoring would be of increasing importance in health care.

In the Netherlands new developments in medical technology are usually evaluated by an independent Health Council. A group of selected temporary advisers prepare a document about the state of the scientific and technological development, the desirability of implementation into the health care system, a cost–benefit analysis and possible ethical, psychological and social implications. In 1977 the Health Council advised the Ministry of Health about the various aspects of biochemical and cytogenetic diagnosis, genetic counselling and prenatal diagnosis [1]. These recommendations were accepted by the government and the major political parties. Between 1977 and 1979 representatives of the health insurers and various medical and laboratory experts negotiated about the future organisation and funding of clinical genetics. An agreement in 1979 formed the basis for the development of seven regional centres of clinical genetics and for the organisation and funding of the various activities up to the present time.

Funding and Legal Background

The overall expenditure on science and technology in the Netherlands amounted to 2.3% of the gross national product in 1990. About 58% of this expenditure is by private institutions (mainly the four large multinationals) and 42% by the government. A total of about Dfl. 1000 million (£350 million) is spent on medical research. Nearly half of this is funded by the Ministry of Education and Science through the eight medical faculties and university hospitals.

New developments usually start as a research effort by individual university departments and to a lesser extent in institutions for medical research outside the university. Only about £40 million for medical research is spent through the Ministry of Health. Heads of department are free to select their own priorities in the case of university funding and they or their coworkers have to compete with other projects if they want additional funding from outside sources. Because of the limited funds and relatively hard competition new developments such as the application of DNA technology, new biochemical methods or experimenting with ultrasound diagnosis are most likely to be initiated by groups that already have a strong position in clinical and basic research.

The situation in the provision of health care services is quite different. Most diagnostic and therapeutic activities are funded by semigovernmental and private health insurance institutions. People with a lower or median income (about two-thirds of the population) have an obligatory insurance with one of the semigovernmental institutions (ziekenfondsen) and the premium is paid partly by the employee and partly by the employer. About one-third of Dutch citizens have a health insurance with one of the private insurance companies. Within a few years this difference will disappear and all citizens will have the same type of health insurance. The premium will then largely be income dependent.

In case of negotiations about the funding of new activities, such as diagnostic chromosome analysis, biochemical tests, DNA studies, genetic counselling, or prenatal diagnosis the experts involved have to agree with representatives of the two categories of health insurers about the indications and the way of funding. If an agreement is reached it has to be approved by a Central (national) Council for Tariffs in Health Care (Centraal Orgaan Tarieven Gezondheidszorg) and by the Minister of Health.

In 1979 tariffs for prenatal and postnatal chromosome analysis, various types of biochemical tests for the postnatal diagnosis of inborn errors of metabolism and for amniocentesis were established (Table 22.1). Initially the funding of genetic counselling and postnatal enzyme tests was based on lump sum subsidies to certain centres of clinical genetics. When sufficient experience had been gained, from 1985 on tariffs were also established for these activities (Table 22.1). Lump sum funding to a limited number of centres is possible via a special joint fund of all health insurers and under the General Law Special Costs in Health Care (Algemene Wet Bijzondere Ziektekosten). This way of funding has been chosen in instances where the daily practice of new activities in genetics still had to be evaluated but where funding by the health care system was nevertheless considered necessary. A recent example is the AWBZ lump sum funding of four centres for diagnostic DNA tests for the period 1988–1991. The centres to be funded are selected on the basis of quality, experience, geographic location and

Table 22.1. Tariffs for different genetic services in the Netherlands

	Dutch guilders	English pound (rate 3.50)
Postnatal chromosome analysis	1112	318
Prenatal chromosome analysis (incl. genetic counselling)	1212	346
Biochemical diagnosis of genetic disease		
Metabolites in blood/urine	tests vary from 12 to 34	average expenditure per patient 300
(Enzyme) protein test on cell material	tests vary from 100 to 1120	average expenditure per patient 300
DNA diagnostic tests (lump sum funding)	vary from 520 000 to 910 000 per centre	vary from 150 000 to 260 000 per centre
Genetic counselling in complex situations	1700	486
Amniocentesis	400	114
Chorion sampling	536	153
Prenatal AFP test (incl. genetic counselling)	198	56
Prenatal biochemical diagnosis + cell bank (lum sum funding of one centre)	ca. 500 000	143 000

expected demand. Once sufficient experience has been obtained with DNA diagnosis it is likely that one or more tariffs will be established and that all seven centres of clinical genetics will receive governmental recognition to perform such tests (see also next section).

The advantage of a tariff per activity is that a laboratory or clinical department can respond to a growing demand by more activities and that the increasing income will allow more personnel to deal with the increasing workload. Of course, the health insurers must have a means of controlling growth and therefore the indications for genetic counselling, prenatal monitoring and cytogenetic, biochemical and DNA tests have been outlined clearly both for the experts and for the medical staff of the insurance companies. For each activity, be it amniocentesis, chorion sampling, laboratory test or genetic counselling, the expert in the clinical genetics centre has to ask the permission of the local health insurer. Permission and hence payment will only be obtained if the activity is performed in one of the seven certified centres of clinical genetics and if a correct indication is used. In addition, each centre provides an annual report with data on the number of persons investigated and the number and percentage of patients detected; these reports are sent to the National Council of Health Insurance and to the Ministry of Health. These data allow a yearly evaluation of the activities per centre and of the overall growth of human genetics services in the country.

An important legal restriction to unlimited growth is the so-called Law of Special Hospital Services (Wet Bijzondere Ziekenhuisvoorzieningen). This law provides the Minister of Health with the power to restrict certain activities, such as heart or liver transplantation, special kinds of radiotherapy, and also some of the genetic services, to a limited number of centres. A centre that wants to be funded has to be approved by the Ministry of Health and again, quality, annual activities, regional location and expected demand and efficacy are major factors in the decision of approval.

For all genetic services it has been decided that seven regional centres for clinical genetics are sufficient to meet with the demand (about 15×10^6 population and 180 000 annual births). Each of the centres has facilities for prenatal and postnatal cytogenetics, postnatal biochemical diagnosis of genetic metabolic disease, genetic counselling in complex situations (see next section), amniocentesis and chorion villus sampling. Prenatal biochemical diagnosis and an associated cell bank are centralised in one centre (Department of Cell Biology and Genetics, Rotterdam) and DNA diagnostic tests are thus far limited to the centres in Groningen, Leiden, Nijmegen and Rotterdam.

An exceptional situation exists for the obstetricians involved in prenatal diagnosis. Contrary to the laboratory analysis, the sampling of fetal material is not legally restricted. As a consequence every obstetrician/gynaecologist is in principle allowed to perform amniocentesis or chorion sampling, but the laboratories of the seven clinical genetics centres have decided to accept material only from obstetricians who they feel have sufficient expertise. In a small proportion of cases this leads to referral of fetal material to laboratories outside the Netherlands. This procedure is sometimes also followed in cases where an obstetrician knows that the insurer will not fund an activity because it is not within the agreed indications (for instance women below the age limit of 36 years for prenatal chromosome analysis) and where the patient still wants prenatal monitoring.

Within the Netherlands no private payment is accepted by members of the clinical genetics centres, but as long as there are no legal restrictions on the sampling of fetal material, referral of patients to private clinicians or laboratories in foreign countries cannot be prevented. In view of the increasing internationalisation and the closer collaboration between European countries it would seem useful to establish better contacts between the various Ministries of Health and between legislators and experts involved in genetic services. The experiences in different countries could then be exchanged, the most effective and reliable procedures might be adopted and the national legislations compared and adjusted.

In the Netherlands genetic services are considered an essential part of health care and in November 1984 the Ministry of Health, on the advice of the National Council of Health Insurance, decided that genetic services are a legal right for every Dutch citizen under the Insurance Law (Verstrekkingenbesluit Ziekenfondsverzekering).

During the past decade most efforts have been focused on the organisation, funding and legislation of the provision of genetic services. Recently, however, there has also been a growing concern about possible negative aspects of future advances in genetic technology. Major points are negative discrimination in admittance for insurances, mortgage and certain jobs on the basis of results of genetic testing. The Health Council has prepared an updated document with advice on new developments in genetics and their social and ethical consequences [2]. At the formation of the present government in 1989, the political leaders of the main coalition parties agreed upon the preparation of legislation that would sufficiently protect citizens against such discrimination. Again, it seems worthwhile if representatives of governments, legislators, medical experts and patient–parent organisations from different EC countries would coordinate activities in this area.

Organisation and Results

The tariffs per activity (Table 22.1) are mainly based on calculations of the number and type of personnel needed for a given number of activities (Table 22.2). In addition the costs of space, equipment, consumables and overhead services are also included in the tariff. The different ways the costs of individual genetic services are calculated in various countries and the different funding of health care, inclusion/exclusion of private honorarium of the clinician and the different economic significance of the various currencies make international comparison very difficult.

The number of activities varies per centre depending on the population it serves, the referral for specific activities based on special interests of the department, and its competence as judged by referring doctors. Also, the strength of related disciplines such as molecular genetics, paediatrics, obstetrics, neurology, haematology, oncology and others, play a role.

Table 22.3 summarises the annual activities in our own centre which has a total of 90–100 staff one third of which is academic. The activities in the Rotterdam centre, which is the largest in the Netherlands, are performed at an annual expense of about Dfl. 8 million (£2.3 million). The centre has subdivisions for

Table 22.2. Staff members for different genetic services

Postnatal chromosome analysis including genetic counselling	1 clinical geneticist, 1 biologist and 10 technicians[a] and administrative staff per 1200 analyses per year
Prenatal chromosome analysis including genetic counselling	same + 2 extra technicians[a] per 1200 analyses per year
Metabolite analysis	1 chemist, 1 senior laboratory assistant, 4 technicians, 0.5 administrative (ca. 1000 new patients per year)
Postnatal enzyme protein analysis	1 biochemist, 2 technicians[a], 0.5 administrative
DNA analysis	1 molecular biologist, 2–4 technicians depending on number of patients investigated + 0.5 administrative
Genetic counselling	3 clinical geneticists, 1 psychologist, 1 genetic associate, 2 administrative for 500–600 complex counsellings
Amniocentesis, CVS	3 obstetricians, 4 technical/administrative per 3000 pregnancies monitored
Ultrasound monitoring for fetal abnormality	No funding from health insurance
Prenatal biochemical diagnosis + cell bank	1 biochemist, 4 technicians and 0.5 administrative

[a] Includes salary for necessary photographers.

Table 22.3. Annual activities in the Rotterdam Regional Centre for Clinical Genetics[a]

	No. of individuals tested	No. of anomalies detected	Percentage anomalies
Postnatal chromosome analysis			
Congenital abnormalities	1673	228	14
Tumour cytogenetics	231	129	56
Chemical analysis metabolites in blood or/and urine	800–900	20–40	3–5
Postnatal (enzyme) protein assays in cell material	260 Dutch 81 foreign referral	50 41	19 51
Postnatal DNA analysis	559	33 carriers identified	
Genetic counselling in complex situations	525	ca. $\frac{1}{4}$ risk not increased $\frac{1}{2}$ risk 1–15 $\frac{1}{4}$ risk 15–50	
Prenatal diagnosis			
Amniocentesis, CVS	2440	88	3.6
Ultrasound	1634	169 fetal abn.	10

[a] Data for 1988–89 kindly provided by Mrs Professor E. Sachs, Mrs Dr M. Jahoda, Professor J. Wladimiroff, Dr W. Kleijer, Dr O. van Diggelen, Dr J. van Hemel, Mrs Dr D. Halley, Professor M. Niermeijer and Ms M. Veldhuizen.

postnatal cytogenetics, including cancer cytogenetics for diagnostic purposes, metabolite analysis in close association with the department of paediatrics, postnatal enzyme studies, DNA analysis, prenatal diagnosis in close collaboration with the department of obstetrics, and genetic counselling. The payments by the health insurers plus funding from other sources for new developments go to a Regional Foundation of Clinical Genetics which has an independent board, a coordinating director and a small administrative staff for personnel, finance and daily management.

Requests for new staff, equipment or space are sent to the coordinating director and decided by the board which meets several times a year. Most staff members, paid by the Regional Foundation of Clinical Genetics, are administratively appointed in the Department of Clinical Genetics, others in the Department of Obstetrics and Gynaecology, the Department of Paediatrics, all of the University Hospital or in the Departments of Cell Biology, Biochemistry, Genetics or Medical Psychology, all of the Medical Faculty. In this way each person works in the setting that is best suited for his/her specialty and at the same time the Board of the Foundation of Clinical Genetics and the coordinating-director have a final say in the distribution of funding and personnel and they can control priorities, efficacy and coordination.

The other clinical genetics centres have similar services, except those for prenatal biochemical diagnosis which is a highly specialised activity that involves too few cases to warrant more than one centre for the whole country. The overall activities in all seven centres for 1989 are summarised in Table 22.4 as well as an estimate of the number and percentage of patients discovered. The total costs of the genetic services in the Netherlands is of the order of Dfl. 35 million (£10 million) per year; this is less than 0.1% of the total expenditure on health care (about £11 000 million).

Table 22.4. Estimate of annual activities in seven clinical genetics centres in the Netherlands[a] (15 million population, 180 000 births)

	No. of individuals tested	No. of abnormalities detected	Abnormalities (%)
Postnatal chromosome studies	ca. 6500	ca. 1250	19
Chemical analysis metabolites	ca. 4000	ca. 220	5
Postnatal enzyme studies on cell material	ca. 1000	ca. 210	20
DNA analysis	ca. 2700	255 carriers 63 patients	
Genetic counselling in complex situations	ca. 2500	about $\frac{1}{4}$ no increased risk about $\frac{1}{2}$ risk 1–15% about $\frac{1}{4}$ risk 15–50%	
Prenatal monitoring by amniocentesis or chorion villus analysis	ca. 7000	ca. 200 affected fetuses	2.9

[a] Data from the seven Dutch centres in Groningen, Nijmegen, Utrecht, Amsterdam, Leiden, Rotterdam and Maastricht. In addition subcentres in Eindhoven and Enschede perform chromosome analysis and obstetricians in Arnhem, Eindhoven and Enschede perform amniocentesis and CVS.

The relatively high proportion of chromosomal abnormalities found is likely to be due to a strong selection by the referring clinicians; at the same time a one-year trial with loosening of the firm indications showed that hardly any extra chromosomal aberrations were detected. In most other countries the proportion of chromosomal abnormalities found is lower which results in a heavy workload without the benefit of a higher detection rate.

The investigation for abnormal metabolites in blood and urine of patients with an unexplained psychomotor retardation or other symptoms and signs that might suggest a genetic metabolic disease (see Chapter 14) requires heavy investment in equipment and in personnel. Most centres investigate between 600 and 800 new "patients" each year and find 4% – 5% of them to suffer from a genetic metabolic disease. Some of these patients require repeated chemical tests to evaluate their clinical situation and the effect of treatment.

The efficacy of direct (enzyme) protein assays on cell material is greater and most centres find 10%–20% genetic (enzyme) protein defects in the material that is referred to them. Again, this depends very much on the expertise of the referring clinician and on regular contacts between clinician and laboratory.

DNA analysis, either in the context of gene tracking or demonstration of a gene mutation (see Chapters 10 and 24), is performed in four clinical genetics centres only. Between those centres a division of labour has been agreed mostly on the basis of a special interest of the laboratory into certain diseases or chromosomal regions, whereas the prevalence of a disease also plays a role. Although the overall number of DNA tests was nearly 3000 in 1989, the output in terms of detected carriers or prenatal diagnosis is still modest. In addition to those two criteria the exclusion of carriership is, of course, also very important in genetic counselling and reproductive decisionmaking.

Genetic counselling by clinical geneticists in one of the regional centres is limited to complex situations. Examples of these are index patients with clinical features that might be associated with different genetic syndromes and varying recurrence risks or situations where in one family patients with different congenital disorders occur. Each centre handles about 300–600 requests for complex genetic counselling. The pattern of referral and the main reasons for anxiety are summarised in Table 22.5

Table 22.5. Referral for genetic counselling in complex situations to the Rotterdam Centre of Clinical Genetics[a]

Referral by		Reason/anxiety	
38%	General practitioner	40%	Previous affected child
31%	Clinical specialist in Rotterdam university hospital	29%	Possible genetic disorder in the family
24%	Clinical specialist in non-university hospital	16%	One of the parents affected by congenital anomaly
3%	In relation to family studies	1%	Consanguinity
2%	Self-referral	14%	Combination of two or more of reasons mentioned above
1%	Others (midwives, other centres)		

[a] Data kindly provided by Professor M. Niermeijer, Professor D. Lindhout and Ms M. Veldhuizen.

Genetic counselling in the centres is provided by an experienced clinical geneticist with the help of a genetic associate, sometimes a psychologist or social worker, and administrative staff. Each counselee usually has two visits of one hour and at the end of the counselling a written summary is sent to them and, of course, to the referring clinician(s). Those counselees who want further support usually find their way in the health care system after guidance by a psychologist or social worker who is especially trained in clinical genetics.

Prenatal diagnosis is now performed in all seven centres and appointments for women with a possible increased risk are first made in the outpatient clinic of the Department of Obstetrics. In the university hospitals the selection of pregnant women is usually made by the obstetrician and a consulting clinical geneticist.

Such collaboration does not exist in smaller peripheral hospitals where an individual obstetrician takes care of the intake without the support of a clinical genetics department. The latter situation exists for 14% of all amniocenteses and chorion samplings in the Netherlands. As is shown in Table 22.6, 86% of the prenatal diagnoses are established in one of the seven clinical genetics centres; the additional 14% are performed by collaboration between individual obstetricians and a laboratory belonging to a clinical genetics centre [3].

Table 22.6. Prenatal diagnoses in the various centres in the Netherlands[a]

	1988 No. pregnant women investigated	% Abnormal fetuses affected	1989 No. pregnant women investigated
State University Amsterdam	1115	2.4	1312
Free University Amsterdam	304	2.3	399
Groningen	707	2.8	802
Leiden	486	5.5	628
Maastricht	360	2.5	415
Nijmegen	50	–	192
Rotterdam	2287	2.4	2304
Utrecht	573	3.0	738
Subcentres			
Arnhem	655	2.4	701
Eindhoven	–	–	196
Enschede	201	3.5	268
Total	6738	2.9	8045

[a] Data collected by the Working Group on Prenatal Diagnosis by the Dutch Society Obstetricians and Gynecologists and published by Dr M. Kloosterman, Ned Tijdschr Obstetrie en Gynecologie 1990; 103: August.

Table 22.7. Increased referral for prenatal diagnosis (Rotterdam 1970–1990)
  1970: start of programme   1980: 20% of women >38 years
  1976: 6% of women >38 years   1990: 60% of women >36 years

	Number of pregnant women investigated during past 10 years							
	1980	1981	1982	1984	1985	1986	1987	1989
Amniocentesis	900	1062	1220	1471	1679	1767	1641	1239
Transcervical CVS	–	–	–	205	319	490	491	90
Transabdominal CVS	–	–	–	–	–	–	216	947
Ultrasound examination fetal anomaly	dozens	many dozens	few hundreds	570	776	873	1020	1634

Data kindly provided by Professor J. Wladimiroff and Mrs Dr M. Jahoda.

During the last few years chorion villus sampling (CVS) is replacing amniocentesis more and more (Table 22.7). In the beginning CVS was used mainly in pregnancies at higher risk, i.e. carriers of a chromosome translocation or couples with a child with a genetic metabolic disease [4]. Gradually it became clear, however, that chromosome analysis of direct chorion villus preparations also gives a reliable result in most instances. In the case of doubt a chorion cell culture, always started in parallel, can be continued for cytogenetic analysis. When a clear diagnostic result is obtained with a direct preparation our group does not proceed with the culture and provides the result after a few days. Thus far we have not had any diagnostic errors in about 4000 cases of CVS with a 100% follow-up either by investigation of fetal material after induced abortion or clinical investigation of the newborn (Jahoda et al. personal communication). Since 70% of these prenatal diagnoses concerned women at advanced age (>36 years) and another 17% women with a previous affected child, we believe that direct chromosome preparation of chorionic villi is also a suitable approach towards prenatal diagnosis in pregnancies at low risk, provided the correct clinical and laboratory procedures are being used. The latter also implies that the age of the pregnant woman is taken into account when the time of chorion sampling is determined [5]. Recent studies have shown that in older women a delay of chorion sampling until 12 weeks considerably reduces the risk of fetal loss [6].

Prenatal biochemical diagnosis can be performed for about 100 different genetic metabolic diseases [7,8]. The vast majority of these can be analysed directly on uncultured chorion villus samples; in a few instances chorion cell culture is required to attain sufficient reliability (*see* Chapter 14). This activity is centralised in Rotterdam and in 1989 nearly 120 prenatal biochemical diagnoses for 40 different genetic metabolic diseases were established. Of these diagnoses 55% were based on biochemical assays of chorion villus samples and 45% concerned cultured amniotic fluid cells. About half the samples were referred by other centres in the Netherlands and the other half were sent by colleagues in other countries. Since prenatal biochemical diagnoses are based on comparative (enyme) protein assays it is mandatory that cultured fibroblasts from an index patient in the family and, if possible, from the heterozygous parents, are available at the time of prenatal diagnosis. Financial support by the health insurers guarantees the service for cultivation and storage of cell material referred by clinicians in the Netherlands. A subsidy from the EC was helpful to also store and register a relatively large number of human mutant cell strains from abroad. At present the Rotterdam cell bank has a total of about 5000 cell lines, one-third of which have biochemically defined single gene mutations. This cell collection is of importance both for (prenatal) diagnosis and for research purposes.

In contrast to the well-developed organisation and funding of clinical genetics services that of ultrasonographic evaluation of pregnancies at risk has so far not received support by the government and the health insurers. A (low) tariff for routine ultrasound examination exists, but there is no funding for the equipment, space and trained personnel needed for the ultrasound detection of fetal abnormalities. One reason is that the responsible authorities are not convinced of the important role of ultrasound in prenatal management and another reason is the fear that "the sky will be the limit" both in costs and in (ethical) limits of what one would define as a fetal abnormality. It is to be hoped that the responsible experts will ultimately be able to convince the authorities that the development of ultrasound is as important to clinical genetics and prenatal management as cytogenetics, biochemistry or DNA analysis and that the basic moral dilemmas are not different from those in other areas of prenatal diagnosis.

Fear and Reality in Public Acceptance

After initial enthusiasm about the new perspectives of genetic technology in early diagnosis and prevention of congenital malformation and genetic disease there are several indications of increasing fear of possible negative aspects. A top politician like Jacques Attali, adviser to the French president and chairman of the Central European Bank, lists "genetic manipulation" among the five major concerns for the future, together with pollution, North–South divide, drugs and arms proliferation. In Denmark the national board of science and technology recently organised a hearing in parliament about the desirability of continuing governmental support to the European programme on human gene mapping and further development of prenatal diagnosis [9]. In Germany legislation limits certain aspects of reproductive technology and genetic research. In the USA there is a strong lobby of the pro-life movement and several states have recently adopted legislation prohibiting abortion in state hospitals. Many governments in Western countries prepare legislation to ensure the privacy of individuals who have undergone genetic testing, to prohibit the (anonymous) use of patient material for other purposes than the initial diagnostic aim and to protect individuals against the use of genetic data by third parties such as insurance companies [10] or employers (*see* Chapter 21).

The cost-effectiveness of prenatal diagnosis and more in general of genetics services has been demonstrated in many studies. Also, it is generally felt by parents of a handicapped child that an early diagnosis followed by genetic counselling means an important psychological contribution. Finally, couples suspecting an increased risk of abnormal offspring consider the genetic counselling as a prerequisite for an informed choice between various options (*see* Chapter 19).

In addition, programmes of carrier screening, genetic counselling and prenatal diagnosis have markedly reduced the incidence and prevalence of Tay–Sachs disease by 95% among Ashkenazy Jews in the USA and of β-thalassaemia by 70%–90% in various Mediterranean areas [11, MM Kaback, personal communication].

Despite all these positive aspects many politicians, professional medical organisations and the general public express a non-defined fear for future developments in genetics. This seems to be based on a mixture of complex issues. Some people point to the eugenics movements in the 1930s and the holocaust during the Second World War [12]. Others have fundamental objections against terminating pregnancies for whatever reason, including the diagnosis of an affected fetus. Still others have no moral objections, but are afraid that minor fetal defects will also become a reason for abortion and they do not approve of attributing "a value to a fetus". Many people are afraid of the idea of early knowledge by DNA testing about the manifestion of a certain disease several decades later or information about increased susceptibility for a disease which might focus people too much on their own health and lifestyle. Also the combination of methods used in in vitro fertilisation and gene transfer causes concern.

It is the combination of all these different fears together with insufficient knowledge about the real perspectives and limitations in human genetics that cause political and public restraint.

The experts involved in genetic services, prenatal care and reproductive technology should acknowledge the existence of fears and try to explain the positive aims of the technologies and activities. Prenatal diagnosis is not adequately defined by "search and destroy" as its opponents do, but it also means "dare to have a pregnancy". Recent follow-up studies on the reproductive behaviour of couples at increased risk of affected offspring clearly show that 50% of the couples at high risk (>15%) refrain from pregnancy [13]. This figure corresponds to that in earlier studies of parents at risk for a child with Duchenne muscular dystrophy [14] or with cystic fibrosis [15]. However, if the option of prenatal diagnosis was available to a similar group of counselees, only 30% of the high-risk couples were deterred from pregnancy [13].

Hence, the number of children born thanks to prenatal monitoring is probably higher than the number of fetuses aborted because of an abnormal result.

Instead of spending time outlining possible negative effects of future genetic technology we might as well prepare ourselves and the health care system for the enormous tasks awaiting. At present early diagnosis and genetic counselling is mainly related to diseases in childhood. As is shown in Table 22.3 this implies for one centre annual numbers of hundreds to thousands of tests and several hundreds of complex genetic counsellings. In the future it is likely that the genetic factors involved in various forms of cancer, cardiovascular diseases, psychiatric and neurological disorders will be identified. Also the relationships between these genetic factors and various environmental factors involved in multifactorial diseases will become more clear. This will result in an enormous increase in the demand for genetic counselling. Close relatives i.e. children, brothers, sisters of affected people will ask advice about their genetic risk and about measures to be taken to improve their health and life expectancy.

If all patients who died of cancer or cardiovascular diseases before 65 years of age, and all patients who were diagnosed to have other diseases with a genetic component, had only one relative who would ask for counselling, the numbers involved would be of the order of 100 000 (Table 22.8). In reality these numbers will be higher because most affected people have more than one close relative who might be concerned about the recurrence risk (*see* Chapter 23).

Table 22.8. Clinical genetics now and in the future

Present activities mainly related to disease in childhood		Future counselling related to multifactorial disease in adulthood[a]	
Annual postnatal chromosome analyses	6000	Because of early death (<65 years) of close relative:	
Annual biochemical assays	5000	Cardiovascular disease	8 000
		Cancer	10 000
Annual DNA analyses	3000	Other disease with genetic component	4 000
Genetic counselling in complex situations	3000	Close relative declared unfit because of:	
Prenatal diagnoses	8000	Mental disease/problems	125 000
		Rheumatoid arthritis	54 000

[a] Data from the Central Office for Statistics, The Hague (1989)

Of course, it will take time before the human gene mapping projects pay off in terms of understanding the genetic background of the major multifactorial diseases. Also, it takes time before a new test is accepted by the majority of the people eligible to undergo this test. This is clearly shown by the uptake rate of prenatal chromosome analysis by pregnant women at advanced age. In the Netherlands this programme started in 1970 and after 10 years not more than 20% of the pregnant women of 38 years and older underwent amniocentesis. However, during the last decade this percentage has gone up to 60% (*see* Table 22.7). Similar trends are observed in other West European and Scandinavian countries [16].

This increase in acceptance is not only due to improvements in technology (introduction of CVS and direct cytogenetic biochemical and DNA analysis), but also by the effects of repeated information, better education in genetics in schools and the support by parent and patient organisations. Finally, it may well be that gradual changes in society play the most important role in the acceptance of new medical technology. This is shown by the great difference in acceptance of genetics services between different ethnic groups within one country and by the enormous differences between countries. The challenges for the future are enormous both in the development of new strategies for early diagnosis, therapy and prevention of disease and in achieving social acceptance [17]. The latter implies that governments should be reserved in legislation about individual decisions about life and death. Professional health workers should be (more) careful in introducing new technologies that are only halfway in terms of diagnostic reliability or therapeutic success. The public should accept that individuals are different in disposition, background and life situation and consequently should be offered the freedom to react in different ways to "the same" situation and to solve "the same" problem in a different manner (*see* Chapter 21).

References

1. Health Council, Advice on "Genetic Counselling". The Hague, 1977.
2. Health Council, Advice on "Human Genetics and Society. The Hague, December 1989.
3. Prenatale Diagnostiek. Thermanummer. Ned Ver Obstetr Gynaecol 103 (6).
4. Sach ES, Jahoda MGJ, Kleijer WJ, Pijpers L, Galjaard H. Impact of first trimester chromosome, DNA and metabolic studies on pregnancies at high risk; experience with 1000 cases. Am J Med Genet 1990; 29:293–303.
5. Jahoda MGJ, Pijpers L, Vosters RPL, Wladimiroff JW, Reuss A, Sachs ES. Role of maternal age in assessment of risk of abortion after prenatal diagnosis during first trimester. Br Med J 1987; 295:1237.
6. Cohen-Overbeek TE, Hop WC, den Ouden M, Pijpers L, Jahoda MGJ, Wladimiroff JW. Spontaneous abortion rate and advanced maternal age; consequences for prenatal diagnosis. Lancet 1990; i:27–9.
7. Galjaard H. Early diagnosis and prevention of genetic disease. In: Cosmi E, Di Renzo R, eds. Reviews in perinatal medicine. New York: Alan Liss, 1989; vol 6: 133–72.
8. Kleijer WJ. Prenatal diagnosis. In: Fernandes J, Saudubray JM, Tada K, eds. Inborn metabolic diseases. Berlin: Springer-Verlag 1990; 638–95.
9. Consensus Conference on the application of knowledge gained from mapping the human genome. Teknologi Naevnet (Danish Board of Technology). Copenhagen: November 1989.
10. Annals of life insurance medicine. Special edition. In: Hefti ML, ed. Proceedings of the 16th international congress of Life Assurance Medicine, The Hague 1989. Karlsruhe: Verlag Versicherungswirtschaft 1990; 9.
11. Cao A. Results of programmes for antenatal detection of thalassemia in reducing the incidence of the disorder. Blood Rev 1987; 1:169–76.
12. Müller-Hill B. Murderous Science (translated by George Fraser). Oxford: Oxford University Press, 1988.

13. Frets PG, Duivenvoorden HJ, Verhage F, Niermeijer MF, vdBerge SMM, Galjaard H. Factors influencing the reproductive decision after genetic counselling. Am J Med Genet 1990; 35:496–502.
14. Emery AEH, Watt MS, Clard ER. The effect of genetic counselling in Duchenne muscular dystrophy. Clin Genet 1972; 3:147–50.
15. Kaback MM, Zippin P, Boyd R et al. Attitudes toward prenatal diagnosis of cystic fibrosis among parents of affected children. In: Lawson D, ed. Cystic fibrosis; horizons. New York: John Wiley, 1984; 16–28.
16. Galjaard H. Interaction between technical and social developments in human genetics. In: Doxiadis S, ed. Early influences shaping the individual. New York: Plenum, 1989; 55–66.
17. Galjaard H. Challenges for the future. In: Emery AEH, Rimoin D, eds. Principles and practice of medical genetics, 2nd edn. Edinburgh: Churchill Livingstone, 1990; Vol 2:2025–35.

Chapter 23

Genetic Services

R. Harris

Obstetricians and Medical Genetics

It is very appropriate that this review of genetic services should appear in a book produced by the Royal College of Obstetricians and Gynaecologists. Obstetricians work closely with genetic services, and their care of the fetus and discerning use of prenatal diagnosis are important in the prevention of handicapping disorders. Prenatal diagnosis of many disorders is possible only because obstetric skill and the use of ultrasound have made amniocentesis and chorion villus biopsy safe procedures; the development of diagnostic ultrasound also owes much to obstetricians. In vitro fertilisation, preimplantation diagnosis and embryo research have begun as a result of the cooperation between obstetricians, biologists and geneticists. Obstetricians are also more generally involved in genetics because women in antenatal clinics frequently ask about genetic disorders and their prevention, and this involvement is likely to increase with population screening for carriers of cystic fibrosis and other common genetic disorders. Pragmatically, obstetricians have been the usual target of litigation resulting from alleged negligence involving the failure to diagnose genetic disorders in utero.

The Burden of Genetic Disease

Monogenic disorders occur in 1.3% of newborns, and pathological chromosome anomalies in 0.7%, to which should be added about 2% of newborns with sporadic congenital malformations [1,2]. Many other genetic disorders become apparent later in life and the considerable load of lost and impaired life years due to some monogenic and chromosomal disease is shown in Table 23.1. The total financial burden to individuals, the Health Service and Social Services is very great. However, an even greater load is suggested by Patricia Baird, working with the Department of Health in British Columbia [3], who found that 5.5% of the population developed a genetic disease by age 25, and 60% in a lifetime (she included degenerative diseases of adult life which have a strong genetic predisposition like coronary heart disease, diabetes, cancer and the major psychoses). As a consequence 5% would be a conservative estimate of the proportion of the population who may require to use genetic services.

Table 23.1. Incidence and burden of common important genetic disorders in the UK [19]

Condition	Birth Incidence per 10 000	Unimpaired years	Impaired years (degree of impairment)	Lost life years
Dominant				
Adult polycystic kidney disease[a]	8	30	30 (30%)	10
Huntington's chorea[a]	5	40	10 (50%)	20
Neurofibromatosis[a]	4	20	30 (50%)	20
Retinoblastoma (treated)[a]	3	3	? (20%)	?
Myotonic dystrophy[a]	2	40	10 (50%)	20
Tuberous sclerosis	1	5	45 (80%)	20
Familial adenomatous polyposis[a]	1	20	30 (20%)	20
Autosomal recessive				
Cystic fibrosis[a]	5	2	10 (50%)	30
Phenylketonuria (treated)[a]	1	60	10 (10%)	0
Neurogenic muscle atrophy[a]	1	1	4 (90%)	65
Early onset blindness	1	5	70 (50%)	0
Non-specific mental retardation[b]	5	0	50 (90%)	20
Sickle cell disease[a]	0.5[c]	30	20 (20%)	20
Thalassaemia major[a]	1[c]	0	35 (20%)	35
Tay–Sachs disease[a]	0.4[d]	0	3 (90%)	67
X-linked recessives				
Duchenne muscular dystrophy[a]	2	4	16 (60%)	50
Haemophilia A[a]	1	0	60 (20%)	10
X-linked mental retardation	10	0	50 (80%)	20
Chromosome abnormalities				
Down's syndrome	12	0	35 (80%)	35
Autosomal structural aneuploidy	5	0	20 (95%)	50

[a]DNA markers available.
[b]Includes polygenic forms.
[c]50 per 10 000 in the ethnic group most affected.
[d]4 per 10 000 in the Jewish population.

The Impact of Medical Genetics

The practical relevance of new prenatal and postnatal DNA tests, including predictive tests for late onset disease, has already greatly improved the management of a number of genetic disorders. These include cystic fibrosis, thalassaemia, Huntington's chorea and Duchenne muscular dystrophy. The international programme to map the human genome will, possibly within the decade, result in every disease gene becoming available for direct study, for population screening and eventually for treatment by genetic replacement or by the use of the products of cloned human genes (see e.g. [4,5]). These advances have important medical, social and political implications [6].

Increasing public awareness and media coverage of sensational genetic advances or ethical dilemmas have already increased demands made on clinicians for explanation of often complex genetic problems and techniques. Practitioners need to know which diseases can be diagnosed by new methods if they are to offer a high quality service and avoid potential medicolegal pitfalls, particularly those that may arise if they have not offered appropriate tests to the parents of infants with genetic disabilities. Ethically there is a need to ensure that individuals and families are well informed about the options available to them.

Regional Genetic Centres

Genetic centres have been established in all Health Service regions in England and Wales, with appropriate arrangements in Scotland and Northern Ireland, to offer specialist counselling and chromosome diagnosis. Obstetricians have often been among the strongest supporters of genetic centres and the Royal College of Obstetricians and Gynaecologists was one of the four medical royal colleges whose 1986 Joint Statement [7] to the Department of Health was influential in obtaining official support for the effective planning of genetic centres. In discussing the "Relation to Obstetric Services" the royal colleges' Joint Statement noted that:

> The combined skills of obstetric ultrasound and clinical genetics allow the prenatal diagnosis of an increasing number of genetic disorders and fetal abnormalities. The obstetrician shares with the family practitioner the continuing care of the pregnancy and deals with the anxieties of the mother and any complications that may arise. Effective communication here, as elsewhere, is essential. It is unsatisfactory and sometimes impossible to offer prenatal diagnosis to women who are seen for the first time when they are already pregnant and a high priority is attached to ascertaining such individuals and ensuring that family studies are completed before pregnancy ... The exigencies of the situation now encourage the closest possible liaison between all the clinicians concerned and the genetic laboratories ... (and) ... those who plan Regional Genetic Services (should) appreciate the ... benefits of close collaboration between clinical geneticists and their scientific colleagues ... to ensure that the full benefits of recent advances are available to patients and their families throughout the country.

The prompt and positive reply [8] from the Department of Health noted:

> The Joint Statement has served to draw wider attention to this expanding area of medicine ... (and) ... the rapid progress being made in genetics, its influence on medical practice and the potential benefits for patients. (There is a) ... need for co-ordinated planning and organization besides appropriate education and training if the new developments are to be given effect ... (A DHSS funded study of DNA methods) ... will ... assist the Department to develop policy and planning guidelines to help health authorities in their consideration of service provision and its organization including both clinical and laboratory elements ... (and will) ... make available to health authorities details of the costs and benefits (including both the tangible and intangible elements) of these services.

In 1989 the Minister for Health replied [9] to the President of the Clinical Genetic Society, and reinforced the support of the Department for regional genetic centres at the time of the introduction of major Health Service reforms. She wrote:

> I recognise that the remarkable and continuing advances being made in clinical genetics offer major opportunities for enhancing care in preparation for parenthood and in preventative medicine generally. I have been especially impressed by the priority given by members of the Society to counselling of individuals and their families, who are – or believe they are – at risk, to enable them to make well informed choices on parenthood. I welcome, too, the initiatives towards setting up effective audit of these services ... we recognise that there will be some services which, because of the pattern of demand and need for concentration of specialised skills and resources, are best provided on a regional basis. It seems that for the foreseeable future clinical genetics comes into this category, on account of the special features to which you draw attention – the centralisation of specialist laboratory facilities and their close links with clinical services and the need for well maintained registers which enable relatives, who may benefit from surveillance, to be offered advice and counselling. As you know, the Department has undertaken a survey of the genetic services currently available and planned for the future. In the light of this survey we shall be considering the need for central guidance on the range and organization of services necessary to provide even access for all NHS users. I think that, whatever views are taken on detailed arrangements, a regional organization will be favoured.

The Functions of Regional Genetic Centres

Clinical genetics is a largely outpatient specialty; diagnostic problems and genetic counselling are the two main clinical activities. Problems referred for diagnosis may involve any genetic or dysmorphic disorder, and these are often too rare for the experience of non-genetic specialists. Diagnosis requires specialised clinical skills, access to dysmorphology databases, close working relationships with genetic laboratories, and a knowledge of genetic principles, including linkage and Bayesian risk estimation.

Genetic counselling is a procedure comparable to surgical operations and to medications in its potential for healing, but also for disaster if ill-informed or coercive. Its consequences may extend through several lifetimes. With the exception of obstetricians, clinical geneticists are unique among clinicians in having many healthy clients (see below) and are also more concerned with families (as distinct from simply individual patients) than any other group, with the possible exception of general practitioners. Research and development are an essential part of the practice and development of medical genetics; they may involve high technology, including molecular biology, the definition of rare clinical syndromes or the less tangible areas of counselling, ethics and outcome measures. There is a major commitment to undergraduate and postgraduate education because of the rapidity of new developments.

Many clinical geneticists hold joint clinics with consultants in other specialties, for example for prenatal diagnosis, handicapped children, neuromuscular disorders, genetic eye and hearing problems and for cancer families. Clinical genetics also has an unusual role in coordinating other services for individuals and families when multi-system genetic disease creates the need for advice from many specialties and support from social services. The facilities provided by a well set up Regional Genetic Centre are summarised in Table 23.2

Table 23.2. Facilities found in a well set up specialist genetic service

---
Regional organisation of genetic centre with integration of clinical, cytogenetic and molecular laboratory genetics

Adequate medical and non-medical clinical staffing

A genetic family register for the support of genetic families especially for those relatives at high risk of genetic disorders

Appropriate outpatient accommodation: many patients will have had recent bereavements or handicapped children and may be distressed to be accommodated where there are normal children or pregnant women

Adequate secretarial and clerical support for contacting extensive families, searching many hospital records, reports to doctors and patients

Academic links with a university department of human or medical genetics to facilitate research and teaching

A specialised genetic library and appropriate computer hardware and specialist databases for clinical audit, recording extensive family studies and diagnosing more than 3000 genetic disorders

Effective links with related specialties, including obstetrics and paediatrics

---

The relationship of clinical geneticists to chromosome and molecular genetics laboratories is close and cooperative rather than managerial. The present state of chromosome services is described in the report of the Association of Clinical Cytogeneticists [10] and molecular genetic services are reviewed in Chapter 24.

Healthy but At Risk of Genetic Disease: Genetic Family Registers

Clinical geneticists are particularly concerned with healthy people who are at risk of developing genetic disease, of producing affected offspring, or both. Between

2% and 3% of all couples are known to be at high and recurrent risk of having a child with an inherited disorder [11] and many other relatives are anxious about their own risks of developing serious genetic disorders, although most will not do so. Genetic family registers exist in some regions to cater for families with serious dominant or X-linked diseases. Although they were originally designed as a way of preventing genetic disease [12], in Manchester it has been found [13] that registers are also the most effective way of providing counselling, support and the benefits of technical advances to individuals, particularly young adults as they enter reproductive years. A special aim is to ensure that young women are counselled and their carrier status determined in advance to avoid panic testing of them and their relatives during pregnancy. The introduction of registers and DNA tests has had the welcome effect of showing that many close relatives of affected individuals are not carriers, thus relieving years of anxiety for themselves and future children. When individuals are shown to be carriers the same DNA tests usually permit prenatal diagnosis.

The register staff are in annual contact with individuals on the register, and each client has the telephone number and name of a member of the team. All entries to the register are voluntary, subject to informed consent and to full confidentiality; information is not released to third persons without permission.

In Manchester a genetic register has been in existence for ten years and includes over a thousand patients with specific genetic disorders (Fig. 23.1). There are now over 5000 relatives at high risk on the register (Fig. 23.2), representing an average of four relatives for each patient. For some diseases, including Huntington disease and hereditary cancers, the ratio is even higher with between eight and ten high-risk relatives per patient. Relatives are reviewed each year and this is an increasing responsibility because we have ascertained only a

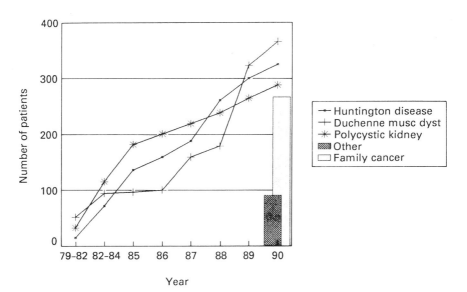

Fig. 23.1. Manchester Genetic Family Register showing numbers of patients with specific inherited diseases.

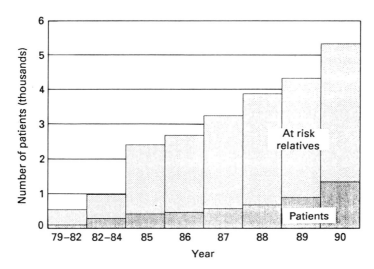

Fig. 23.2. Manchester Genetic Family Register showing relationship between numbers of patients and relatives known to be at high risk.

fraction of the families with the diseases which we currently register, and there are at least 3000 genetic disorders [14], although not all require or would be suitable for the register approach. The register workload falls particularly on our senior clinical medical officer and our genetic nurses and genetic associates, requiring much correspondence and many domiciliary visits.

Are Genetic Services Adequate for the Tasks? Clinical Genetic Manpower

The Royal College of Physicians of London has surveyed [15] clinical manpower in genetic centres. There has been a considerable increase in the number of consultant posts in Britain, from 20 in 1983 to 48 in 1990, and all English regions have funded at least one NHS consultant post. However the projection [16] of one consultant whole time equivalent (WTE) per million population has been achieved in only two regions (East Anglian and Oxford). There are about 50 WTE trainees and other medical staff and in all there are 1.6 WTE medical staff per million population. There are also 69.80 WTE clinical co-workers, representing 1.22 genetic nurses, genetic associates and others per million population in Britain.

Including a few non-clinical academics, but excluding laboratory scientists, the total number of specialised clinical genetic staff in Britain is only 165 WTE to cope with the 5% of the population [3] (and many more relatives) that require genetic services. This represents about 17000 genetic patients, plus their relatives, for each genetic worker! Given this potential workload the recommendations of the Royal College of Physicians Clinical Genetics Committee [15] are remarkably modest (see below).

If There are Too Few Genetic Specialists can Doctors in General Cope with These Challenges?

The numbers involved make it inevitable that clinicians in general will have to adapt to the new technology and to the clinical changes that will follow. Obstetricians already play a large part in providing genetic services and many other doctors, most notably paediatricians, routinely deal with the genetic aspects of clinical problems within their specialty. However, there are few data on the effectiveness of the present ad hoc arrangements and there is reason to doubt the general level of genetic literacy among doctors. A recent survey by the Royal College of Physicians [17] showed that medical students are not always taught genetics adequately, and what they are taught frequently has little relevance to what patients need. General postgraduate medical education in genetics is also rudimentary, although this is not yet adequately documented (A.W. Johnston, personal communication). It cannot be assumed, therefore, that doctors in general are well equipped to deal with the increasing complexity of genetics, or with the service needs arising from the clinical usefulness of genetic techniques.

What can be Done?

The Royal College of Physicians' Committee on Clinical Genetics recommended [15] that more specialist consultant clinical geneticists should be trained, with at lest 77 new posts in Britain. The Committee also supported the need for many more non-medical genetic counsellors to be trained by genetic specialists to work in the community and in antenatal and other clinics. However, it is unlikely that there will ever be enough genetic specialists and the ability of all specialties to deal with genetic problems must, therefore, be reinforced. In addition to more effective undergraduate teaching and intensive postgraduate genetic training for doctors in general, a Confidential Enquiry into Recurring Genetic Disorders has been established.

Confidential Enquiry into Recurring Genetic Disorders

The enquiry is multispecialty and was initiated by the Royal College of Physicians (London) with the support of the Royal College of Obstetricians, British Paediatric Association, Royal College of Midwives, Health Visitors Association, Clinical Genetic Society and others. The object of the genetic enquiry is overtly educational, aiming to improve the quality of medical records and hence good clinical practice in many specialties that deal with genetic disorders. The enquiry is confidential and non-censorious.

The enquiry will review "memorable events" involving marker or sentinel disorders, defined as those that might have been predicted by maternal age, family history or commonly available screening tests. The initial markers selected are Down's syndrome, neural tube defect, haemophilia, thalassaemia, cystic fibrosis and multiple endocrine neoplasia type 2 (MEN2, an example of a late onset disease in which prevention among relatives is a high priority). There will be local scrutiny of records by the clinicians who are responsible for the care of the patient/family, using questionnaires prepared by a multispecialty planning group. The confidential enquiry is likely to have a rapidly beneficial effect, judging by the success of the Confidential Enquiry into Maternal Deaths organised by the RCOG [18].

Acknowledgments. I am grateful to Dr Anwar Khan for data on the Manchester Genetics Register and to colleagues in medical genetics in Manchester and elsewhere for continuing collaboration.

References

1. Carter CO. Monogenic disorders. J Med Genet 1977; 14:316–20.
2. Polani PE. The impact of genetics on medicine: the Harveian Oration of 1988. London: The Royal College of Physicians, 1990.
3. Baird PA, Anderson TW, Newcombe HB, Lowry RB. Genetic disorders in children and young adults: a population study. Am J Hum Genet 1988; 42:677–93.
4. Bodmer WF. Human genetics: the molecular challenge. Cold Spring Harbor Symposia on Quantitative Biology, Volume LI. Cold Spring Harbor Laboratory, 1986.
5. Bodmer WF. The human genome sequence and the analysis of multifactorial traits. Molecular approaches to human polygenic disease. Chichester: Wiley (Ciba Foundation Symposium 130), 1987; 215–28.
6. Bodmer WF. Social and political implications of the mapping of the human genome. Science in parliament 1990; 47.
7. Joint Statement on Genetic Centres from Royal Colleges of Physicians (London), Pathologists, Obstetricians and Gynaecologists and General Practitioners. Unpublished (available from Hon. Secretary, Committee on Clinical Genetics RCP), 1986.
8. Reply from DHSS to Joint Statement. Available from Hon. Secretary Royal College of Physicians Committee on Clinical Genetics.
9. Reply to President of Clinical Genetics Society from Mrs Virginia Bottomley, Minister for Health. Available from Hon. Secretary Royal College of Physicians Committee on Clinical Genetics.
10. Staff structure and services in cytogenetic laboratories, England, Scotland and Wales. A report. Association of Clinical Cytogeneticists, December 1989.
11. Prenatal diagnosis and genetic screening. The Royal College of Physicians of London, 1989.
12. Emery AEH, Brough C, Crawfurd M, Harper P, Harris R, Oakshott J. A report on genetic registers. Based on the Report of the Clinical Genetics Society. J Med Genet 1978; 15:435–42.
13. Read AP, Kerzin-Storrar L, Mountford RC, Elles RG, Harris R. A register-based system for gene tracking in Duchenne muscular dystrophy. J Med Genet 1986; 23:581–6.
14. McKusick VA. Mendelian inheritance in man. Ninth edn. Baltimore and London: Johns Hopkins University Press, 1990.
15. Harris R. Clinical genetic services in 1990 and beyond; report of a survey of genetic manpower in Britain. J R Coll Physicians, in press.
16. Harris R (Chairman), Emery AEH, Johnston AW, Pembrey ME, Winter R, Insley J. Role and training of clinical geneticists. Report of the Clinical Genetics Society Working Party. London: Eugenics Society. Suppl no. 5, 1983.
17. Teaching genetics to medical students. London: Royal College of Physicians of London, 1990.
18. Department of Health and Social Security. United Kingdom reports on confidential enquiries into maternal deaths 1985–87 and 1988–90.
19. Department of Health and Social Security. On the State of the Public Health for the Year 1986. London: HMSO, 1987; 12.

Chapter 24

National Coordination of Molecular Genetic Services

A. P. Read

Virologists, pathologists, oncologists, immunologists and biochemists all use molecular genetic techniques as one of their investigative tools. Here, however, I shall restrict myself to molecular genetic investigations performed in a clinical genetics setting, but including tests for any clinical genetic purpose and not just antenatal diagnosis. These services are at present applicable only to diseases with a simple mendelian mode of inheritance.

Mendelian diseases are numerous but individually all rare, and this is the factor which underlies all discussion of national coordination of molecular genetic services. Over 4000 mendelian phenotypes (not all are diseases) are listed in McKusick's catalogue [1], but only one significant mendelian disease affects more than one in 1000 of the UK population. Table 24.1 shows the incidences of the commonest mendelian diseases in the UK. It follows that for any particular disease a laboratory can generate a large workload only by drawing on a large population. But a large population will include cases of very many different diseases, too many for a single laboratory to cope with. This is the case for coordination.

To discuss coordination I shall first discuss how many diseases a laboratory might reasonably tackle simultaneously and whether there is a minimum viable workload for a single disease. I shall then try to assess how far these figures will remain valid in the face of rapid technical and scientific change. Finally, I will consider some possible arrangements for meeting the genetic needs of large populations, and describe the progress to date in the UK towards a nationally coordinated service.

Table 24.1. The commonest serious mendelian diseases in the UK

Disease	Inheritance[a]	Incidence in UK, per 1000
Familial hypercholesterolaemia	AD	2
Polycystic kidney disease	AD	1
Fragile-X mental retardation	XL	1
Cystic fibrosis	AR	0.5
Huntington's disease	AD	0.5
Neurofibromatosis I	AD	0.4
Duchenne muscular dystrophy	XL	0.3
Myotonic dystrophy	AD	0.2
Haemophilia A	XL	0.2
Phenylketonuria	AR	0.1
Congenital adrenal hyperplasia	AR	0.1
Familial polyposis coli	AD	0.1

[a]AD, autosomal dominant; AR autosomal recessive; XL, X-linked.

The Case for Specialisation

It could be argued that DNA tests, unlike enzyme assays, are all the same. If a laboratory is competent at DNA extraction, restriction digestion, electrophoresis, Southern blotting and hybridisation, then it can apply any probe to any digest. Similarly, if it is competent at preparing primers and running polymerase chain reactions (PCR), then it can amplify and study any DNA sequence, given only a printout of the primer sequences. Indeed this is at least partly true, and explains the ability of laboratories to switch their attention rapidly from one disease to another. Nevertheless, no laboratory would use a probe or polymorphism for diagnosis without several practice runs on known material. Probes and primers have quirks, produce artefacts, come wrongly labelled or in the wrong vector, and in general require familiarity before they can be used with confidence.

Thus a laboratory incurs an overhead when it takes on a new disease, which will deter it from offering its population the full range of available tests. A further argument for specialisation is the rapid advance in scientific knowledge. Laboratories need to have the latest probes and recombination data for each disease, and to know the current catalogue of mutations and polymorphisms in cloned disease genes. They cannot maintain this awareness across the whole range of diseases, and the increasing accessibility of information from the public databases is no substitute for knowledge in depth based on experience.

Similarly, a certain continuing workload is required for efficiency. This is partly because of general economies of scale, particularly for procedures using radio-labelling. ^{32}P has a half-life of 14 days, so labelled nucleotide needs using within a week of receipt, and the minimum batch size which can be ordered is sufficient for many tests. But for individual tests low throughput also brings inefficiencies. There are three reasons for this. First, familiarity with the reagents and interpretation needs regular reinforcement. Second, probes, primers and enzymes may deteriorate in storage, and need testing if they have not been used within the last month or so. Third, many laboratory procedures, but particularly Southern blotting, use batches of 15–40 samples. It is almost as much work on Southern blots to test two DNA samples for a particular polymorphism as 20.

These considerations suggest that a laboratory using radiolabelling and Southern blots is unlikely to be able to provide an efficient service for more than a dozen or so different genetic problems, or for problems which generate less than one set of work per month. A laboratory relying on PCR and on non-radioactive labelling methods could perhaps handle twice the diversity and half the workload, but the calculations below show that this is still incompatible with supplying a full genetic service to any population.

Gene Tracking, Direct Diagnosis and Laboratory Workload

I suggested above that a laboratory needs to perform at least one "set of work" per month on a genetic problem to keep efficient. This is not necessarily the same as accepting one referral per month. In considering workload, an essential distinction is between problems where direct dagnosis is possible and those where gene tracking must be used. Diagnosis by either method will use essentially the same laboratory techniques and skills (Southern blotting and hybridisation with labelled probes, or some version of the PCR reaction). However, they produce different work patterns, they require different relationships between the laboratory and the referring clinician, and they may result in very different laboratory workloads per referral.

In direct diagnosis the patient's DNA is tested for the presence or absence of a pathogenic change. This is similar to most other biochemical tests: a sample is sent, a test is performed, and a report is issued. From the laboratory's standpoint the case is then closed. Direct diagnosis is possible once a gene has been cloned and sequenced and the pathogenic mutations defined.

Gene tracking is used when we know the chromosomal location but not the DNA sequence of the pathogenic gene. With many diseases this is a passing phase, during the stage where the gene has been mapped but not yet cloned. This stage can last several years (seven years so far for Huntington's disease). During this time gene tracking is the only way to meet the urgent demands of patients for antenatal or presymptomatic diagnosis. Once the gene is cloned a direct test is usually preferable, but sometimes a known gene has so many different possible mutations that it is impracticable on a service basis to check by direct tests for all of them, so gene tracking is still used. Non-deletion mutations in the dystrophin gene would be an example of the latter category.

Gene tracking concentrates on the parent(s) who may or may not have transmitted the disease gene to the patient. Polymorphic DNA markers are used to distinguish the two homologues of the relevant chromosomal segment in this parent, and then other family members are studied to work out which of the markers is associated with the disease-bearing homologue. Knowing this, we can then test the patient's DNA to see whether he or she inherited the high-risk or low-risk homologue.

For gene tracking we study families rather than individuals. The results on an isolated individual are meaningless; it is the pattern of transmission in the family which makes the diagnosis. At the very least, the laboratory needs to have a set of family samples and the pedigree. Which family members require sampling depends on the results with the probes. If one person turns out to be uninformative or recombinant, other family members may need to be sampled. Gene tracking can be enormously laborious in large families with X-linked or dominant diseases where the available probes are not highly informative or where recombination between the probe and disease locus is relatively common. Autosomal recessive diseases are much simpler. Gene tracking for these normally requires only four samples (the parents, the previous affected child and the fetus) and one or two informative probes.

By contrast, direct diagnosis requires only a single DNA sample, regardless of the mode of inheritance. It may still require many tests. Each test confirms the presence or absence of one specific mutation. Over 60 different mutations are known in the CFTR gene, any one of which can cause cystic fibrosis. Confirming the presence or absence of any mutation in the CFTR gene would be a very different task from checking for one specific mutation. Gene tracking can only be applied to families with a history of the disease, but once direct diagnosis becomes possible, people with no family history ask for tests to confirm that they are not carriers of a disease gene. This obliges the laboratory to test for many possible mutations.

Changes Due to Scientific and Technical Advances

Detailed descriptions have been published [2,3] of the workloads of several laboratories. Unfortunately in a rapidly changing area these historical accounts are of limited use for planning future services. It is essential to try to look ahead, even though attempts to predict the nature or speed of future advances are guaranteed to fail. However, we can perhaps see two trends and an end-point. One trend is the increasing use of PCR techniques. One effect of this has been to make new things possible, such as genotyping single cells. But its major impact has been to make difficult jobs easy. Direct detection of mutations has been a prime example of this. Investigations which formerly required construction and screening of genomic libraries can now be performed, 30 at a time, in a day.

This is one cause of the second trend, which is an increase in direct diagnosis at the expense of gene tracking. For dominant or X-linked diseases the transition from gene tracking to direct diagnosis will mean a substantial reduction in workload per referral. In contrast, the workload for autosomal recessive diseases can only grow. They are numerous, serious and mostly as yet unmapped. Gene tracking for these diseases is uncomplicated, while direct diagnosis is paradoxically more laborious at present, because many different possible mutations must be checked.

Perhaps the gene tracking workload will be sustained by newly mapped dominant or X-linked diseases coming along to replace those now cloned, but I think a decrease is predictable. Natural selection ensures that serious dominant diseases are rare, except for late-onset diseases which do not impair reproduction. Dominant diseases with long survival produce good large pedigrees with many living affected members. Hence they are particularly amenable to mapping by linkage analysis, and most of the major ones have already been mapped. The implication is that as direct diagnosis reduces the laboratory gene tracking workload for Huntington's disease, fragile-X, polycystic kidney disease, myotonic dystrophy and familial cancers, there is not a large workload from as yet unmapped dominant diseases waiting to take its place. Also, present experience is clear that there is little demand for antenatal (as distinct from presymptomatic) diagnosis of late-onset diseases, however devastating the disease.

The shift of emphasis away from gene tracking and Southern blotting towards direct diagnosis and PCR has considerable implications for the organisation of molecular genetic services. Southern blotting and hybridising with radiolabelled probes are relatively difficult techniques which are not amenable to packaging in kit form, and require a general biochemical competence. Interpretation of the results depends on accurate diagnosis; it requires the use of the clinical notes and a close working relationship between the clinician and the scientists. Gene tracking also usually implies DNA banking because results may be updated as new probes are developed (making people more informative or reducing numbers of recombinants), and because in an extended family other members may seek testing later. Thus Southern blotting and gene tracking have encouraged the development of specialised molecular genetic laboratories at a regional level, integrated with the regional clinical genetic services.

Direct testing and PCR technology will change this. PCR is manipulatively much simpler than the methods it replaces. PCR-based test kits may become available for use in side-rooms or district general hospital laboratories. Clinical genetic skills are not needed to interpret the results of direct tests, although they may well be as necessary as ever for counselling the patients. On the other hand, a predictable end-point of present developments is that it will become easy and cheap to determine the sequence of any desired region of any patient's DNA, but only by the use of automated genotyping machines [4] which will require a large throughput to justify their cost. Sequencing would allow all possible mutations in a gene to be checked in a single test and reduce all direct tests to a single procedure.

Frequency and Diversity of Requests for Service

The trend towards direct diagnosis by a single procedure means that we should estimate workloads simply by asking how large a population is needed to generate one referral a month for a particular genetic problem, and what variety of problems such a population is likely to throw up. When a new test becomes possible there is a backlog, but eventually a steady state is reached when the rate of referrals depends mainly on the incidence. Assuming an average UK birth rate of 12 per thousand population per year, and using the rule of thumb that a laboratory needs at least one referral a month to keep efficient, we can very simply estimate the population it must serve (Table 24.2). For a serious disease most cases will lead to referrals; the figures in Table 24.2 assume 50% of cases (new families and new cases in old families) lead to a request for testing.

Table 24.2. Population required to generate a steady-state load of one referral per month

Incidence	Population to generate one referral per month[a]
1 in 1 000	2 million
1 in 10 000	20 million
1 in 100 000	200 million

[a]Assuming 50% of births or diagnoses generate a referral.

These figures are of course only the crudest of estimates. They ignore all the special factors existing with every disease and above all, they refer to a hypothetical "steady state" which in practice will probably never be reached. Scientific advances change patterns of referral, and even in families already referred new probes generate requests for retesting. Much depends on the assiduity of the clinicians in referring cases and the attitude of the clients to offers of testing. However, taken in conjunction with the incidences in Table 24.1, they do suggest that only the very commonest genetic diseases will generate a viable long-term laboratory workload of direct tests within the 2 million–5 million population of a typical single Health Service region. We can guess that the specialist molecular genetics laboratory of the future will need to serve a population of 5 million–50 million for each disease.

Over 200 of the 4000 known mendelian conditions are already amenable to DNA testing [5], and the number is growing monthly. Even the rarer ones will have a measurable incidence in a population of 50 million. Even if a laboratory could handle 50 diseases, it could not hope to offer a comprehensive DNA diagnostic service to any population. Specialisation and sharing are unavoidable.

Present Arrangements

At present almost every Health Service region has one or more molecular genetics laboratories. These function as a branch of the regional clinical genetics service, sharing premises with the clinicians and cytogeneticists. They have usually been established on soft money and reflect the determination of the local clinicians and scientists not to be left out of an exciting fast-developing area. Haemoglobinopathies stand apart, being largely dealt with by the National Reference Laboratory at Oxford. In addition to these service laboratories, University or Research Council research laboratories often pass through a phase of offering a limited diagnostic service for a disease they have newly mapped or cloned.

Most of the service laboratories offer testing for Duchenne and Becker muscular dystrophy, cystic fibrosis and Huntington disease; many offer myotonic dystrophy or fragile-X. Many offer a service for rarer diseases where they have a research interest. These might be individual diseases, diseases of one organ (eye diseases, for example) or diseases mapping to one chromosome, usually the X.

Some coordination has already taken place. The four Scottish laboratories (in Aberdeen, Dundee, Edinburgh and Glasgow) agreed from the outset to serve their 5-million population as a consortium. The arrangements have been described by Brock [6]. Briefly, each centre receives and processes all samples originating from its geographic area, then forwards the DNA to whichever laboratory handles that disease. A somewhat similar arrangement has been proposed for nine laboratories covering a population of 16 million in the West of Britain. Here, each laboratory offers a service for Duchenne and Becker muscular dystrophy, Huntington's disease, cystic fibrosis and fragile-X. For each other disease one main and one backup laboratory is nominated. Laboratories in other areas have discussed similar arrangements but not produced working arrangements.

Models for Future Coordination

Given that coordination is necessary, three models can be proposed: supraregional disease-based consortia, a national disease-based consortium or a national chromosome-based consortium. The merits and problems of each will be discussed, but it is important to remember that they are not mutually exclusive. Different genetic problems may need different consortium arrangements.

Supraregional Disease-Based Consortia

The advantages are ease of organisation, ease of communication, and continuity (since the Scottish and West of Britain consortia have already been set up). Unfortunately these consortia emerged from local initiatives rather than a national strategy to divide the country into viable groupings. The laboratories which are not in these consortia do not naturally group into their own supraregional consortia. Their population size is rather arbitrary (5 million for Scotland, 16 million for the West of Britain). In favour of supraregional groupings, some diseases will generate too little work for a regional laboratory and too much for one national laboratory.

A National Disease-Based Consortium

The main advantages are clarity of purpose and generality. Depending on the workload a disease generates, a variable number of laboratories could offer a service, so that regional, supraregional and national arrangements could all coexist and evolve flexibly within the one system. The organising body for a national consortium would also be well placed to negotiate pan-European arrangements for the rarest diseases.

A National Chromosome-Based Consortium

An alternative model would concentrate not on diseases but on chromosomal regions. In gene tracking (and most diseases have been through a gene tracking stage, even those where direct diagnosis is now usually possible), the markers used track a chromosomal segment rather than a disease gene. Many laboratories have linkage research projects, which give them expertise in tracking particular chromosomal segments. Thus in this model for a consortium, laboratories would offer a national service for markers in one or more specific chromosomal regions.

A particular advantage of this model is that it would facilitate the development of molecular cytogenetics. It is often desirable to define translocation or deletion breakpoints at the molecular level against the genetic map of the region. Technically this is not particularly difficult, but it is laborious. Each breakpoint is different, so a different set of probes must be obtained, grown up and tested for each case. As a result, these cases are usually only investigated adequately if somebody in the laboratory needs to do a small research project. Molecular cytogenetics on a service basis scarcely exists. Only a national consortium seems likely to make it possible.

A second advantage of a chromosome-based consortium is that it might lead to a simpler division of the work. Even when all genes have been mapped, it is hard to imagine that there would be a laboratory assigned to each of the thousands of mendelian diseases. Many diseases would be like "orphan" drugs: important to a few patients, but too rare to be part of a cost-effective service. Assigning consortium work on a chromosomal basis would allow a logical and finite partitioning of the total work. If, say, the human genome were partitioned into 200 chromosomal regions covering one or two G-bands, each region might on average contain a few hundred genes, leading eventually to perhaps a few dozen diseases. Most of these diseases would be so rare that developing a direct test might never be worth the effort, but they could all be tracked by the same dozen or so markers.

Finally, since many chromosomal locations will remain unclaimed for several years to come, a chromosomal consortium offers emerging laboratories a route into consortium work, with the hope that the region picked might turn out to contain some interesting genes.

Despite these theoretical virtues, it is generally recognised that a consortium cannot be organised solely on a chromosomal basis. Perhaps a pointer to future developments is the present state of work on the X chromosome. More diseases are mapped to the X than to any other chromosome, because mapping a disease known to be X-linked is much easier than mapping a disease which could be on any autosome. Laboratories often combine expertise on a limited set of X-linked diseases with more general familiarity with markers from one band or arm of the X chromosome.

Progress Towards Establishing a National Consortium

Discussions within the profession have led to agreement among clinicians and scientists that a national consortium should be established, organised by the

Clinical Molecular Genetics Society which is the professional body of the scientists. Initially the consortium will be a purely voluntary grouping. Laboratories can offer a consortium service for diseases, for chromosomal locations, or for any combination of these. The only condition is that the laboratory must agree to audit of its consortium work. At first the audit will be purely quantitative, recording the amount of consortium work performed.

Without financial rewards or sanctions, the organisers will not be in a position to impose any pattern of consortium work. However, if political changes in the Health Service lead to near-market conditions for laboratory services, the audit would pick out the laboratories, diseases and chromosome regions where there is a significant volume of consortium work. Hopefully the consortium could also become the vehicle for organising arrangements to protect the "orphan" diseases and regions, where there is no cost-effective workload.

References

1. McKusick VA. Mendelian inheritance in man, 9th edn. Baltimore: Johns Hopkins, 1990.
2. Harris R, Elles R, Craufurd et al. Molecular genetics in the National Health Service in Britain. J Med Genet 1989; 26:219–25.
3. Rona RJ, Swan AV, Beech R et al. Demand for DNA probe testing in three genetic centres in Britain (August 1986 to July 1987). J Med Genet 1989; 26:226–36.
4. Landegren U, Kaiser R, Caskey CT, Hood L. DNA diagnostics – molecular techniques and automation. Science 1988; 242:229–37.
5. Cooper DN, Schmidtke J. Diagnosis of genetic disease using recombinant DNA, 2nd edn. Hum Genet 1989; 83:307–34.
6. Brock DJH. A consortium approach to molecular genetic services. J Med Genet 1990; 27:8–13.

Discussion

Donnai: Dr Read suggested that to do polymerase chain reaction (PCR) it is not necessary to be in the same place as the clinical geneticists, but one could construct an argument for being in the same place. The molecular laboratory gets a sample of blood and somebody needs to have examined the patient and decided that the blood sample should be taken, and someone must have taken a family history to work out who else in the family may be at risk. Although the same accuracy of diagnosis as is needed for linkage analysis does not necessarily apply, if the service is to be delivered efficiently to the patients who require it someone must take family trees, work out who is at risk and look out for possible clinical heterogeneity.

Read: It is a question of which bits of the work get done centrally. Take, for example, the X-linked immunodeficiencies. If one had a patient for whom gene tracking needed to be done, probably the matter would be referred to Professor Pembrey's laboratory at the Institute of Child Health. Does one then tell the laboratory to test with specified probes in order to track this part of the X-chromosome, or does one say that this patient seems to have an X-linked immunodeficiency, pass on the clinical notes and ask the laboratory to try to work out which X-linked immunodeficiency it is and then use appropriate tracking to get the carrier risk?

My feeling is that, although it would be extremely inefficient to centralise the diagnostic side, for those tests which are fairly straightforward it is hard to resist centralising the DNA testing.

I very much hope that DNA laboratories will continue to be associated with clinical genetics centres. We are all part of the same endeavour. But the DNA testing for a disease by direct mutation testing does not necessarily benefit from being done in the clinical genetic centre that first ascertains the family.

Donnai: I agree. But anybody who has taken part in a multicentre study learns whom they can trust in terms of diagnosis and reliability of information, and presumably the reputation and accuracy of the laboratory depends on their results, and on the accuracy of what comes in.

R. Harris: It is important to point out that Sir David Wetherall set up a haemoglobin reference centre where the laboratory work is virtually error-free. It takes in material from every region in Britain for prenatal diagnosis of haemoglobin disorders, and one of the stimuli to the Confidential Inquiry was Sir David's extreme anxiety because the samples came in with inadequate clinical information, and he had no idea what was being done with the results in some centres or what the patients were being told.

The message from that must be that there has to be close association with people who understand genetics until the professions generally in medicine are sufficiently well informed to handle this problem. It is something of a race to get everybody up to a certain genetic level, but we have to be very cautious about the uses that will be made of excellent lab work if people do not understand what information to send, and what to give patients in the way of counselling at the other end.

Williamson: I disagree with Dr Read on only one of his points but it is important as far as I am concerned because I should like to see it appear in the final recommendations. He said we were discussing a service, and not research. In this field we are talking about something that is moving so quickly that much of what we do is to turn research into service. This is something which we in the UK are peculiarly bad at funding. Getting money from the Medical Research Council to do what might be called pure research is not that difficult in molecular genetics. Getting money for service per se may be difficult but we know how to do it. But getting money to turn research results into service is nearly impossible at the moment, and it is something that should be a specific recommendation, that facilitative funding should be available, not just in this field, although this field is a particularly acute example because it is moving so quickly.

The second point is a question to Professor Harris. He slipped in a phrase about community education in genetics. So far as we are concerned in the cystic fibrosis area, a lot of what we are talking about is educating the community – educating the people whom we go and deliver things to. Assuming that we cannot convince Professor Galjaard to move to the UK and do a television programme every week of the year, how do we actually do something on community education in genetics?

R. Harris: I believe I was instrumental in suggesting that GIG (Genetics Interest Group) be formed, because I felt that getting together all the charities interested in genetic handicap and disorders would facilitate the wider educational requirement. One of their aims is to produce literature for general practitioners. I have always suggested that they ought to produce literature for the general public which the GP should also get, and then one could be sure that everybody will understand it.

GIG has a big role in this but the media, television and the newspapers are doing quite a lot and will do a lot more.

Williamson: We have already talked about this and I think one of the things we should be doing professionally is to look at the rewriting of the school curriculum. We should really be looking at how to introduce genetics, and clinical concepts generally, into primary schools rather than secondary schools. It is children who are 5, 6 and 7 years old who should be getting these concepts.

Pembrey: Genetics is in the National Curriculum and it is pretty good. The genetic component is built in from age 5 years onwards and that is a very reasonable starting point. I know it is also being taught in special schools with learning difficulties, or there have been attempts to try and put it in.

Drife: Is the programme on Dutch Television specifically to do with genetics? Is it a regular feature on television?

Galjaard: Not as such. I do a lot of television outside of genetics.

In our curriculum, by the way, there is quite a lot of education. Not from five years on, but you need that in the UK because otherwise you are not ready for screening at 20. We screen later so we can have our teaching programme later. It is in the secondary schools.

Lilford: One of the things that needs to be taught in schools is not just the genetics but also the concept of probability as mathematics teaching. Where counselling fails, it is not so much that it fails because the genetics fail; it fails because people have great difficulty in understanding probability.

But what is the state of the art advice for people who are related to sufferers of cystic fibrosis (CF)? There are a few places doing screening for CF, and everybody is diagnosing and looking for a prognosis of CF when couples have had a child, but should one be offering the service if one gets a cousin, or a sister or a brother in one's clinic?

Pembrey: This interests me very much. One thing that did upset me was that when the Cystic Fibrosis Trust advertised screening programmes, they made the statement that they thought that involving other family members was already part of the service. It has not yet been evaluated and this sort of cascade counselling, of approaching relatives where the mutation is known, raises a number of very difficult questions which have not been properly assessed and evaluated at the present time.

The first thing is how to approach those relatives. We have been testing the efficacy of going back to the parents and saying that we knew what mutation they had got, and would they know of any relatives, and would they be prepared to make contact? Of 190 that have gone out we have had 30 returns. So it is not self-evident. And we have also had some very confusing problems; a couple being literally dragged up by an uncle. They were perfectly happy with their 1 in 320 risk but they said they would have the test anyway to satisfy him, and that was because he had had a telephone call from some lab without having any counselling at all – just saying that he had a risk of the 508 deletion. His reaction was that he was not at risk because he had completed his family, but that they should see his children.

It has not been evaluated and yet it seems very efficient, having got the mutation. This comes down to mutation detection as well. No matter how good the machines, if one member of the family is known to have a particular mutation, those machines will not read and diagnose which mutation is in each of the other family members. This comes back to family medicine. Gene tracking is not so different from mutation detection. We can diagnose in one which mutation it is, and then we will let people know that that is a mutation for the other relatives.

Lilford: There is a subsidiary on that. Depending how far back we take this concept, we could almost replace the whole notion of screening, presumably. If we went back beyond second degree relatives, then I presume it would take out practically all the known genes in the community.

Pembrey: For rare recessives, cascade counselling, as I call it, is probably the most efficient way to do it. One cannot say that if a recessive disease is sufficiently uncommon for population screening to have no or negligible impact, in that case we should forget about it. Cascade counselling where the lab will only ever analyse and will only counsel people who are at 50% risk or at 25% risk because of missing out generations, is a very efficient way of spreading the genetic service.

Brock: In comparatively small countries – and Scotland is a small country – one begins to recognise surnames quite quickly, and these surnames are at risk of diseases like cystic fibrosis because these are the diseases that segregate in those families.

But I wanted to make another point, a different one about organisation of molecular genetic services. The Scottish Consortium has been mentioned as a model, but the Scottish Consortium now feels that it got it wrong, and it is a pity that it should be cited constantly as a model. Where it got it wrong is on the major diseases; it got it right on the minor diseases. What we are arguing about at the moment in the Consortium is what are major and what are minor diseases. We have decided that cystic fibrosis, Huntington's, muscular dystrophies, possibly fragile-X in the future, and possibly haemophilia (unless the haematologists insist on doing it) will need to be looked at in each centre, and thereafter, having got a nucleus of disorders which each centre will cover, one would then go to one of the different models which is proposed.

Read: The other point about the Scottish Consortium is that what got it going was money, and this is the problem with national coordination within the rest of Britain – we are not at the moment talking money.

Coming back to what Professor Williamson said, we are in agreement about the development of the service from research. We saw the Consortium as being perhaps a tool to help that, because what happens is that a laboratory maps or clones the gene for a rare disease, and for a time does research and collects in patients, and so they offer a service. Then after a time the grants run out, their interest moves on and that comes to a stop. If there were a national consortium which enabled the lab to be labelled as the lab which does such-and-such a disease, then with luck in the future that can be linked to funding arrangements. But we do have the problem that we are trying to run this with no teeth of any sort and it is quite a tribute to the amicability which seems to rule in clinical genetics that clinicians and the laboratories are agreed that we should embark on this.

Campbell: Can I raise something much more basic, the crisis in the provision of cytogenetic services in the UK? In our own department we perform some 600 chorion biopsies a year but not one of these is analysed in the regional centre. Many of these are from our own region and some from our own district. Professor Harris has shown that the South-East Thames is one of the better provided of the London health regions in terms of genetic services, and yet not one of our chorion biopsy samples can be processed by the regional genetic centre.

We are also doing biopsies for South-West Thames, whose provision of genetic services is diabolically low, and women are having to pay for the laboratory analysis. The Health Service is totally failing to provide cytogenetic services in the London Region, and we should put this on record. It is a desperate state of affairs. While all the discussions here are important for the future, it is important we get the basics right before we launch into new initiatives.

Chapter 25

Provision of Service: The Obstetrician's View

R. H. T. Ward

Although the last decade has seen major technical advances in the field of prenatal diagnosis, the delivery of the service remains uneven across the UK. In 1983 the RCOG Prenatal Diagnosis Study Group concluded their report with a comprehensive list of recommendations [1] and many of these are highlighted again in reports from both the King's Fund Forum [2] and the RCP [3].

Current Situation

Wide variations in the level and type of service within the maternity services were found in the National Audit Report [4]; for example in the provision of the number of maternity beds which varied from 31 to 51 births per bed. Not surprisingly variations were also evident in the different criteria used in deciding which antenatal screening tests should be offered. In Scotland there was greater uniformity in that maternal serum alphafetoprotein (MSAFP) screening was available to all, but was only offered in England and Wales to 66% of the pregnant population. The differences were less marked in the provision of ultrasound (100% in Scotland and 95% in England and Wales), but in this audit there is no indication about the timing of the examination. Similar criteria are different for the provision for amniocentesis in women over 35 years of age – being 73% in Scotland and only 62% in England. In the same national audit report, four regions were asked to identify their objectives for prenatal screening and diagnosis. These objectives varied from "none" to offering a test for Down's syndrome in all pregnancies in women over 35 years old and in another region testing for Down's syndrome in 80% of women over 34 years old. The report also commented that in England and Wales there was no detailed guidance or recommendations as to the development of services to support and counsel patients about the purpose and possible outcome of screening. Furthermore, none of the regions monitored performance against their objectives.

We have only to look at the numbers of malformed babies at birth between 1977 and 1988 (Table 25.1) to appreciate that whereas the numbers of some types of malformations are falling, others notably exomphalos, cardiovascular abnormalities and Down's syndrome have not. In order to appreciate the scale of the problem for individual hospitals, Table 25.2 gives the incidence of these congenital abnormalities in five regions of England. In principal the whole population is at risk. However, traditionally we have been limited to screening and testing certain high risk groups such as older women; but fetal anomalies other than chromosomal are not more common with advancing maternal age (see Table 25.3). We also offer testing to women known to be translocation carriers (or their partners), those women who have had a previous Down's syndrome pregnancy or other fetal abnormality particularly a neural tube defect (NTD). It is thought that the MSAFP screening programme has been responsible for much of the decline in NTD yet the number of babies born with spina bifida remains depressingly high. Some districts that decided against MSAFP screening for neural tube defects may now well consider its introduction in detecting those pregnancies with a greater risk of carrying a Down's syndrome fetus [5]. In addition obstetricians have been slow to appreciate that MSAFP screening for NTD has a "spin off" in detecting a proportion of "normal" singleton pregnan-

Table 25.1. Malformations in babies in England and Wales

	1977	1980	1983	1986	1988
Anencephaly	568	342	114	52	41
Spina bifida	881	756	422	267	157
Hydrocephalus	259	222	194	138	137
Cardiovascular	649	866	995	882	726
Exomphalos	170	150	176	127	147
Down's syndrome	425	481	497	445	428

Source: OPCS Monitor [44].

Table 25.2. Congenital malformations by areas of residence

	Total births	Anencephaly	Spina bifida	CNS malformation	Hydrocephaly	Down's syndrome
East Anglia	26 702	2	3	15	4	11
NW Thames	50 876	2	12	41	9	38
NE Thames	56 784	5	11	39	9	37
SE Thames	51 682	2	9	28	5	33
Trent	61 921	5	11	44	16	40

From OPCS [44].

Table 25.3. Malformed babies in relation to mother's age (1986, England and Wales)

	Total	Age (years)			
		29+ under	30–34	35–39	40+ over
Anencephaly	52	34	10	6	2
Spina bifida	187	119	55	13	—
Hydrocephalus	138	99	21	17	1
CNS	637	465	120	47	5
Cardiovascular	882	625	167	77	13
Exomphalos	127	97	17	12	1
Down's syndrome	435	207	126	77	25

OPCS Monitor [44]

cies with a high perinatal loss rate as well as increased rate of spontaneous abortion [6,7].

However, other factors have led to difficulties in providing a satisfactory service. Inevitably the family doctor and those working within the community play a vital role but this has hardly been exploited [3]. The obstetrician is often limited in counselling patients because they are referred "too late" in pregnancy. This may be a deliberate ploy with couples who would wish to avoid prenatal diagnosis, but is more often because the patients themselves lack the necessary information and the traditional habit of referring patients to hospital after the second missed period.

Nevertheless the community has taken on a major role in establishing prenatal counselling and diagnosis for certain ethnic groups, particularly those at risk of thalassaemia [8] enabling couples to revert to their normal reproductive patterns. This could provide the model for carrier screening in the future for cystic fibrosis.

Obstetricians are also entirely dependent on neonatal and genetic colleagues for identifying high risk couples and many of these specialist centres have developed close links with individual obstetric units thus providing for a regional and often national service (usually unfunded).

Future Provisions

Ultrasound

The role of ultrasound is of paramount importance in modern obstetric practice. Its value in detecting fetal abnormalities has recently been shown in Helsinki [9] with the perinatal mortality halved in the 16–20 weeks scanned group; the difference in perinatal mortality being largely the contribution of congenital abnormalities in the control group. Ultrasound is increasingly used to diagnosis spina bifida [10,11] and possibly Down's syndrome [12–14] but it is essential for the smooth running of a MSAFP screening programme at 16–18 weeks. Whilst a 19 week scan seems desirable to detect fetal abnormalities an initial ultrasound examination before the booking visit is increasingly recognised to be of general value but will become essential if screening biochemical tests become applicable to the first trimester [15]. In north London about 10% of our patients refuse a booking ultrasound, but only 2.5% of women of 35 years and over. However, the vast majority of those who have had an ultrasound at booking invariably find it "exciting, enjoyable or informative". Very few respond negatively and none regarded the examination as disappointing or "a waste of time" [16]. An initial ultrasound examination before the first visit prevents the booking of patients with non-viable pregnancies and should detect multiple pregnancies (of particular importance in the context of prenatal diagnosis). Accurate dating at an early stage permits a date to be given for a 16 weeks' MSAFP estimation and a 19 week scan. However, with improvements in ultrasound it is probable that more fetal abnormalities will be detectable at the initial scan, and selective scanning using a vaginal probe may also be of value [17]. The major argument in favour of a booking scan is the continuing problem of "uncertain dates" which in one community is as high as 49% (McNay, personal communication). As biochemical screening is likely to become routinely offered, accurate dating is essential. There is natural concern that the cost and the level of existing ultrasound facilities may preclude offering all (rather than selected high risk patients) a second scan at 18–20 weeks. However, since introducing such a two-stage scan procedure locally, the number of scans per patient has actually decreased due to a fall in the number of late pregnancy scans.

Although some regions have designated obstetric ultrasound referral centres, many do not. The case for such a regional referral pattern is overwhelming not only to provide a service for patients but for the training and the continuing audit of such a service.

Amniocentesis and Chorion Villus Sampling (CVS)

The main group of patients who have had mid-trimester amniocentesis has been the older mother. In the immediate years to come we can expect an increasing number of births to women over 35 years old. Since Holloway and Brock [18] highlighted a rise of 65% in the age group between the years 1977 and 1985, the trend continues but still the number is considerably lower than the early 1960s (Table 25.4), so that the average age of motherhood in 1989 is 27.4 years, the highest average since 1961 [19] and projected to rise to 28.3 years by 1995. On the other hand, the statistics for therapeutic abortion show that fewer women in this age group are terminating their pregnancies (Table 25.5). Nevertheless the

Table 25.4. Live births by age of mother: England and Wales

	No. of births (thousands)			
	1961	1971	1981	1988
Ages 30–34	152.3	109.6	126.6	141.0
Ages 35–39	77.5	45.2	34.2	47.6
Ages 40 + over	23.3	12.7	6.9	9.0
All ages	811.3	783.2	634.5	693.6

From Population Trends [19].

abortion figure for 1988 shows that 1732 women had abortions on the grounds of Clause 4, representing just over 1% of the 168298 abortions [20].

Table 25.5. Termination of pregnancy in England and Wales by age group

Age (years)	% of all conceptions			
	1977	1982	1986	1987
30–34	15.3	14.3	13.3	13.6
35–39	32.5	30.8	25.4	24.8
40+	52.9	50.2	46.0	45.2
All	15.2	17.0	18.0	18.9

From OPCS [44].

Counselling in Early Pregnancy

However, not all older women will wish to have prenatal diagnosis and fetal karyotyping, although some may still want a MSAFP or a 19 weeks scan. It is essential to document clearly the agreed plan for each patient and when these tests are to be done whilst also indicating on the notes when patients do not wish to have the test, to avoid the mother being subjected to persistent (although well meaning) offers for prenatal diagnosis. Such documentation is also important medicolegally. In order to give couples more time to take decisions about prenatal diagnosis, we established in 1985 an early pregnancy genetic clinic for the older mother [21,22]. This coincided with the offer of first trimester chorionic villus sampling (CVS) for karyotyping and resulted in 100% offer rate of prenatal diagnosis to women aged 38 years and over at their expected date of delivery. Inevitably a number of non-viable pregnancies (18%) were found in the clinic with a further 9% aborting spontaneously before the diagnostic procedure. These patients require sympathetic counselling. We offered them all a D and C and although 82% accepted, when they were asked at a later date, one-third of them wished they had allowed a spontaneous abortion to occur. Whatever the management, these patients must be told that the operation itself is unlikely to establish the reason for the abortion.

For this type of prenatal genetic clinic to function efficiently, there needs to be a direct telephone line to a designated coordinator. Ideally patients and partners should be seen and scanned before the tenth week to allow a decision about which test they might prefer. The information imparted at the counselling should be backed up with suitable explanatory leaflets. Traditionally a doctor has offered such counselling but there seems no reason why a trained non-medical person would not be satisfactory in this role.

Women's choice of prenatal tests

In a study when CVS was first on offer for fetal karyotyping 131 women were given a questionnaire before they were counselled and another after being counselled [16]. The results are shown in Table 25.6. It was not surprising that many were initially unsure (61%) but the fact that 15 women finally decided on "no test" shows that even with the option available of an early test and a rapid result, a number of women for religious or ethical reasons will wish to avoid prenatal diagnosis.

Table 25.6. Specific test preference in 131 women aged 38 or over at EDD

Test	Precounselling	Postcounselling	P value
Amniocentesis	12 (9.2%)	58 (44.3%)	<0.001
CVS	31 (23.7%)	51 (38.9%)	<0.01
None	8 (6.1%)	15 (11.5%)	<0.5
Don't know	80 (61.0%)	7 (5.3%)	<0.001

From Knott [16].

In another questionnaire 128 patients out of a group of 158 were asked what their perception was about the risk of a miscarriage, of having a baby with Down's syndrome and what the risks were for pregnancy loss after CVS or amniocentesis [16]. Patients were not expected to give exact answers and a reasonable range of replies was acceptable. The results are shown in Table 25.7. There were highly significant increases in the number of correct responses to all four questions in the postcounselling as compared with the precounselling questionnaire indicating the positive value of the interview in improving the patient's knowledge.

Table 25.7. Effect of counselling interview in interpretation of risk factors

Risk factor	Miscarriage	Down's syndrome	Amniocentesis	CVS
Precounselling				
Correct	20	42	34	29
Incorrect/don't know	105	84	94	97
Postcounselling				
Correct	50	94	83	83
Incorrect/don't know	75	32	45	43
P value	<0.001	<0.001	<0.001	<0.001

From Knott [16].

Patient choice of test was assessed by McGovern et al. [23] who found that among women who had undergone prenatal diagnosis by amniocentesis 68% chose that method mostly because of the lower risk of abortion. However 87% of those who had chosen amniocentesis said that if the risk of fetal loss after CVS was equal or less than amniocentesis, they would choose CVS in the future. In a more recent study [24] patients were questioned about their attitudes to testing in a future pregnancy. The questionnaire was sent to 190 patients 10 weeks after CVS and there was an 80% response. The majority 93%, would wish CVS in a future pregnancy and some 97% would accept a risk of miscarriage, twice that quoted for amniocentesis.

Our own experience is of interest in that in 1985, a study showed that the majority of patients (77%) who had either CVS or amniocentesis elected for the test that they already had. However, some would wish to have CVS reflecting in many cases their late referral, i.e. after 11 weeks in the pregnancy tested which effectively precluded CVS which was only done by the transcervical route at that time (Table 25.8) [16]. There were also three patients who would prefer amniocentesis and not CVS in the future but in all three there had been a failure of diagnosis. However, the unexpected finding in this study was that there were 11 patients who would not have either test in the future. None of these who had amniocentesis had had an abnormal fetus and all pregnancies progressed normally although one patient did need a repeat amniocentesis for failure of culture. As the questionnaires were given immediately after the procedure one of the explanations of this high refusal rate for future testing could be that of concern about the possibility of miscarriage or of fetal abnormality, exacerbated by the delay in receiving a result.

Table 25.8. Test preference

Test performed in current pregnancy	Test preference for future pregnancy		
	CVS	Amniocentesis	No test
CVS	51	47	3	1
Amniocentesis	73	14	49	10
Total	124	61	52	11

From Knott [16].

Attitudes to Prenatal Diagnosis

Patients were also asked about their attitudes to five selected features of the diagnosis which was tested on two separate occasions. In fact there was very little difference between the results before and after the test had been performed. In order of importance (ranking) was accuracy, safety and early diagnosis. The least important was pain and embarrassment. It is not surprising that of overriding importance was accuracy. All these patients had been counselled that about 1% of CVS samples would give false positive results but the majority of these would be readily resolved by amniocentesis. Safety was ranked more important than early diagnosis overall but the CVS patients did put a higher priority on early diagnosis compared with those who had amniocentesis. Thus there are a large number of patients for whom the advantage of early diagnosis outweighs the probable increased risk of fetal loss with CVS. This agrees with the findings of McCormack et al. [24] whose larger series confirmed that CVS is an acceptable alternative to amniocentesis for many women in this older age group.

A study by Spencer and Cox [25] further supported patients' acceptability to CVS. The responses of the patients who had undergone CVS indicated a reduction of anxiety up to 10 weeks earlier than those who had amniocentesis. In addition, patients found that CVS was less uncomfortable than amniocentesis.

In fact in our own recent practice, 160 patients of 38 years of age or older who accepted a prenatal test for fetal karyotyping, only 20 had CVS (12.5%) and 10 others preferred to rely on the result of a negative MSAFP estimation. This left 130 electing for a standard amniocentesis. In the same annual audit, whereas an offer of prenatal testing was recorded in 206/218 (94.5%) and accepted in 77.7%, there were eight patients (3.7%) where there was no documentation of an offer of prenatal fetal karyotyping. (Four patients' notes could not be traced.)

Amniocentesis

The technique of the so-called standard amniocentesis (SA) as opposed to early amniocentesis (EA) has changed over the years so that most now use continuous ultrasound surveillance [26]. The other advance has been the use of a smaller gauge needle, although it is surprising to learn that there are still operators who use 20 or even 18 gauge needles. Our own experience [27] in which 513 patients, including nine with twins, underwent amniocentesis confirmed the safety of Romero's technique. There were 12 taps that failed to produce a clear sample and on only two occasions was frank blood obtained. It was also of interest to note that the success of the procedure did not depend on the placental site. Seven pregnancies (1.4%) of those intended to continue were lost following the procedure.

The experience of the operator clearly influences the success of amniocentesis. In a series from a university department, success at the first attempt varied between 94% for the principal author to 69% for the other occasional operators [28]. A clear specimen was obtained in 86% of the pregnancies by the main operator but only in 71% from the others. Despite these findings there was no obvious difference between the pregnancy outcomes, but a study of this size would be unlikely to distinguish small differences in the miscarriage rates.

In the UK the majority of amniocenteses are performed by registrars in training. How often though do they receive specific training in amniocentesis by an experienced consultant and to what extent is this training backed up by continuing supervision?

Early Amniocentesis

As amniocentesis became technically easier it was natural that the timing of the procedure, traditionally at 16 weeks should be advanced. I suspect that early amniocentesis was provided as an alternative by centres and obstetricians in the USA when CVS was not readily available outside university departments yet at a time when patients were pressing for an early test. As amniocentesis was a recognised procedure for insurance purposes it mattered little in this context at what gestational age it was done because certainly in the early days of CVS this was not an allowable insurance procedure. Whereas many obstetricians in the UK had been undertaking amniocentesis more frequently at 14–15 weeks there needs to be considerable caution exercised before early amniocentesis is encouraged [29]. At a recent international workshop (XII Fetoscopy Group, Bologna, October 1990), all centres confirmed that amniocentesis was routinely performed between 14 and 16 weeks. Only one (in France) undertook the procedure earlier.

Very early amniocentesis at 8–11 weeks is feasible but karyotyping is only possible in 68% of cases [30]. However, the same authors had a 100% success rate of karotyping specimens of amniotic fluid taken between 12 and 14 weeks. Hackett et al. [31] have suggested a low failure rate of 2.8% to obtain fluid at the first attempt between 11 and 14 weeks. In all 105 samples obtained, a fetal karyotype was possible. Two of the 105 pregnancies had been lost at the time of the report but only 62 of the babies had been delivered.

Complications of Amniocentesis

The theoretical concern of fetal lung development being arrested following amniocentesis was addressed in a monkey model by Hislop and Fairweather [32]. Their conclusion was that there was arrest of fetal lung development in those pregnancies subjected to amniocentesis compared with the control population and this was so even if no fluid was removed. A recent report [33] does much to allay anxieties with regard to children's development (whose mothers had amniocentesis) but ear infections and middle ear impedance abnormalities appeared to be more common than in the control group. However, with early amniocentesis and the suggestion that at 10 weeks 10 ml of amniotic fluid requires to be removed (approximately one-third of the total amniotic fluid volume), the situation is obviously different to that pertaining at 16 weeks when this quantity of fluid would represent about 5% of the total volume. Meade and Grant [29] have suggested that early amniocentesis should be the subject of a randomised trial and could usefully be compared with transabdominal CVS. Such a study would compare the two procedures, not only for safety but for the quality of karyotyping, technical ease and the associated pregnancy loss. A potential long-term follow-up of babies should be built into the study. Another aspect of the study might be the comparative costs of prenatal diagnosis. It is well known that obtaining a karyotype from chorionic villi is more labour intensive than from amniotic fluid and it is likely that, should there be a sudden demand for early diagnosis using CVS, many regions could not either expand their laboratory or find obstetric personnel to satisfy this demand.

Safety and Timing of CVS

The safety of CVS has been the subject of three randomised studies, in Canada [34], USA [35] and the UK (yet to be published). In all three, pregnancy outcome was compared with pregnancies in which standard amniocentesis was undertaken. The Canadian study, the first to be published, recruited women from 11 centres. Each centre had "reasonable experience" of CVS both in the cytogenetic laboratory and with the transcervical CVS technique. Nevertheless one of the striking findings of this study was that 9.9% of women who had had a CVS needed an amniocentesis for either obstetric (6.8%) or laboratory (3.1%) reasons. The obstetric "failure rate" is considerably higher than reported by experienced operators and compares with a 1.8% figure in the NIH study. Such a finding might suggest that the Canadian centres were considerably "newer" to the technique of CVS than those in the seven American centres. However, in an up-date of the

Canadian result [36] the excess of total loss in the CVS group was 0.6% over that of amniocentesis, whereas in the NIH study the adjusted difference between the two was 0.8%. Should there be similar findings in the MRC study there is likely to be an increased demand for CVS in the UK. It is important to appreciate that apart from the increased provision of cytogenetic and obstetric time, there will need to be a greater input into counselling patients particularly as the risk of spontaneous abortion is so much higher in the first rather than in the second trimester of pregnancy.

Whilst planning ahead for a standard amniocentesis presents few problems, this is not the case for CVS as patients expect the procedure to be arranged as soon as possible. Which CVS technique is "best" is now largely academic, as both approaches are as successful as regards achieving a diagnosis and safety. Bleeding following transcervical CVS may occur in as many as 6.0% of patients (as compared with 1.9% for transabodiminal CVS) but these patients do not have a worse pregnancy outcome (Brambati, XII Fetoscopy Group, Bologna, 1990). As regards losses following CVS both techniques have been shown to have acceptably low abortion or perinatal loss rates (Table 25.9).

Table 25.9. Pregnancy outcome in randomised CVS studies

Author	Number of patients	Transabdominal	Transcervical
Brambati	1194		
Total pregnancy loss		16.5%	15.5%
Excluding TOP		4.9%	6.4%
Jackson <28 weeks losses	3608	2.4%	2.6%
Wapner <28 weeks losses	1691	2.4%	1.6%

Data presented at XII Fetoscopy Group, October (1990).

In fact there may be some value in waiting until 12 weeks before proceeding to CVS in the older patient with a low risk. This approach has been advocated by Jahoda et al. [37] to avoid the higher spontaneous abortion loss before 12 weeks. In their series the post-procedural loss was 2.5% after a 12 weeks' transabdominal CVS compared with a 5.8% loss when performed earlier. In a more recent paper on transabdominal CVS, Saura et al. [38] concluded that the procedure should not be performed before week 12. Delaying CVS until this time might well be acceptable to the older women with a comparatively small risk of fetal chromosomal abnormality, but the same cannot be said for those with a higher risk. Indeed it is this particular group of patients, such as those at risk of thalassaemia [39] who should be offered CVS at an early stage.

MSAFP Screening

The only reliable test for spina bifida is the measurement of amniotic AFP levels, although increasingly it can be detected by ultrasound. In fact many units are now confident of excluding the lesion ultrasonically without proceeding to a diagnostic amniocentesis and measurement of amniotic fluid AFP. This may be a satisfactory policy in certain centres but not one that can yet be advocated nationally.

Whatever the individual policy, there must be a rapid recall system with immediate counselling and offer of ultrasound and possibly amniocentesis for couples with a raised MSAFP level. When both an MSAFP estimation and 19-week scan are offered routinely, patients are frequently prewarned that they will not be contacted if the results are normal. If these tests are offered to patients with the aim of detecting fetal abnormalities, it is not surprising that patients remain anxious until their next visit at 28 or 32 weeks. Fearn et al. [40] found that the anxiety for 2–3 weeks after the MSAFP test was greatly influenced by whether the patient was given a definitive normal result or whether she was told to assume that the result was normal if she did not hear from the clinic. Marteau [41] stresses the psychological importance of giving a negative result by telephone backed up with a confirmatory letter. My experience suggests that patients like to hear the result first hand. An antenatal visit combined with a 19-week scan in addition to allowing various uncertainties of the scan to be discussed, such as "low lying placenta" can also be used to check there have been no omissions at booking. One of these is failing to screen ethnic groups who are at risk of a haemoglobinopathy – a problem highlighted in the RCP report [3].

It seems likely that MSAFP and other biochemical screening tests will identify many more pregnancies at risk for Down's syndrome in the younger age group. On the other hand there is evidence to suggest that the older mother (especially after long periods of infertility or IVF) may elect biochemical screening before deciding on whether to proceed to an amniocentesis.

When amniocentesis is performed for fetal karyotyping, it is still wise to do a preprocedure MSAFP test. Should results subsequently be high, in the absence of any fetal abnormality, the pregnancy should be regarded as high risk and the antenatal care modified. On the other hand if there is a culture failure, the result of the MSAFP can be of great help in advising the couple whether they should proceed to either repeat amniocentesis or another procedure such as a fetal blood sample.

Other Diagnostic Tests

There are occasions when fetal blood sampling and late "CVS" [42] may be indicated. Either can be useful for the late booker or "failed amniocentess" or when an ultrasonic fetal abnormality has been diagnosed [43]. In the latter situation the additional information of the fetal karyotype is important in a couple's decision-making even if the pregnancy is too far advanced for legal therapeutic abortion. Fetal blood sampling is a technique that should be confined to a few centres. Even in the hands of an expert the loss rate is greater than with amniocentesis – depending on the method used and the indication (Table 25.10). The majority of the patients in Ansaklis' series were at risk of thalassaemia.

Delivery of Service

The scope for improving prenatal diagnostic services is considerable even with existing facilities. If each unit had a defined policy which can be supported by the community and genetic services, it is probable that more couples would be offered screening and testing. It is the offer rate that is important and obviously the acceptance rate will vary with population. To ensure a 100% offer rate requires cooperation between family doctor and hospital, and an efficient method for counselling early in pregnancy for certain well-defined groups such as the older mother.

Table 25.10. Fetal loss rate following diagnostic fetal blood sampling

	Number of patients	Fetal loss (%)
Fetoscopy 1987–90	147	3.6
Cordocentesis	1265	1.41
Cardiocentesis	168	3.6

Ansakalis, XII Fetoscopy Group (1990).

Ultrasound examination and biochemical screening should be available to all with an efficient system of recall as well as of informing the patient of normal results. Regional referral centres should be established to offer tertiary scanning, and provision of obstetric techniques such as fetal blood sampling. Such centres should work in conjunction with regional genetic services and should be able to call upon the expertise of a perinatal pathologist. Finally, such centres should provide training in all aspects of prenatal diagnosis and be responsible for an annual audit.

Acknowledgement. I am particularly grateful to Mr Peter Knott for his permission to include much of his research which culminated in a MD thesis.

References

1. Rodeck CH, Nicolaides KH (eds.) Proceedings eleventh study group of the Royal College of Obstetricians and Gynaecologists. Prenatal Diagnosis. 1983; 397–8.
2. King's Fund Forum. Screening for fetal and genetic abnormality. Lancet 1987; ii:1408.
3. Royal College of Physicians of London. Prenatal diagnosis and genetic screening, community service and implications. A report of the Royal College of Physicians, 1989.
4. National Audit Office. Maternity services. London: HMSO, 1990.
5. Merkatz IR, Nitowsky HM, Macri JN, Johnson WE. An association between low maternal serum alpha-fetoprotein and fetal chromosome abnormalities. Am J Obstet Gynecol 1984; 148:886–946.
6. Brock DJH, Barron L, Duncan P, Scrimgeour JB, Watt M. Significance of elevated mid-trimester maternal plasma-alpha-fetoprotein values. Lancet 1979; i:1281–2.
7. Milunsky A, Jick SS, Bruell CL, et al. Predictive values, relative risks, and overall benefits of high and low maternal serum α-fetoprotein screening in singleton pregnancies: new epidemiologic data. Am J Obstet Gynecol 1989; 161:291–7.
8. Modell W, Ward RHT, Fairweather DVI. Effect of introducing antenatal diagnosis on the reproductive behaviour of families at risk for thalassaemia major. Br Med J 1980; 2:737.
9. Saari-Kemppainen A, Karjalainen O, Ylostalo P, Heinonen OP. Ultrasound screening and perinatal mortality. Controlled trial of systematic one stage screening in pregnancy. Lancet 1990; 336:387–91.
10. Campbell J, Gilbert WM, Nicolaides KH, Campbell S. Ultrasound screening for spina bifida: cranial and cerebral signs in a high-risk population. Obstet Gynecol 1987; 70:247–50.
11. Van den Hof M, Nicolaides KH, Campbell J, Campbell S. Evaluation of the lemon and banana signs in one hundred thirty fetuses with open spina bifida. Am J Obstet Gynecol 1990; 162:322–7.

12. Benacerraf BR, Barss VA, Laboda LA. A sonographic sign for the detection in the second trimester of the fetus with Down Syndrome. Am J Obstet Gynecol 1985; 151:1078–9.
13. Lynch L, Berkowitz GS, Chitkara U, Wilkins IA, Mehalek KE, Berkowitz RL. Ultrasound detection of Down syndrome: is it really possible? Obstet Gynecol 1989; 73:267–70.
14. Tongue M, Rodeck C. Is ultrasound of any value in screening in Down's syndrome? Br J Obstet Gynaecol 1989, 96:1369–71.
15. Brock DJH, Barron K. Holloway S, Liston WA, Hillier SG, Seppala M. First-trimester maternal serum biochemical indicators in Down syndrome. Prenat Diagn 1990; 10:245–51.
16. Knott PD. The introduction of chorionic villus sampling as a screening test for Down's syndrome. MD Thesis, London University, 1989.
17. Rottem S, Bronshtein M, Thaler I, Brandes JM. First trimester transvaginal sonographic diagnosis of fetal anomalies. Lancet 1989; i:444–5.
18. Holloway S, Brock DJH. Changes in maternal age distribution and their possible impact on demand for prenatal diagnostic services. Br Med J 1988; 296:978–82.
19. Population Trends Health and Personal Social Services Statistics. London: HMSO, 1989.
20. Abortion Statistics, OPCS, London: HMSO.
21. Knott PD, Penketh JA, Lucas MK. Uptake of amniocentesis in women aged 38 years or more by the time of the expected date of delivery: a two-year retrospective study. Br J Obstet Gynaecol 1986; 93:1246–50.
22. Knott PD, Ward RHT, Lucas MK. Effect of chorionic villus sampling and early pregnancy counselling on uptake of prenatal diagnosis. Br Med J 1986; 293:471–4.
23. McGovern MM, Goldberg JD, Desnick RJ. Acceptability of chorionic villi sampling for prenatal diagnosis. Am J Obstet Gynecol 1986; 155:25–9.
24. McCormack MJ, Rylance ME, Mackenzie WE, Newton J. Patients' attitudes following chorionic villus sampling. Prenat Diagn 1990; 10:253–5.
25. Spencer JW, Cox DN. Emotional responses of pregnant women to chorionic villi sampling or amniocentesis. Am J Obstet Gynecol 1987; 157:1155–60.
26. Romero R, Jeanty P, Reece EA et al. Sonographically monitored amniocentesis to decrease intraoperative complications. Obstet Gynecol 1985; 6:426–30.
27. Knott PD, Colley NV. Amniocentesis with continuous ultrasound surveillance: reduced complications with greater sampling success. J Obstet Gynecol 1989; 10:77.
28. Wiener JJ, Farrow A, Farrow SC. Audit of amniocentesis from a district general hospital: is it worth it? Br Med J 1990; 300:1243–5.
29. Meade TW, Grant AW. Early amniocentesis: a cytogenic evaluation. Br Med J 1989; 299:623.
30. Rooney De, McLachlan N, Smith J et al. Early amniocentesis: a cytogenetic evaluation. Br Med J 1989; 229:25.
31. Hackett GA, Smith JH, Rebellow MT et al. Early amniocentesis at 11 to 14 weeks gestation for the diagnosis of fetal chromosomal abnormality – a clinical evaluation. Prenat Diagn 1990; in press.
32. Hislop A, Fairweather D. Amniocentesis and lung growth: an animal experiment with clinical implications. Lancet 1982; ii:1271–2.
33. Finegan J-AK, Quarrington BJ, Hughes HE et al. Child outcome following mid-trimester amniocentesis: development, behaviour, and physical status at age 4 years. Br J Obstet Gynaecol 1990; 97:32–40.
34. Canadian Collaborative CVS Amniocentesis Clinical Trial Group. Multicentre randomized clinical trial of chorion villus sampling and amniocentesis. Lancet 1989; i:1–6.
35. Rhoads GG, Jackson LG, Schlesselman SE et al. The safety of chorionic villus sampling for early prenatal diagnosis of cytogenetic abnormalities. N Engl J Med 1989; 320:609–17.
36. Hamerton JL. Canadian collaborative CVS Amniocentesis Clinical Trial Group. Up-date Proceedings of the fifth international early fetal diagnosis meeting, Prague, 1990.
37. Jahoda MGJ, Sachs ES. Experiences with 3000 CVS procedures in Rotterdam. In: proceedings of the fifth international meeting on early fetal diagnosis, Prague, 1990.
38. Saura R, Longy M, Horovitz J et al. Risks of transabdominal chorionic villus sampling before the 12th week of amenorrhoea. Prenat Diagn 1990; 10:461–7.
39. Ward RHT, Petrou M, Modell BM et al. Chorionic villus sampling in a high-risk population – 4 years experience. Br J Obstet Gynaecol 1988; 95:1030–5.
40. Fearn J, Hibbard BM, Laurence KM, Roberts A, Robinson JO. Screening for neural-tube defects and maternal anxiety. Br J Obstet Gynaecol 1982; 89:218–21.
41. Marteau TM. Reducing the psychological costs. Br Med J 1990; 301:26–8.
42. Holgreve W, Miny P, Schloo R. "Late CVS" international registry compilation of data from 24 centres. Prenat Diagn 1990; 10:159–67.

43. Palmer CG, Miles JH, Howard-Peebles, PN, Magenis RE, Patil S, Friedman JM. Fetal karyotype following ascertainment of fetal anomalies by ultrasound. Prenat Diagn 1987; 7:551–5.
44. OPCS monitor. London: HMSO, 1987.

Discussion

Rodeck: I am always impressed at hearing my genetic friends talking about their splendid regional organisations of genetic services. Deficient though they may be in some areas, it is possible to organise those services on a regional basis. It is much, much more difficult to provide basic prenatal diagnosis because women go to antenatal clinics in district general hospitals. One of the most important things that we could do is to recommend that there should be specialist obstetricians who have responsibility for prenatal diagnosis. This is something that the College, the regions and the government have never taken on board. New techniques cannot be introduced and grafted on to existing antenatal clinics which churn on and on. We need to have specialists who will take responsibility for these things at consultant level. That is quite separate from having regional fetal medicine centres which should work in close collaboration with, and preferably on the same site as the genetic centres. Professor Galjaard made that point with his umbrella organisation which we rarely have in the UK.

Allan: Can I clarify the financial implications of karyotyping? We have heard that the take-up rate for chorionic villus sampling (CVS) is likely to be much higher in the older age woman. The main delay as I understand it in the processing of an amniocentesis test is administrative and financial rather than growing the cells. If we cannot afford to have an efficient amniocentesis service, how can we possibly afford to have something much more intensive to a wider population in terms of CVS?

Ward: Professor Gosden could probably answer better than I could, but my understanding from our people is that it is because all samples are given the full cytogenetic analysis.

In the case of low-risk individuals, that is, say, less than 1%, would it be acceptable to do an aneuploidy count on a direct sample? If it were accepted that a direct sample analysis was sufficient for low-risk samples, (possibly holding something in reserve to be used if one had an abnormal result) it would transform the economics of the situation. But this is very contentious within the cytogenetics community.

Gosden: There is really a major problem about whether or not the profession wants to take on board AFP screening to get to more people, or whether to set a maternal age limit. The vast majority wanting CVS are the older mothers and those with a previous affected child, and they are a low-risk group. There is a high risk of false positive results. I do not think there is any risk of false negative results in chorion. If the count is abnormal for certain things like trisomy 18, 45X – almost any of the other trisomies other than 21 – the back-up sample would need to be put into play. But one has to ask the basic question: should this low-risk group, say between 35 and 40, be offered CVS? Should CVS be restricted to the over-40s? They would have a slightly higher risk, and a greater risk of needing termination. Should age limits be adjusted even more and the 35–40 year olds be persuaded to go for serum screening and amniocentesis?

Ward: I did not advocate CVS for the low-risk group. I was making the point that our demand has dropped 12.5% for several reasons: the laboratory in fact saying we must limit the numbers. With the number of other people coming through whose indication for CVS is clearly there, I would agree with what Professor Gosden is suggesting. We should be aiming for a cut-off, perhaps at 40 and over, or the woman who has a family already and is unexpectedly pregnant. They might wish to take up CVS, rather than the infertile woman who has been desperate for a baby and who may well wish to have biochemical testing before going on further.

Gosden: Even for amniocentesis we have got to ask whether for the low-risk patients we should do three cells only. We have got major questions to try and answer. We could offer many more patients amniocentesis.

Lilford: It is perhaps worth summarising the three reasons why CVS has arguments against it at low genetic risk. The first, if we do direct and culture, is cost. The second is that the slightly higher miscarriage rate compared to amniocentesis becomes more important as genetic risk drops. The third is that the chance of getting an ambiguous result relative to Down's risk is much higher.

I used to advocate only doing a direct preparation as a Down's exclusion test, but it is very hard to persuade cytogenetic laboratories to do what they would see as lowering their standards. I disagree with Professor Gosden. There is a risk of a false negative result if they only go for trophoblast. I am doing a medicolegal report at the moment on a patient where they missed a Down's and they had only done a direct.

Nevin: It is a matter of organisation. Our laboratory has not been involved in CVS at all; we have gone for early amniocenteses. We have one cytogeneticist who copes with 700 amniocenteses per year. On each amniocentesis we do a full-banded preparation for a total of 12 cells and the karyotype is available, within 9 to 10 days.

Campbell: I should like to challenge the concept that we need regional centres to do cytogenetics. If 5% of our patients need an amniocentesis or a chorion biopsy, that is a large number of samples wending their way from far parts of, say, South-East Thames, all the way to the centre of London. Why can districts not organise cytogenetic services? They organise so many areas – biochemical, chemical pathology, morbid anatomy, etc. – why can they not organise cytogenetics? When the district hospitals opt out of that particular service, they downgrade it in importance.

I am not talking about genetic counselling, but taking the laboratory side, and each district doing its own thing, or maybe a group of districts. I do not see for cytogenetics the need for a centralised regional service.

Gosden: As Professor Nevin is saying, a well set up laboratory that is efficient and has a good policy and goes for efficiency can go through many more samples with a very much higher degree of accuracy than is generally reported. The difficulty, if it is devolved to numbers of small centres, is that they do not necessarily have the expertise and it is usually less efficient to be in a small lab because one person is having to do such things as media making, which can become more efficient if the service is centralised. But we have the example of a small service which can be super efficient and so it is difficult to reconcile the two views.

Marteau: Can I ask for the source of Dr Allan's data that most women prefer CVS instead of amniocentesis?

Allan: There is a lot of evidence that the take up rate of an earlier diagnostic test like CVS is higher even in the ordinary mothers.

Marteau: Even with Peter Knott's data one still did not get the majority opting for CVS, and that was someone who would be counselling very much in favour. I am not sure that the evidence is there.

Lilford: There are three studies that show that if one suggests a woman with a 1 in 100 risk for Down's – leaving out false positives and false negatives, which we should not do in real life – and asks about relative miscarriage rates, all three studies [1–3] show that women are prepared to run up to about a doubling of procedure-related risk in order to get an earlier test.

Nicolaides: The conversion of research into clinical practice, is something that is very much related to amniocentesis. This is a very established technique, and for many years we have been counselling – without any evidence – that the procedure-related mortality was 0.5%. It is only two or three years ago that there was only one prospective study, by a leading expert who demonstrated clearly that the procedure-related mortality is at least 1%.

On the same basis I would discuss the conversion of the data from an interested group on a research basis into clinical practice. We were saying that it takes 48 hours to get the results of a fetal blood karyotyping. Yes, when it was of research interest! At the moment at King's College Hospital it takes one to two weeks to get fetal blood karyotyping. And "CVS takes from a few hours to two or three days": at the moment it takes us three weeks to get CVS results.

Amniocentesis in private laboratories in the United States and in research interested groups in Britain takes 10 days. At King's College Hospital it takes four weeks if we are lucky, usually five weeks to get the results. It is extremely important that there is contact between the laboratories and the patients. I spend approximately 20%–25% of my time answering the phone to justify to the patients why we do not yet have their results. If the system required the laboratory technicians to answer the phone, perhaps they could understand the urgency of doing so.

The second point is that if the laboratories are integrated in institutions where they come in touch with patients, and they understand research, it becomes an interesting aspect of their job, rather than that boring thing of looking under a microscope and counting chromosomes.

Atkins: Regarding the regional cytogenetics laboratory in the Northern Region our laboratory has metastasised, so there is a regional laboratory in two places. It is not the district laboratory although it happens to be in our district and they pay us so much for the space and so on. The result is that the specimens do not have to go vast distances.

As far as funding it is concerned, the proposal that has been put forward is that the region be top sliced, that districts be top sliced, and they then pay so much per maternity.

Nevin: To come back to Mr Nicolaides's point, in Northern Ireland we have one genetic obstetrics centre. All the amniocenteses and genetic ultrasound are done in the one hospital and the patients travel to the hospital, so there is an advantage in terms of that. All the amniocenteses are done on the one day and are batch processed through the laboratory.

We have always given the patient a card with a senior cytogeneticist's number in the laboratory and the patient is told to phone at 12 days for the result. If it is an adverse result then a clinical geneticist is available to cope with that patient on the telephone, but normal results are given out by the laboratory so that the obstetrician is not bothered about giving those results, although the results will filter through to him in time.

References

1. Thornton JG, Lilford RJ. Genetic counselling. Br Med J 1988; 296:933–4.
2. Knott PD, Ward RH, Lucas MK. Effect of chorionic villus sampling and early pregnancy counselling on uptake of prenatal diagnosis. Br Med J 1986; 293:479–80.
3. Bryce RL, Bradley MT, McCormick SM. To what extent would women prefer chorionic villus sampling to amniocentesis for prenatal diagnosis? Paediatr Perinatol Epidemiol 1989; 3:137–45.

Conclusions and Recommendations

1. There is growing public demand for prenatal diagnosis and screening. Families are increasingly looking to obstetric and genetic services for help towards the birth of healthy children.

2. We strongly support previous statements from medical Royal Colleges and the Department of Health that genetic services should be regionally organised, with integration of their clinical and laboratory arms. There is a special need to provide training for the clinical genetic co-workers, whose background may be in science, nursing or social work, and who provide an important link between genetic centres, antenatal clinics practising prenatal diagnosis and community services. Specialist fetal medicine centres are appropriate at regional level, ideally on the same site as regional genetic centres to facilitate close working.

3. Prenatal diagnosis (PND) is now a major part of all antenatal care. The obstetric element cannot be centralised to the same degree as genetic services organised at regional level, but has to be incorporated into the routine antenatal clinic. Effective delivery of PND cannot be achieved until each district general hospital has a consultant obstetrician with a special interest in PND, to take responsibility (in collaboration with appropriate colleagues) for the organisation of ultrasound, invasive procedures, patient information and professional education.

4. We emphasise the need for better medical undergraduate and postgraduate, as well as lay, education in genetics to increase the level of understanding of the expanding range of options available to individuals, couples and families at risk of inherited and congenital disorders.

5. Systems for monitoring outcomes and quality of care, including clinical audit, should be an inherent part of prenatal diagnostic services.

6. There is a need for adequate registers of congenital malformations for the effective audit of screening programmes. Current systems are unsatisfactory because information is incomplete. For many conditions there is merit in the organisation of malformation registers at regional level but interlinked at national level. Experience with regional registers suggests that they provide an effective tool for monitoring outcome. For some specific conditions additional national registers are more appropriate.

7. All terminations of pregnancy for fetal abnormality should be followed by appropriate laboratory investigations and the fetus should be fully examined by a specialist with specific expertise in fetal pathology. The practice of storing samples for future diagnosis (possibly by DNA analysis) is encouraged.

8. The Royal College of Obstetricians and Gynaecologists has previously recommended routine ultrasound scanning at 16–18 weeks' gestation. The Study Group considered that from an ultrasonographic point of view the optimum time is 18–20 weeks' gestation. There was support for an additional simpler early scan for dating and detection of major malformations and this would improve interpretation of maternal serum alphafetoprotrein and other biochemical tests.

9. Fetal anomaly scanning should have clearly defined objectives. A standardised examination of fetal anatomy should involve the fetal heart (four chamber view), brain, spine, kidneys, abdominal wall, diaphragm and limbs. It is estimated that this approach could identify up to 70% of serious structural anomalies present in the newborn. There are established training programmes in obstetric ultrasound, and fetal anomaly scanning should be performed only by appropriately trained personnel.

10. There is a need to continue research, including that into ultrasound markers of Down's syndrome, and the place of a routine 30 week scan to identify abnormalities of the urinary tract with a view to postnatal follow-up and treatment. Further evaluation is needed of Doppler ultrasound and magnetic resonance imaging in the detection of fetal abnormality.

11. Significant chromosome abnormalities are found in 1 in 150 newborn babies. For pregnancies in which the genetic risk is high there is a need for early testing which is accurate and gives rapid results. The two approaches to chorionic villus biopsy (transabdominal and transcervical) seem equally successful, whereas the relative merits of early amniocentesis require further investigation. Further conclusions should await publication of the Medical Research Council's trial comparing chorion villus sampling with amniocentesis.

12. Two-thirds of babies with Down's syndrome are born to women below the age of 35 years, yet investigations are usually available only to those women of 35 years or over. The birth prevalence of Down's syndrome has not fallen over the past 15 years. There was considerable support in the Study Group for offering serum biochemical screening to all pregnant women: this has the potential to place over 60% of all affected pregnancies into a high risk category which should be investigated further. Any such screening programme should be set up with due regard to the organisational, training and counselling aspects.

13. Adequate information should be given to women eligible for screening or diagnostic tests for fetal abnormality. Genetic counselling should be available for those at high risk and where possible counselling and preliminary investigations should be completed before pregnancy. Records should be made in the patient's

notes of diagnostic tests offered, as well as explanations given, and of the decision made by the patient.

14. All diagnostic results should be transmitted to the patient as soon as possible. Genetic counselling should be available when an abnormal result is given, especially for those women with a result which is difficult to interpret or where the outcome is variable. If a woman seeks termination of pregnancy for fetal abnormality, appropriate support, obstetric follow-up and genetic counselling should be available to her (and to her partner where relevant).

15. The use of DNA-linked markers has facilitated prenatal diagnosis for many single gene disorders. Care should be exercised because of the possibility that apparently identical disorders may result from mutations at different chromosomal sites.

16. Direct gene analysis is now possible on DNA from a single cell. The extreme sensitivity of the polymerase chain reaction (PCR) technique when applied to single cells requires stringent precautions against contamination. Though not yet generally available, PCR offers the possibility of rapid preimplantation diagnosis of genetic defects by embryo biopsy; and perhaps even from the first polar body of an unfertilised egg or from fetal cells in the maternal circulation. Whilst preimplantation diagnosis is a means of avoiding abortion of an affected embryo, the low availability and high failure rate of in vitro fertilisation techniques may limit its acceptability.

17. Fetal blood sampling has a diagnostic role especially in growth-retarded infants and those shown on ultrasound scans to have anomalies. It also has a therapeutic role in treatment of infants affected by red blood cell isoimmunisation, hydrops due to infection and some other haematological disorders. Procedures for draining hydrothorax and dilated renal tracts in utero may have benefit in some cases. Pretreatment evaluation is crucial.

18. Fragile-X mental retardation poses particular diagnostic problems. The prenatal diagnostic method of choice is chorion villus sampling with cytogenetic analysis in a regional centre. If positive, appropriate action is taken. If negative, referral for fetal blood sampling should be considered. In the future, when suitable linked probes are available, prenatal diagnosis and carrier detection should be carried out by DNA testing.

19. Co-operation and collaboration between clinicians, molecular scientists and biochemists are important for research and development. Effective clinical management of individual patients requires early and continuing co-operation between professionals in all the disciplines involved. Discussions about the introduction of new techniques should involve ethicists as well as other professionals to ensure that methods of screening and treatment are not only effective but also ethical.

Subject Index

Abdominal wall defects 20–2, 29
Abnormal fetal karyotype, Doppler umbilical studies 116–18
Abnormality, definition of 281–2
Abortion. *See* Termination of pregnancy
Accuracy problems 82–4, 93, 164–5
Acetylcholinesterace 57
--Acetyl-α-galactosaminidase deficiency 194
Adenosine deaminase 142
AIDs counselling 292
Alloimmune thrombocytopenia 211–12, 220–21
Alphafetoprotein. *See* Maternal serum alphafetoprotein (MSAFP) screening
Alzheimer's disease 283
Amino acid sequencing 194
Amiodarone 218
Amniocentesis 15, 26, 52, 64, 77, 79, 81–8, 244, 246, 260–1, 285, 349, 350
  complications of 343
  diagnosis of spina bifida 48–50
  discussion 91–5
  early 87–8, 93, 94, 342–3, 349
  factors influencing uptake of 244
  failed 206
  first trimester 83
  future provision 338–44
  standard technique 342
  traditional 86–7
Amniotic fluid supernatant, chemical analysis of 187
Anencephalus 7
Anencephaly 4, 5, 8, 24, 33, 46, 124
  diagnosis of 49
  incidence of 25
Aneuploidy 120
Angiotensin infusion sensitivity test (AIST) 117
Antenatal treatment 19–20, 29
Aortic arch anomalies 109
Aortic atresia 106
Aortic override 107

Aortic stenosis 240
APRT (adenine phosphoribosyltransferase) 142
Ashkenazi Jews 184
Ataxia telangiectasia 155
Atrial septal defects (ASDs) 124
Atrioventricular septal defect 104–5
Audit 29, 122
Autoimmune thrombocytopenia 211–12
Autopsy examination 15, 28
Autosomal dominant disorders 131

Balloon valvuloplasty 240
Beckwith–Weidemann syndrome (BWS) 155
Beta-thalassaemia 184
Bilateral hydronephrosis 224
Bilateral renal agenesis 18
Biochemical diagnosis 137, 306
Biochemical disorders 183–98
  discussion 197–8
  postnatal diagnosis 185–7
  prenatal diagnosis 187–91
Biochemical microassay 142–3
Biopsy 139–40
Birth management 22
Bladder outlet obstruction 19, 20
Blastocysts 140
Blood disorders 202–3
Bloom's syndrome 155
British Paediatric Association 318

Caesarean section 20, 21, 40
Canadian randomised trial 158, 161
Canadian trial on CVS 76, 87, 92
Cancer 308
Cardiac anomalies 38
Cardiac arrhythmia 19, 20
Cardiac disease 39
Cardiac malformations 102–9

Cardiac ultrasound scanning 97–111
  abnormalities of connection 103–7
  abnormalities other than connections 108–9
  discussion 122–5
  four-chamber screening 98–100
  great arteries 102–3
  Guy's Hospital 109–10
  levels of 97
Cardiovascular diseases 308
Care effectiveness 122
Cascade counselling 332
Central nervous system (CNS) defects 15, 25
Chorionic villus biopsy (CVB) 15, 26, 262
Chorionic villus sampling (CVS) 156, 159, 219, 244, 246, 305, 348–40
  abnormal results in 161
  biochemical assays 188
  Canadian trial on 76, 87, 92
  costs 88
  discussion 91–5, 286
  early and late 154
  failure rate 343
  fetal karyotyping 153–67
  first trimester 77, 83, 154, 157, 160
  fragile-X syndrome 170–81
  future provision 338–44
  labour intensive 164
  late 163, 345
  misdiagnosis 161
  mosaicism in 157–60
  MRC trial 73–7, 93
  pregnancy outcome 344
  safety and timing of 343–4
  second trimester 163–4
  third trimester 163–4
  transabdominal 77, 79, 81, 83–7, 89, 91–4
  transcervical 79–81, 83–7, 89, 91–3
Chromatographic analyses 185
Chromosome abnormalities 4, 15, 16, 26–8, 82–4, 125, 153–5, 158–60, 164, 207–8, 354
Chromosome analysis 137, 306, 309, 313
Chromosome imprinting 164
Chromosome rearrangements 261–62
Chromosome translocation 306
Chylothorax 222
Cleavage stage embryos 139–40
Clinical Genetic Society 318
Clinical intuition 291
Clinical Molecular Genetics Society 329
Collaborative Survey of Perinatal Mortality 16
Community education 330–1
Confidential enquiry into recurring genetic disorders 318–19
Confidentiality 292
Confined placental karyotypes 160
Congenital abnormalities 222
  by areas of residence 336
  classification 4–9
  implications of 3
  in babies in England and Wales 336
  in relation to mother's age 337
  losses due to 35
  national voluntary reporting scheme for 27
  prevalence of 3–5
  recognition of risk 3–4
  registers of 354
  trends in 3–11
  see also under specific abnormalities
Congenital heart abnormalities 5
Congenital heart disease (CHD) 97, 99–101, 109–10, 119
Congenital thrombocytopenia 210–12
Consent for treatment 286–7, 290
Cordocentesis 201–15
  indications 202–12
  risks 202
  technique 201–2
  see also under specific conditions
Cost–benefit analysis 68, 270–3
Cost-effectiveness analysis 270, 271, 307
Counselling 56, 253–65, 313
  after prenatal diagnosis 259
  and ultrasound examination 291
  at time of testing 258
  discussion 265–6, 289
  effect in interpretation of risk factors 340
  in early pregnancy 339
  termination of pregnancy 263
  see also Genetic counselling
Couple screening 64, 68
Cystic fibrosis 33, 59–70, 144, 148, 149, 184, 313, 319, 330, 331
  cost–benefit analysis 68
  detection of alleles 61
  discussion 67–70
  gene cloning and identification 60
  gene locus 61
  heterozygote screening 62, 63, 65
  life expectation of 59
  moderate risk problem 63
  national survey of 59
  pilot trials of different models of screening 65
  prevalence of 59
  social and organisational considerations 65–6
Cystic fibrosis transmembrane conductance regulator (CFTR) 61, 133, 324
Cystic hygromatous changes 28
Cystinosis 194
Cytogenetic analysis 82–3, 206
Cytogenetic disorder 155
Cytogenetic services
  laboratory organisation 349–51
  provision of 349
Cytogenetic studies 141
Cytomegalovirus 204

Decision analysis 265
Delivery management 20–2

Subject Index

De novo chromosome rearrangements 261
Diagnostic dilemmas 161–2
Diaphragmatic hernia 22, 39, 40, 164, 222
Diazepam 232
Digoxin 217
Direct diagnosis 324, 325
Direct gene analysis 355
Discounting effect 293
DNA analysis 33, 55, 94, 129–37, 143–5, 164, 181, 190, 194, 300, 302, 304, 355
  see also Recombinant DNA technology
DNA markers 129, 132, 134, 177, 355
DNA probes 161, 162
DNA segments 144
DNA sequence 130–36, 143, 322
DNA technology 194
DNA tests 313, 316, 322, 326, 330, 355
Doctor–patient relationship 290, 292
Dominant diseases 324–5
Doppler flow velocity waveforms 114–16
Doppler studies 208, 210
Doppler ultrasound 113–25
Doppler umbilical studies 114
  abnormal fetal karyotype 116–18
  structural fetal abnormalities 118–19
  total obstetric population 119–20
Double inlet connection 105
Double outlet right ventricle 107
Down's syndrome 4, 27, 69, 125, 141, 155, 156, 250, 285, 319, 336, 350, 354
  discussion 54–7
  economic efficiency studies 274–5
  impact of antenatal diagnosis on prevalence rates 9
  monitoring of screening programmes 45–7
  prevalence of 8–9, 47
  role of maternal age 52–3
  screening for 50–3
  trends in prevalence rate 8–9
Duchenne muscular dystrophy 33, 134, 144, 276, 313
Dysplastic kidney 23

Echo planar imaging (EPI) 234–6
Echocardiogram 97, 100–1
Echocardiography 123
Economic aspects
  discussion 293
  prenatal diagnosis and screening 269–78
Economic efficiency studies 270, 274
Educational dimensions 281
Embryo assessment 140–1
Embryo recovery 138
Embryo transfer 138
Encephalocele 4, 24
Environmental abnormalities 4
Enzyme assays 82
Epispadias 31

Ethical aspects 271, 279–93, 313
  discussion 289–93
European Registration of Congenital Abnormalities and Twins (EUROCAT) 7, 9
Extrauterine surgery 39

Family history
  at risk 315–17
  high risk 256, 266
  increased risk 257
  low risk 257
  registers 315–17
Fanconi's anaemia 155
Fetal abnormalities 232
  case histories 29–30
  Northern Regional survey 13–34
  prenatal diagnosis of 110–11
  scanning 354
Fetal anaemia 209–10
Fetal blood gases 208
Fetal blood sampling 201, 218, 239, 345, 355
Fetal blood transfusion 201, 219–20
Fetal growth retardation 18
Fetal heart rate (FHR) monitoring 208, 210
Fetal hydrothorax 222–3
Fetal imaging, limitations of 232–3
Fetal immobilisation 231
Fetal infection 203–5
Fetal karyotyping 139, 140, 164, 205–12
  chorionic villus sampling 153–67
  possible sources of diagnostic error 155–61
  risks involved 154–5
Fetal malformation 207–8
Fetal surgery 222
Fetal vesico-amniotic shunt 20
Fetocide 93
Fetofetal transfusion syndrome (FFTS) 221
Flecainide 20, 218
Four-chamber screening. See Cardiac ultrasound scanning
Fragile-X syndrome 169–81, 206, 355
  chorionic villus sampling (CVS) 170–81
  discussion 176–81
  problems encountered during study 176
  protocol for 179
  results of study 172–5
FRAXA 169–81
Friedreich's ataxia 33
FUdR 170, 172, 173, 177

Gametes 139
Gender selection, motives for 281
Gene mapping projects 130, 308
Gene probe diagnosis 82
Gene-specific probe 130
Gene tracking 129, 130, 323–4
  limitations of 133–4

Genetic analysis 139
Genetic centres 313–15
Genetic counselling 68, 183, 184, 187, 207, 249, 255–6, 304, 308, 315, 354–5
 definition 255–6
 discussion 265, 289
 information from third parties 288
 see also Counselling
Genetic disease 183, 184, 186, 190, 313, 317
 classification 4–9
 confidential enquiry 318–19
 healthy but at risk of 315–17
 incidence and burden of 311–12
 interventions to prevent 284
 prevalence of 3–5
 recognition of risk 3–4
 risk of developing 315–17
 see also under specific diseases
Genetic factors in human disease and congenital abnormalities 4
Genetic finger-printing 83
Genetic heterogeneity 131
Genetic metabolic disease 184–6, 188, 189, 191, 306
Genetic prediction 129
Genetic registers 315–17
Genetic risk 92, 94
Genetic services 311–19
 adequacy of 317
 arrangements 326–7
 case for specialisation 322–3
 clinical manpower 317–18
 facilities found in 315
 future coordination models 327
 impact of 313
 laboratory workload 323–4
 national chromosome-based consortium 328–9
 national coordination of 321–33
 national disease-based consortium 327
 Netherlands. See Netherlands
 obstetricians in 311, 335–51
 recommendations 353
 regional organisation of 348
 requests for service 325–6
 scientific and technical advances 324–5
 specialist consultant clinical geneticists 318
 steady-state load 326
 supraregional disease-based consortia 327
Genetics education 330–31
Genic disorders 4
Genitourinary (GU) anomalies 31
Genome mapping project 280, 286, 313
GIG (Genetics Interest Group) 331
GPI-deficiency 143
Great arteries
 imaging 102–3
 transposed 106
Great Ormond Street Hospital 111
Guy's Hospital 109–10

Haemoglobin reference centre 330
Haemoglobinopathies 326
Haemolytic anaemia 143
Haemophilia 319
Handicap
 attitudes to 250, 283, 292
 discussion 265, 290
Happiness argument 290, 292
Health professionals in prenatal testing 249
Health Visitors Association 318
Heart disease 97
Heteozygote screening 59–70
Hexosaminidase A assay 184
Hexosaminidase-deficiency 143
High-performance liquid chromatography (HPLC) 185–6
Hirschprung's disease 30
HLA antigens 83
HPRT (hypoxanthine phosphoribosyltransferase) 142
HPRT-deficiency 142, 143
Human chorionic gonadotrophin (hCG) 51, 53, 55
Huntington's chorea 284–5, 292, 313
Hydatidiform moles 233
Hydrocephalus 24, 25, 39
Hydronephrosis 23, 40
Hypospadias 31
Hysterotomy 201, 222

IgG antibodies 211–12
Immunoelectron-microscopical studies 192
Infertility treatment 138–9, 144
Inguinal hernia 31
Inherited diseases 316
In situ hybridisation 136, 141–42
International Human Genome Project 129
Intracranial haemorrhage (ICH) 220–1
Intraperitoneal transfusion (IPT) 220
Intrathoracic lesions 222–3
Intrauterine diagnosis 110
Intrauterine growth retardation (IUGR) 232
Intrauterine therapy 217–27
Intravascular transfusion (IVT) 219–20
In utero treatment 39
Invasive diagnostic procedures in first trimester 79–95
In vitro fertilisation 138–9, 144, 190

Karyotyping procedures. See Fetal karyotyping
Klinefelter's syndrome 179

Law of Special Hospital Services 300
Legal aspects 282, 292
Lesch-Nyhan syndrome 142, 143
Lethal abnormality 18–19
Linkage analysis 129–36
Locus heterogeneity 130–3

Subject Index

Long-term disability 24–6, 29, 38
Low birth weight (LBW) infants 218
Lung hypoplasia 224

Magnetic resonance imaging (MRI) 229–40
  assessment of pelvic size 233
  background 229
  discussion 238–40
  early work 230
  fetal imaging 230–1
  principle of 229
  see also under specific techniques and applications
Martin–Bell syndrome 169–81
Maternal cell contamination 161
Maternal serum alphafetoprotein (MSAFP) screening 6, 13, 24, 25, 32, 33, 42, 45, 47, 50, 51, 53, 55–7, 65, 69, 244, 246, 249, 266, 275, 335–7, 339, 344–5, 348
  ultrasound dating scan to interpret 48
Maternal serum biochemistry 206–7
Medical genetics. See Genetics services
Medical Research Council (MRC) European trial 73–7, 93
  background 73–4
  BBC television "Horizon" programme 75
  eligibility of participating centre 74
  eligibility of pregnant women 74
  encouragement for future trials 75
  involvement of special interest groups 74–5
  main objectives 74
  progress of 74–6
  recruitment of other European centres 75–6
Mendelian disorders 183, 185, 194, 195, 321, 322
Menkes' disease 93
Menkes' syndrome 82
Metabolic disorders 203
Methotrexate (MTX) 170, 172, 174, 177
Microvillar enzyme testing (MVT) 64
Miscarriage risk 82, 88, 92, 93
Mitral atresia 104
Molecular genetic techniques 321
Monogenic disorders 129, 131, 311
Mosaicism 93, 157–60, 162–3, 206, 262
Mouse model systems 143
Mullerian abnormalities 234
Multifactorial disorders 4
Multiple endocrine neoplasia type 2 (MEN2) 319
Multiple pregnancy 83, 93
  diagnosis of 39
Muscular dystrophy 283–4
Mutation detection 129–30, 134–6

National Audit Office Survey 56, 335
National Perinatal Epidemiology Unit 74
Neonatal surgery, identification of conditions amenable to 40

Netherlands genetic services 297–310
  annual activities 302, 303
  fear and reality in public acceptance 307–9
  funding of genetic services 298–301
  legal restrictions 300–1
  organisation and results 301–6
  requests for new staff, equipment or space 303
  staff members 302
  tariffs for prenatal and postnatal chromosome analysis 299–301
Neural tube defects 4–8, 31–3, 65, 238–9, 263, 319, 336
  discussion 54–7
  economic efficiency studies 274
  impact of antenatal screening and diagnosis 6–8
  incidence of 24–6
  monitoring of screening programmes 45–7
  neonatal survivors 26
  prevalence of 46
  screening for 48
  trend in prevalence rates 6
Neurological disorders 308
Northern Regional fetal abnormality survey 13–34
  background 13–14
  case histories 29–30
  difficulties 17
  discussion 28–34
  input 16
  methods 14–17
  output 17
  results 17–28
  teething problems 17
  validation 15
North-West Thames Regional survey 31

Obstetricians and genetic services 311, 335–51
Obstructive uropathy 39, 223–4
Office of Population Censuses and Surveys (OPCS) 15, 27, 31, 46, 47
Oligohydramnios 18, 117, 224
Osteogenesis imperfecta (OI) 132–3

Pallister–Killian syndrome (PKS) 164
Parental decisions 4
Parvovirus B19 infection 205
Patient/health professional relationship 286–8
Pelvic masses 234
Pelvi-ureteric obstruction 23
Perinatal alloimmune thrombocytopenia (PAIT) 220–21
Perinatal mortality and morbidity, reduction of 38–9
Peritoneal dialysis 20
Placental biopsy 163–4, 206
Placental localisation 233
Placental mosaicism 93, 160
PNP (purine nucleotide phophorylase) 142

Polar body 139, 140, 144, 149
Polyhydramnios 116, 117
Polymerase chain reaction (PCR) 61, 82, 134–5, 143, 162, 322–5, 329, 355
Population screening 17–26
Posterior urethral valves (PUV) 223–4
Post-mortem examination 3
Post-neonatal care 23–4
Prednisolone 212
Preimplantation diagnosis 137, 142–4
  discussion 148–9
Preimplantation embryos 138–9
Prenatal diagnosis and screening 110–11, 282
  abnormal results
    difficult to interpret 260–2
    easy to interpret 259–60
  access to information 286–8
  attitudes to 341–2
  choices available 269–70, 279
  current situation 335–7
  cut-off risk 276
  discussion 265–6
  economic aspects of 269–78
  future developments 307–9, 337–46
  groups presenting for 256–8
  growing public demand for 269
  improvement of delivery of service 345–6
  issues involved 282–8
  privacy and third parties access to information 280–1
  provision of services 335–51
  psychological implications 242–54
  recommendations 353
  regional organisation 348
  resource consequences for other conditions 276
  revealing information to patients 386
  women's choice of 339–41
  see also Counselling; Family history; Genetic services; and under specific aspects, techniques and conditions
Prenatal karyotyping. See Fetal karyotyping
Prenatal testing
  attitudes of society towards 250–1
  availability of 243
  factors influencing use of 243
  impact of 246–8
  recommendations to improve staff delivery of 250
  recommendations to increase informed uptake of 245–6
  recommendations to reduce adverse psychological consequences of 248–9
Preventive health care services
  efficiency of 271–2
  options available 272
  status quo option 272
Procainamide 218
Profound fetal compromise 119
Propranolol 218
Protective protein 192

Protein assays 187, 304
Protein defects 183, 184, 186, 187, 194
  analysis of 191–2
Prune Belly syndrome 19
Pseudomosaicism 206
Psychiatric disorders 308
Psychological implications of prenatal diagnosis 243–54
Pulmonary atresia 105–6
Pulmonary cystic dysplasia 22
Pulmonary hypoplasia 223
Pulmonary stenosis 124
Pulmonary venous drainage 103
Pulsatility index 114

Quality-adjusted life-years gained (QALYs) 273, 276
Quality of life 273

Radioimmunolabelling of cultured skin fibroblasts 184
Radioimmunoprecipitation studies 193
Recombinant DNA technology 162, 185
Red cell alloimmunisation 218–20
Reflux 23, 40
Renal abnormality 39
Renal agenesis 18, 39
Renal cysts 224
Renal damage 224
Renal disease 40, 54
Renal dysgenesis 18
Renal dysplasia 223–4
Renal function 224
Renal parenchyma 224
Renal polycystic disease 39
Renal scans 40
Renal scarring 23
Research and development 355
Resistance index 116, 117, 121
Resource considerations 270
Return on investment 293
Risk factors and risk assessment 3–4, 63, 82, 88, 92–4, 154–5, 164–5, 202, 256, 257, 266, 276, 315–17, 340
Rotterdam Regional Centre for Clinical Genetics 302
Royal College of Midwives 318
Royal College of Obstetricians and Gynaecologists (RCOG) 40, 73, 311, 318, 335, 354
Royal College of Physicians 317, 318
Royal College of Radiologists 40
Rubella 203–5

Safety evaluation 84–9
Salla disease 194
Scottish Consortium 332
Selective fetocide 83
Selective screening 26–7
Septic abortion 86

Septicaemia 86
Service provision. *See* Genetic services
Severe combined immunodeficiency disease (SCID) 142
Sex chromosome aneuploidies 260–1
Sex diagnosis 144
Short-limbed dwarfism 18
Shunting procedures 39
Sialic acid storage disorders 194
Sickle cell anaemia 145
Single cell biopsy 286
Single gene defects 82, 144–5
Skeletal dysplasias 18
Skin biopsy 184
Skin fibroblasts 186
Small for gestation fetuses 208
Social dimensions 281
Sonar examination 13
South-East Thames 333
South-West Thames 333
Southern blotting 323
Spina bifida 4, 5, 7, 8, 24, 38, 46, 55, 122, 239, 246
  diagnosis of 48–50
Structural fetal abnormalities, Doppler umbilical studies 118–19
Suffering, obligation to prevent 285–6
Supraventricular tachycardia 20, 217–18
Systolic/diastolic ratio 114, 119, 120

Tachyarrhythmias 217–18
Tay–Sachs disease 68, 143, 184, 276
Termination of pregnancy 17, 18, 26–9, 38, 92, 137, 141, 248, 249, 271
  by age group 339
  discussion 292
  fetal examination after 262–3
  follow-up 354
  issues involved 282–4
  psychological sequelae 263
  subsequent pregnancy after 263
  support for family after 263
Testicular abnormalities 31
Testicular cancer 31
Tetrahydrofolate reductase 170
Thalassaemia 33, 68, 184, 313, 319
Thalassaemia major 276
Thalidomide disaster 27
Thrombocytopenia 210–21
Thymidine 170
Thymidylate synthetase 170
Tissue characterisation 113
Toxoplasmosis 204
Tricuspid atresia 103
Triple X syndrome 179, 180
Triploidy 116, 158
Trisomy 13 116, 158
Trisomy 18 116, 117, 158, 164
Trisomy 21 26, 27, 30, 116, 118, 120, 141, 158, 206

Trophoblast cells 93
Trust element 290, 291
Tuberose sclerosis (TS) 130
Turner's syndrome 260
Twins 7, 9, 92

Ultrasound 23, 24, 27–8, 35, 45, 207–8, 222, 223, 244, 246, 254
  and counselling 291
  cost effectiveness of screening programme 41
  cross-utility studies 54
  dating scan to interpret AFP values 48
  departments 110
  diagnosis of spina bifida 48–50
  discussion 54–7
  frequency of visualisation of individual structures 37
  future provision 337–8
  identification of anomalies by 36–7
  objectives of screening programme 38, 42
  organisation of screening programme 40–1
  randomised trials 54
  staffing requirements 40
  structures visible by 36
  timing of screening scan 41
  training programmes 40
  use in amniocentesis 81
  see also Cardiac ultrasound scanning; Doppler ultrasound; Prenatal diagnosis and scanning
Umbilical artery
  Doppler ultrasound 114
  flow velocity waveform 116
Unconjugated oestriol (uE$_3$) 51, 53
Undescended testicle 31
Ureteric obstruction 23
Ureterovesical obstruction 23
Urethral valves 19, 39
Urological abnormality 23
Uterine lavage 138–9

Valve stenosis 108
Valvular dysplasia 108–9
Varicella-zoster virus 204–5
Venous-atrial junction 103
Ventricular septal defect 108
Ventriculo-arterial junction 105–7
Verapamil 218
Vesicoamniotic shunting 224

X-inactivation 156
X-linked disorders 129, 131, 133, 140, 144, 155, 169, 324, 329–30

Y chromosome 141–2, 143

Zona pellucida 139–40, 144